Alfred Hudson, Sir John Eric Erichsen

Lectures on the Study of Fever

Alfred Hudson, Sir John Eric Erichsen

Lectures on the Study of Fever

ISBN/EAN: 9783337418854

Printed in Europe, USA, Canada, Australia, Japan

Cover: Foto ©berggeist007 / pixelio.de

More available books at **www.hansebooks.com**

LECTURES

ON THE

STUDY OF FEVER

BY

ALFRED HUDSON, M.D., M.R.I.A.,
PHYSICIAN TO THE MEATH HOSPITAL.

PHILADELPHIA:
HENRY C. LEA.
1869.

PHILADELPHIA:
COLLINS, PRINTER, 705 JAYNE STREET.

TO

WILLIAM STOKES, M.D., F.R.S., M.R.I.A.,

PRESIDENT OF THE KING AND QUEEN'S COLLEGE OF PHYSICIANS,

REGIUS PROFESSOR OF PHYSIC IN THE UNIVERSITY OF DUBLIN,

PHYSICIAN IN ORDINARY TO THE QUEEN IN IRELAND,

PHYSICIAN TO THE MEATH HOSPITAL,

This Volume is Dedicated

BY

HIS FRIEND AND COLLEAGUE,

THE AUTHOR.

PREFACE.

It has long appeared to me that the student usually has a less clear idea of fever than of any disease which he meets with in the wards of the hospital. The difficulty he experiences may arise partly from the want of correspondence of the particular case before him with the description of some form of fever from which he has obtained his *ideal*, and partly from his not possessing the key, so to speak, to its solution in a knowledge of fever in the abstract, of its laws, and of the phenomena which are common to all types of the disease. It is true he has in the great work of Murchison the fullest and most comprehensive descriptions of *every* form of continued fever, but I cannot but regard these distinct treatises (for such they are) as much more likely to be useful as a future work of reference, than to serve as a guide to the early study of a complex and difficult subject like fever.

The object I have had in view in delivering the ensuing Lectures to the students of the Meath Hospital, and in now publishing them, is to furnish the student with a guide to his bedside analysis of each case, by treating of febrile phenomena in succession; first, generally or abstractedly, and secondly, in their relation to each form of the disease. Thus forming in his mind an *ideal* of fever, such as he may readily apply to the case before him, and which he may certainly find to conform to that case, be it of what species, or how complicated soever it may. Should it appear that I have in any degree succeeded in this task, I shall have gained my object, and the student will be assisted, not only in his bedside study of the disease, but also, I trust, in the prosecution of his more extended investigations into its nature and relations, as treated of in the numerous and valuable works which exist on the subject.

In order to avoid prolixity I have omitted the details of some illustrative cases, and have transferred these, and several published observations which I have thought likely to be useful to the student, to an Appendix. I have also reserved for this an examination of some of the arguments on the controverted question of identity or non-identity of fever poisons.

It gives me much pleasure to acknowledge my obligations to my friend Dr. Sinclair, for his supervision of this work when passing through the press, and for many important corrections and emendations suggested by him. I am also indebted to his kindness for the valuable index appended to this volume.

CONTENTS.

	PAGE
INTRODUCTION	17—33

LECTURE I.

Fever considered as one of the group of morbid poisons—Mode of action on the blood, nervous system, and molecular nutrition of the body—Congestive form of Fever—Stage of reaction—Elimination—Crisis—Recapitulation 34

LECTURE II.

Predisposition, nature of—General and Special—Degrees of susceptibility—Predisposing causes—Influences of idiosyncrasy—Recapitulation . 45

LECTURE III.

Exciting causes—Contagion, miasma, and epidemic influence—Spontaneous generation of contagion by ochlesis—Different forms of disease produced by each of the exciting causes—Recapitulation . . . 57

LECTURE IV.

Pathology and Symptomatology—Phenomena set up by the action of the fever poison—Primary or secondary—External phenomena—Temperature—colour, expression, eruptions, in typhus, typhoid, and relapsing fevers—External characters of different complications . . . 69

LECTURE V.

Examination of the pulmonary and circulating systems in succession, and in their mutual relations—Condition of the left ventricle—Observations of Dr. Stokes—Capillary circulation—Respiration—Anatomical changes in the respiratory organs—Physical signs 80

LECTURE VI.

Peculiar form of pneumonia, described by Dr. Stokes—Connection between nervous excitement and pneumonia—Condition of the right cavities of the heart in fever—Examples—Recapitulation 91

LECTURE VII.

Condition of the digestive organs in fever—Different states of tongue—Anorexia, thirst—Value of vomiting as a symptom—Jaundice—Diarrhœa—Intestinal hemorrhage—Peritonitis—Anatomical lesions in different forms of fever 101

LECTURE VIII.

Changes in different secretions—Importance of the urinary secretion—Its relation to molecular nutrition—Characters of febrile urine—Deviations from this type—Uræmia, from retention, from deficient elimination, from diseased kidney—Recapitulation 112

LECTURE IX.

Difficulty of Diagnosis of cerebro-spinal lesions in fever—Functions deranged—Modes of derangement—Delirium, watchfulness, stupor—Derangements of motor function, of sensation, of the special senses.—Anatomical lesions—Symptoms of active congestion: of cerebro-spinal arachnitis—Examples 123

LECTURE X.

Cerebro-spinal lesions, continued—Influence of pre-existing states of nutrition of the nervous centres:—of accidental causes:—of previous disease, or injury—Supervention of symptoms of anæmia or exhaustion upon those of active congestion—Difficulty of differential diagnosis—Nervous derangements of crisis, sleeplessness, convulsions, paralysis . . 137

LECTURE XI.

Epileptiform convulsions after crisis—Influence of depressing emotions—Examples—Influence of secondary blood contaminations—Peculiar symptoms of uræmic poisoning—Recapitulation. . . . 149

LECTURE XII.

Diagnosis of fevers as an essential disease—Fever paroxysm—Differences of the mode of invasion—Progress, duration, and mode of termination of each form of fever—Pathological changes in each—Characteristic differences between typhus and typhoid fevers 162

LECTURE XIII.

Rational prognosis.—The knowledge of tendencies derived from the condition and circumstances of the patient—history of the disease—existing symptoms—complications—effects of treatment.—Prognostic signs at the period of crisis 187

LECTURE XIV.

Principles which should regulate the general management of a case of Fever—Possibility of arresting fever by a shock—Means employed for this purpose—Regulation of the surrounding conditions, food, sleep, &c.—Special indications of treatment in abdominal and thoracic complications 206

LECTURE XV.

Treatment of urinary complication—Of cerebro-spinal lesions—Management of the period of crisis—Danger of abuse, or too long continuance of stimulants—Treatment of complications occurring at this period—Management of early convalescence—Recapitulation . . . 221

APPENDIX 245

LECTURES ON THE STUDY OF FEVER.

INTRODUCTION.

In prefixing to the following Lectures on Fever, some portion of an introductory lecture which I delivered at the opening of the session of 1864-5, on the study of clinical medicine generally, I do so because I consider it of importance that the student should apply the *method* which I have here endeavoured to illustrate and explain, to his bedside study of fever in particular. If, to the more advanced student, my illustrations of this method should appear trite and perhaps tedious, I hope he will bear with me, in consideration of the necessity that they should be understood by those for whose instruction they are intended.

Method[1] is defined by Coleridge, in his admirable Essay, to be "a progressive transition from one step to another." "Without method," says this great thinker, "all things in us and about us are a chaos, and so long as the mind is entirely passive—so long as there is an habitual submission of the understanding to mere events and images as such, without any attempt to classify and arrange them, so long the chaos must continue. There may be transition, but there cannot be progress; there may be sensation, but there cannot be thought; for the total absence of method renders thinking impracticable, as we find that partial defects of method render thinking a trouble and fatigue; but as soon as the mind becomes accustomed to contemplate not things only, but likewise relations of things, there is immediate need of some path or way of transit from one to the other of the things related—there must be some

[1] "On Method, Encyclopedia Metropolitana."

law of agreement, of contrast, between them—there must be some mode of comparison; in short, there must be method. We may, therefore, assert that the relations of things form the prime objects, or, so to speak, the materials of method; and that the contemplation of these relations is the indispensable condition of thinking methodically. Of these relations of things we distinguish two principal kinds. One of them is the relation by which we understand that a thing must be; the other, that by which we merely perceive that it is. The one we call the relation of law, using that word in its highest and original sense—namely, that of laying down a rule to which the subjects of the law must necessarily conform; the other we call the relation of theory. Medicine, Chemistry, and Physiology are examples of a method founded on this second sort of relation which, as well as the former, always supposes the necessary connection of cause and effect."

It is in accordance with the idea that the relations of things are of two sorts according as they present themselves to the mind as necessary, or merely as the result of observation—(relations, in short, of law or of theory)—that medicine, which deals with the latter, has been generally termed a science of observation.

The definitions of Science are almost as numerous and varied as the works written on different sciences. I shall select two as obviously admitting medicine within their comprehension.

"Science," says Whewell, "is that precise and comprehensive kind of knowledge which results from the application to facts which are certain and sufficiently numerous, of conceptions clear and distinct in themselves, and so suited to the facts as to produce an exact and uniform accordance; and the construction of science is a process which comprises methods of observation: methods of obtaining clear ideas; and methods of induction." "Science," says Lord Bacon, evidently following the definition of Pliny,[1] "is the interpretation of nature;" "a comparison," says Bain,[2] "that transfixes the mind with the idea of observing, recording, and explaining the facts of the world."

We shall appropriate this definition for clinical medicine, and proceed to illustrate the application of its method at the bedside; premising that the term observation, in medicine, is used, not in the sense of seeing or looking, merely, but *perceiving;* as expressing

---

[1] "Non unius terræ sed totius naturæ interpretes sumus."

[2] Bain on the Senses and the Intellect, page 545.

"insight, rather than sight"—that it is an observation which, to be comprehensive and discriminating, requires that our purpose and object, as well as our mode of observing, should be clearly and fully recognized before we begin to observe. According to this estimate of the scope and comprehension of the term, there is included in the observation of facts a clear conception of their relations to their cause, and to each other; conceptions which will be clear and precise according to the extent and precision of our knowledge, more especially of our knowledge of vital phenomena, and according to the activity of the exercise of our faculties of comparison, reflection, and suggestion.

The primary object of your examination of a patient is to ascertain the nature and extent of the deviations from health in the several functions, and the changes, if any, in the physical condition of the several organs. Your ability to undertake this task presupposes an acquaintance with the body in health in the most comprehensive sense; with the laws of healthy function; and with the physical signs by which you ascertain the healthy condition of the viscera of the chest and abdomen, their bulk, situation, and relation to each other. To these you refer as standards of comparison in your examination of the same organs in disease. This last is a preliminary qualification so easy of acquirement, and, at the same time, so neglected by students, that I may be excused for pressing it upon your attention. The value of all physical signs depends on comparison with healthy standards. Alterations in the shape of one side of the chest, dulness on percussion, absence or change of respiratory sounds, also of the sounds of the heart in different situations, are all determined by comparison; first, with corresponding parts in the same individual, but ultimately, though tacitly and instinctively, as it were, with standards consisting of our memory of the healthy signs. And yet how often does some student press forward to the bedside, that he may listen to diseased phenomena, who has never heard the sounds of the heart or lung in health.

The same rule applies to the derangements of function. To observe these with accuracy, you must be familiarly acquainted with their healthy exercise, and, *ceteris paribus*, the student whose knowledge of physiology is the most perfect will excel as an observer of disease—as a pathologist in the true meaning of the term.

To insure the accuracy of perception necessary for observation, careful training of the senses must be practised. I would therefore advise the junior student to take advantage of the opportunity of doing so afforded to him in the dispensary attached to this hospital; and in his examination of patients, there as well as within its wards, I would draw his attention to the necessity of frequent reference to healthy standards of comparison. The great requisites of this perception are, that it should be accurate, and that it should be vivid. The first you will insure, by repetition, by attention, and by comparison with the perceptions of others, not by merely repeating (because a wrong impression may be, and probably will be, thus confirmed), but also by comparing the impressions of others, your fellow-students and teachers, with your own. It is thus you will attain accuracy and uniformity, both of which are required in facts that are to serve as the basis of observation. The second requisite —or that vividness;—that mental photograph of the case under observation;—without which you cannot acquire present insight into its nature, or retain a future memory of its features—is to be obtained only by concentrating your attention on the case while under observation; by subsequently reflecting on it, and comparing it with the descriptions of others, or with other cases observed by yourself; and so far as memory is concerned, by associating it with something in itself or external to itself. Thus you will retain a recollection of one case of fever by a peculiarity in the eruption; of another, by some unusual nervous symptoms; of another, by a peculiar effect of a remedy, and so on; and if your faculty of attention is fully aroused while the case is under your observation, the slightest link of this description will bring back the whole chain. It is this combination of vivid perception and facile suggestion, which makes experience valuable in our profession, by enabling its possessor to recall and adopt the facts which have fallen under his observation at different periods.

The next step in the method of observation, is the application of the sign perceived, to the conception in our own mind of the condition which produced it. This conception is equally essential as the preceding perception. For example, you have placed your stethoscope over the side of a patient suffering acute rheumatism, and you have heard a loud "to and fro" sound. You simply perceive it. It has taught you nothing. You listen again, and your perception furnishes the idea of the rubbing together of two roughened surfaces. Continued attention makes you aware that this sound coincides with the heart's movements. Then follows the supposition that an exudation of fibrin upon the heart's surface and the inner surface of the pericardium is the cause of the sound. Probably this is so, and if there were no other condition capable of producing the phenomenon, it must be so. But you are prevented coming to so hasty a conclusion by your knowledge of such other conditions; and you remember having perceived a somewhat similar double sound, and felt a similar fremitus or thrill, in a case which *post-mortem* examination showed to be not one of pericarditis, but of diseased aortic valves. Here your conception is corrected by experience, and by your knowledge of pathology, which, moreover, suggests a number of other conditions by which the problem may be determined.

You thus see that the two elements in the first step in diagnosis, are perception of the sign and a clear conception of the condition which causes it; and as the perfection of the first of these depends on attention, or the direction of the will to the act of observation, so that of the other will depend on the mind being furnished with knowledge, and on the due exercise of the faculties of reflection and suggestion. Knowledge of facts gives us the materials for our induction, which will always be complete and accurate in proportion to the amount and importance of these; and the value of facts is best ascertained by constant reflection upon them. "How did you make your wonderful discoveries?" asked some one of Sir Isaac Newton. "I do not know, unless it was by constantly thinking on them," was the reply. Reflection and knowledge are the prerequisites of *suggestion*, and in proportion to the activity of the faculty of suggestion, you will become distinguished by comprehension of view of the features of each case under your observation; by the power of illustrating each by your past experience; and by the fertility of your resources in fulfilling those indications of treatment which, by the same faculty, are educed from the morbid phenomena. Perhaps, more than any other mental faculty it is strengthened by exercise, and there is no fact too common or minute, or apparently irrelevant, for it to appropriate. Thus, with the aid of past experience, it supplies the frequent missing links in the chain of facts; enabling us to decipher the imperfect record with a readiness and accuracy, resembling that with which the practised antiquary deciphers the half-defaced inscriptions of his cherished monuments.

Observation would be incomplete without a record; and, accordingly, the next step in our method is the reporting of our observations. Such reports include the evidence, or data, on which our judgment is founded—in other words, the *facts* perceived are stated—the *diagnosis*, or judgment of the existing facts, and their relations in the most comprehensive sense—the *prognosis*, or judgment of the future; and the indications of *treatment* suggested by a consideration of the whole. This is called, in technical language, case-taking, and is not merely an important, but an essential, portion of the method of clinical study. By it you are trained to habits of close and correct observation. You acquire by it the power of placing your carefully-collected facts in their proper order of sequence; of arranging the events in the history; of describing your own perceptions clearly and methodically; of tracing the development of the disease and its onward progress; and of noting and recording the effects of remedies. You acquire, moreover, familiar knowledge of the terminology of the science of medicine, so as to be able to think in its language, as well as to use it in your descriptions, which are thus rendered not only terse in their expressions, but pictorial in their effect. And this familiarity will enable you to realize the written descriptions of others, and to reproduce, as it were, to your mind, a case faithfully described by the great observers of former times. It is only those who have accustomed themselves to observe and describe, who can thus appreciate and profit by the cases recorded by others.[1]

There are a few simple and obvious rules, which you will do well to bear in mind, in case-taking. First, examine your patient thoroughly before you put pen to paper, and, if possible, let nothing escape your scrutiny. Next, in noting your observations, adopt a systematic method. It is well to commence with the history of the patient, so far as this consists of facts; which, however, you must carefully distinguish from the impressions patients often substitute for them. The patient's narrative will require to be supplemented by questions, suggested partly by what he has stated, and partly by your own knowledge of the disease under which he labours. Intelligent questioning will suggest many important facts to the patient's mind, which otherwise he might not communicate; and, on the other hand, irrelevant and aimless questions, while they betray your ignorance, will weary, disgust,

---

[1] See Mr. Bain's Remarks on "Realizing of Representation or Description."—*The Senses and the Intellect*, p. 604.

and render him incommunicative. Patients are often much more alive to the scope of questions, and the knowledge of the querist, than you may suppose. From the history and *subjective* symptoms, you will obtain a clue to the *objective* symptoms or signs likely to be met, and to the region which demands your more special physical examination—I say more special, for in every case each region and every function should be examined. This physical examination you should proceed to make with great care, and with due regard to delicacy; and your description of the results—in other words, of the physical signs—should be couched in terse and accurate phraseology; in fact, in the terminology of our science, descriptive and technical, but avoiding equally the tediousness of common language, and the error of involving a theory in your technical description. Suppose, for example, you described a murmur at the apex of the heart in a case of diseased mitral valve; you would not do so as if describing some new or unknown phenomenon, which could only be recognized by a lengthened detail of its character, in every-day language. Nor, on the other hand, would you describe it as a mitral or regurgitant murmur; because, however true the fact, this would involve a theory. But you would tersely and accurately describe the sign as a systolic, post-systolic, or diastolic murmur heard at the apex. Thus giving all the conditions required for its identification, and drawing no conclusions from it which could possibly be questioned.

I am sorry to say that students are too apt to overlook this distinction between fact and inference, in their reports of cases, and that the fault is not confined to students, but is to be met in medical writings of reputation. Everything recognized by the senses may have been correctly described; and yet the observation, as a whole, may have been vitiated by the incorrect interpretation of the facts observed. The description of the phenomena has often been vivid and accurate, but erroneous, because from them, the writer has inferred the existence of given disease, and has regarded the phenomena as symptoms or proofs of a condition which in reality had no existence. Examples of this fallacy, as it may be termed, will suggest themselves to you all.

Having ascertained and described the existing condition of the patient, you may have to review portions of his past history, and perhaps that of his family, so far as these may throw light upon the nature and causation of his disease; and you will not fail to notice his personal peculiarities and idiosyncrasies, so far as these bear upon his disease or its treatment. Lastly, you are not to suppose that your duties as a case-taker end here, but you must keep a record of the progress of the disease, to whatever result; of the treatment employed, and of its effects; carefully studying both, if you would become accurate and trustworthy observers.

Having carefully observed and recorded the phenomena of disease, the next step in our method is their explanation. Morbid phenomena may be illustrated and explained by reference to and comparison with those of health, or with other morbid conditions. The causal relation of a morbid sign or symptom, will sometimes be ascertained by reasoning from the diseased to the healthy condition; at others, by the opposite course; and again, sometimes one morbid phenomenon will explain another. Take an illustration of each. We hear a double murmur in the situation of the aortic opening; we find, after death, the valve roughed and inadequate; we have a previous knowledge of the functions of this valve, and of the sounds generated in health; referring to this, we at once explain the alteration in the sound of exodus by the roughened condition of the valve, and the marked change in the second sound by the inadequacy of the valve to prevent regurgitation from the vessel into the ventricle. Here we refer from the diseased to the healthy; but sometimes the contemplation of a healthy phenomenon will suggest the alteration which it must undergo in disease, and so anticipate the observed sign, and at the same time explain it. For example, a former student of this hospital, walking arm in arm with a fellow-student, was struck with the *fremitus*, or thrill, communicated to his arm from the chest of his companion, when speaking. Reasoning on the cause of this phenomenon he supposed that the interposition of a medium, such as air or fluid, between the lung and costal pleura, would arrest the *fremitus*, by interrupting the vibrations causing it. And he also reasoned that an alteration in the structure of the lung, such as condensation or solidification, would not interrupt, but might rather increase it. Experiment verified the anticipation, and proved the value of vocal *fremitus* as a diagnostic sign.[1]

An example of the explanation of one morbid phenomenon by another, is furnished by albuminuria with dropsy. In Bright's

---

[1] This absence of vocal fremitus as a sign of pleuritic effusion, was recognized about the same time, perhaps a little earlier, by M. Reynaud, of Paris.

disease, we know that the albumen found in the urine is furnished by the blood, which is thus drained of one of its most important constituents; becoming poor, thin, and watery, and losing the composition by which the endosmotic current in the extreme venous capillaries is maintained, and favouring the transudation of its serum into the areolar tissue. I need not remind you that this explanation is not hypothetical, but founded on a simple and decisive experiment. Nor is the other great fact in Bright's disease, namely, the imperfect elimination of urea, less capable of being explained by, and, in turn, explaining similar morbid phenomena. Indeed, you will find it frequently recurring in your study of the pathology of different forms of fever. Here, then, one diseased phenomenon illustrates and explains another.

But we constantly meet with groups of phenomena, of which some can be explained by one or other of these modes, while others remain for the present what may be termed "ultimate facts." Thus, in typhus, the remarkable increase of temperature is fully explained by the increased activity of metamorphosis of tissue compared with the standard of health; while we have no similar guide to the explanation of its peculiar eruption, or its tendency to terminate on certain days by crisis; similar phenomena in other diseases so far from illustrating and explaining these, requiring themselves to be explained. You will remark, that in these explanations there is not only an interdependence of theory upon fact, and fact upon theory, but that there is also a mutual interchange: each one becoming the other by turns, and a well-observed fact often in this process explaining many others. Thus, by the recognition of a permanently patent aortic valve, not only is the murmur of regurgitation explained, but the fact of regurgitation anticipates and explains other phenomena—as the increased labour imposed on the ventricle—this in turn leading to anticipation of the existence of hypertrophy or increased muscular growth of this cavity; which again explains the visible increase of pulsation in the larger arteries, as the regurgitation explains the collapse of the pulse. Again, our knowledge leads us to anticipate, that this overworked ventricle will, in time, become weakened and exhausted, and that the murmur of exodus accordingly will become less intense, while the mechanical condition of the valve remaining unaltered, that of regurgitation continues; and observation shows that our anticipations are correct. Take one more illustration of this anticipation of fact by theory. The phenomena of vocalization having been demonstrated by the laryngoscope, the observer anticipates that a lesion of the innervation of one of the vocal cords would modify or destroy the voice. The conditions which would cause this lesion at once suggest themselves, one of these being the stretching or compression of the inferior or recurrent laryngeal nerve by a thoracic aneurism: the recollection of aphonia previously met in cases of aneurism associates itself with this suggestion. He tests the accordance of his anticipation with fact in such a case, and finds the vocal cord paralyzed and no longer corresponding with the opposite one, and thus a new and important sign of aneurism is obtained.

It is of such facts, wedded to their explanations, that Bacon said: " One is worth a thousand ;" that, " being rightly understood, they draw with them whole ranks (*agmina*) of operations." It has been truly said, that the natural philosopher, who cannot, or will not see that it is the "enlightening" fact which really causes all the others to *be* facts in any scientific sense; he who has not the head to comprehend, and the soul to reverence this " parent experiment," to him no auspicious answer will ever be granted by the oracle of nature.

Three things are especially necessary in conducting your explanation, so as to arrive at a safe and sound induction. The first is comprehensiveness. Take care that you have all the facts of the case before you, when you enter on your explanation. Insufficient comprehension is one of the vices of imperfect induction. Want of clear conception is the other.

Secondly, exercise proper caution in the selection of your facts, and in adapting them to your conception. The fact and explanation must be homogeneous, so to speak; otherwise, like ill-fitting pieces of machinery, they will not hold together. You have, we will suppose, a case presenting signs and symptoms of pneumothorax. You know that air and liquid shaken together produce a splashing sound; you resort to succussion, and you hear the sound. The explanation and diagnosis are obvious, for the fact and conception are homogeneous. But suppose you were to anticipate the same result from succussion in a case of empyema—or conversely —suppose that, meeting the sign, you excluded from your explanation the presence of air and diagnosed simple liquid effusion; there would, in either case, be a want of homogeneity, and your explanation would be faulty.[1] Your selection must also include a correct judgment of the comparative value of your facts when regarded in relation to their cause and to each other; a comparison that will enable you to correct your deductions, by distinguishing those "diacritica signa," on which differential diagnosis rests. As in this process, one explanation after another is excluded from your diagnosis, on account of its inconsistency with some fact, theory, or law, the process is usually termed "the method of exclusion."

Lastly, modesty should be associated with your caution. You must not too soon think, that in observing what is, you have discovered what must be. You must not mistake your theory for a law. You must remember that when we use the term law of nature, it is most frequently in a figurative sense; for those constant and essential laws which regulate the processes of life, are precisely those about which our knowledge is the least satisfactory. Neither must you too soon imagine yourselves inventors or discoverers, or cry "Ευρηκα," thinking you have found the "enlightening fact." While you should, by all means, cultivate and develop in yourselves the talent which enables us to call before our minds the many possible explanations, and select the appropriate one—that talent of quick and fertile *suggestion*, which more than any other, characterizes the discoverer of new truths—you should carefully guard against the vanity and the haste which would lead you prematurely to regard yourselves, each, as one of these discoverers.

But we must glance for a moment at a further step in our method, which requires, if possible, a still higher exercise of the above requisites. I mean the anticipation of the results of disease. This anticipation, or prognosis, is usually regarded as the supplement of diagnosis. I should rather say, it is its necessary complement, for no consideration of the phenomena of disease can be complete, that does not include a knowledge of its tendencies, leading to an anticipation of its results, even of its consequences on the future health; nor can any indications of treatment be regarded as comprehensive and safe, which do not pre-suppose a knowledge of the natural, unaided, and unmodified tendencies of the disease which our treatment seeks to combat, to assist, or to modify, as the case

---

[1] I have more than once heard students under examination enumerate succussion amongst the signs of empyema; in fact, it is a common mistake.

may be.[1] I cannot, therefore, but regard prognosis as a most essential, though much neglected, step in the progress of our method.

Prognosis is partly empirical, partly rational. The former was cultivated by Hippocrates, and added to by Galen, Baglivi, Lommius, Stoll, and many other writers. It may be said to consist simply in the observation of sequences, no attempt being made to reason on them; many of them, while admittedly true not being yet explained. Every experienced practitioner must gradually acquire this prognostic skill, which, however, being incommunicable in a great measure, and so leading to nothing beyond itself, contributes nothing to the science. Rational prognosis, on the other hand, arrives at its results by combining the most comprehensive diagnosis of the disease, its nature and tendencies, with an estimate of the resistance of which the patient's constitution is capable; and with a similar estimate of the probable influence of treatment. The terms of this complex problem will show you its great difficulty arising chiefly from the fact that even the very longest experience will not enable us to become so intimately acquainted with the individuality of our patients as to say in what case, or why, one will survive an attack of phthisis for a year, another for twenty years; or why, of two youths suffering from endocarditis from acute rheumatism, eventuating in valvular disease, one should die in six months; and the other live for fifty years. Neither is our knowledge of the influence of treatment so certain or precise as to give us much assistance. The tendency of the disease is somewhat easier of determination. It may be to certain death, in which case the nature of the disease alone determines the prognosis. In other cases, the result may depend upon the balance between the destructive tendency of the disease on the one hand, and the conservative power of the constitution on the other, of which phthisis furnishes a good example.

But the talent of prognosis is more especially called into requisition in a class of cases in which the natural tendency of the disease is to cure or resolution, and yet in which a fatal result is determined by an unusual amount of disease or other causes.

---

[1] "Hippocrates recommends the cultivation of prescience ($\pi\rho\sigma\nu\delta\iota\alpha$) to physicians, for three reasons; first, for the confidence of mankind, which it will conciliate to the physician; then, because it will free the practitioner from all blame if he has announced beforehand the fatal result of diseases; and further, as being a very great instrument in effecting the cure."—*Sydenham Society's Edition*, by Adams.

Fever and pneumonia are examples of this. It is in these and other acute diseases that the most complex and interesting problems in prognosis arise,—in fever more especially; and in directing your attention, during the ensuing session, to this portion of its clinical study you will find that the following conditions will, with others, enter into the determination of the result of any given case, of Fever:—

The dose of the "Fever Poison," as ascertained by its primary action on the blood and nervous system; sometimes killing both, as it were "uno ictu."—The condition of the vital organs at the time of seizure.—The condition of the function of nutrition with reference to the balance between waste and repair of the whole system, or of any particular organ.—The age and occupation of the patient.—The moral, mental, and hygienic conditions of the patient. —The previous medical treatment, if any.—The conformity of the symptoms, as ascertained by daily observation, to the typical symptoms of the disease at the given period; it being an axiom of prognosis, that any exceptional or unusual phenomenon is of evil augury in a case of fever.—The existence of symptoms indicating a tendency to one or other mode of termination at or about the critical period.

You thus see that caution and comprehensiveness are required in estimating tendencies and anticipating results, no less than in explaining phenomena; and in order to guard against presumption, arising from over-confidence in your prognostic powers, it will be well that you bear in mind the axiom that, "Inasmuch as it is impossible to predict anything with certainty so long as we remain unacquainted with any of the conditions of a given event, and ignorant how they may act in a particular instance; and, *as in the living body, no occurrence or effect ever does take place constantly, or in the same manner at all times; we can never with perfect certainty predict any occurrence or effect whatever.*"[1]

The steps in the progressive method which I have recounted, have for their practical object—the crowning step to which they all lead—the discovery of the indications of treatment. In the same manner that the student is apt to confound the fact with the inference, and to state a presumed pathological condition when he should, logically speaking, state only its signs, he sometimes falls into the mistake of confounding the thing to be done with the

[1] Œsterlen, Medical Logic, p. 129.

means by which it may be effected; thus, in both cases, passing over the important intermediate term, and jumping at a foregone conclusion. "Truth," said Curran, in one of his finest speeches, "is to be sought by slow and painful progress, while error is, in its nature, flippant and compendious: hopping with fastidious levity over proofs and arguments, it perches on assertion, which it calls conclusion." This error arises from the innate dislike we all feel to slow and laborious processes of thought, if we can reach the object by a shorter and easier method. Empirical facts are substituted for principles, and instead of reasoning out the indications, the student too often, merely collects formulæ, copying the prescription at the bedside without reference to the views of the prescriber.

I would by no means wish you to understand that your are not to pay the closest attention to the effects of remedies in the cases under your observation, or that you may neglect to acquire an intimate knowledge of medicines and medical appliances in the fullest sense of the term; but this should be done with a constant reference to the indications to be fulfilled. What condition are we called upon to treat? What symptom demands our attention? What function is to be influenced, and in what direction? Is this to be attempted by direct or by indirect means? and with what ultimate object? are questions which, in rational or scientific medicine, must always precede the selection of our remedies.

The student who undertakes the care, observation, and reporting of cases in the hospital, is called upon, as it were, to defend his thesis in the presence of his fellow students. Having read his report, and submitted it to the necessary correction of his teacher, he is called upon for his explanation of the phenomena; for the process of induction (generally by the method of exclusion) by which he arrives at his diagnosis; and he is usually asked to state the indications of treatment which a consideration of the entire case may suggest to his mind. By this method, his faculties are educated in the truest sense (*i.e.*, drawn out), his intellectual powers are awakened and developed, and so trained by the exercise of observation, comparison, reflection, and suggestion, as to form what has been aptly termed "medical mind."

But, gentlemen, I must caution you who are entering on your studies against the mistake of supposing that this progression, however clear and satisfactory, is an easy one, or one to which the method of some sciences you may have studied can be exactly applied. Medicine is not an exact science in any sense of the term; nor is it, as has been said of theology, built up of "truths crystalline in form and character, unchangeable, and unchanged." In estimating its phenomena and their results, there are, as you must have seen, too many probabilities to be balanced—too many relations (in the case of each patient) to be determined—with nature, with other individuals, and with himself. There are, also, too many contradictions to be reconciled; and too many limitations and qualifications to be placed on the meaning of its terms. You cannot march from step to step on a level and measured track, as you can in mathematics and astronomy. You cannot test every step of your progress by analysis, as in the chemical laboratory. Medicine, in its progress, is rather like an advance towards some goal unseen and far distant, along a path as yet but imperfectly indicated, through a region as yet but imperfectly surveyed. The path is devious and uncertain. The disciple is often drawn aside from it by friendly sciences, even when they proffer their assistance as his guides. He meets sometimes with the absence of landmarks; the presence of such as can be but of partial use; and sometimes with false lights which tend to lead him astray. Many an hypothesis, which has a foundation in fact, he must remove from the way, while at the same time, he may appropriate some of its material. Such an one was Broussais' theory of fever; fallacious, but yet according with many facts. Again, he meets with some lofty eminence which he must scale, in the shape of some great pathological problem, which is, or seems to be irreconcilable with his present comprehension of disease, but which, if surmounted, will contribute largely to the advancement of his science, by substituting for shadowy, symptomatic outlines, the solid, definite, visible, and tangible phenomena which it reveals.

Many men and many minds must co-operate in this progress, to the aid of which numerous arts and sciences contribute their assistance. Indeed it may be said, that no advance is now made in any science, more especially in those of biology and chemistry, without its influence being felt in the progress of medicine. Among the appliances they have furnished I may mention the stethoscope, the microscope, the laryngoscope, the opthalmoscope, the speculum, the endoscope, the scales, the thermometer, the entire laboratory. Armed with these, the machinery of knowledge, different minds go forth to labour. There are the pioneers of science, men in advance of their age, but who from this cause fail to secure the co-operation of their fellows, and that accordance of the facts and ideas of others with their own, which is termed consilience in the process of induction. Then there are observers of facts, who furnish the material for the road; and there are the reasoners, who apply to the pathological heights of which I spoke—the ladder, of which Bacon said: "There will be good hope for the sciences then, and not till then, when, by a true scale, or ladder, and by successive steps, following continuously without gaps or breaks, men shall ascend from particulars to the narrow propositions, from those to the intermediate ones, rising in order one above another, and at last to the most general." Standing on these heights of pathology, now and then, a master-mind—a Hunter or a Laennec, a Bright or a Virchow—far-seeing, scans the horizon of truth, and points the way from height to height towards future discovery, realizing in some degree, the noble eulogy on Bacon of the poet Cowley:—

> "From these and all long errors of the way,
>   In which our wandering predecessors went,
> And like the old Hebrews, many years did stray
>   In deserts but of small extent,
> Bacon, like Moses, led us forth at last.
> The barren wilderness he passed;
> Did on the very border stand
> Of the blest promised land,
> And from the mountain-top of his exalted wit,
> Saw it himself, and showed us it."

To those of you, whose ambition is to tread the path of original investigation, medicine offers a rich and tempting field of labour. The subject of fever, for example, to which I am about to invite your attention in the ensuing lectures, is one which has as yet been but imperfectly cultivated. An immense mass of facts and observations has been accumulated, and a great advance has been made towards the elucidation of its natural history, pathology, and classification; but much remains for future labourers to do, and many minds will be employed in the development of its latent processes, and in determining their numerous and complex relations to healthy and to otherwise diseased functions. To this study, I invite you to bring all your powers of patient and accurate observation and close reasoning, and your most extended knowledge of physiology and pathology, in the assurance that you will here find their profitable exercise; and I would also advise you to apply the method which I have endeavoured to illustrate in the foregoing pages to your daily bedside study of this disease.

LECTURE I.

THEORY OF FEVER.

IN recommending you to take advantage of the present opportunity of studying the different forms of fever in our wards, I would observe, that fever is the disease which beyond all others presents the strongest claims to your careful investigation, whether regarded as an object of scientific interest, or of social importance; whether on account of the varied and complex phenomena it offers for our investigation, or the immense influence which it exercises upon the welfare of individuals and of the community.

To the thoughtful student fever presents an object of contemplation around which the inferences from his past experiences and his new acquirements may continually accumulate, and which will thus grow with his daily observations, and ripen with his maturer judgment. More than any other disease, it calls into requisition his knowledge of healthy and perverted functions; it educates his faculty of observation, it cultivates his talent for judging the value of existing signs and symptoms, and of foreseeing the course of the disease, and providing against future contingencies; and, more than any other, it will put to the test his therapeutic resources, his fertility in suggesting expedients, and his tact in the management of the patient, and in regulating the surrounding conditions. In a word, no single disease will teach the student so much that is of value, or so train him for the future practice of his profession, as fever.

But it must be admitted that the difficulty of the study is in proportion to its interest. Of the cases of this complex disease which you will meet in the wards, perhaps no two will be exactly alike, and, at different periods, you will meet with different groups of phenomena, of which the one will throw no light on the study of the other. A similar want of correspondence will exist between your own observed cases, and the descriptions of many writers of monographs on fever, the best of these being descriptions of some epidemic, which perhaps may not correspond in its characteristics with the one under your own observation. In short, if the clearness of our views of a subject is, as we believe, in direct proportion to the comprehensiveness of our knowledge of its nature and relations, there is every reason why fever should be that disease regarding which the most confused ideas are entertained by students in general, and by not a few members of the profession. This will appear from the fact that in no disease has so much controversy existed with reference to its nature, classification, origin, diffusion, pathological anatomy, or treatment, as in fever.

If this be so, the student should endeavour, by all means, to enlarge his view of the subject by looking at it from various stand points, and with regard to its various relations: studying its analogies with other cognate diseases, and comparing the several functions deranged by fever with the same functions in their healthy condition, so as thus to build up synthetically in his mind, a concrete idea of this complex series of phenomena, at the same time that he is engaged in the daily analysis of these phenomena at the bedside.

You will find that by conjoining these two modes of study, each enlightening the other, what at first appears a confused and contradictory subject, will become gradually clear and in harmony with the laws of physiology in health and disease.

With a view to aid you to the extent of my ability in this comprehensive study of fever, I propose to supplement our bedside observations by a few lectures of a *general* character, not strictly clinical, but making occasional reference to such cases as may present themselves in our wards, for the purpose of illustration of the doctrines and practice which I shall advance here.

In this course I shall consider fever, using the term in a general sense, as an object of study regarded from different points of view :—

1st. As one of the group of morbid poisons; identified with them by similarity of origin, and by conformity to certain laws; like them received into the blood, affecting the entire organism, and exerting its influence on different organs in succession; like them having its periods or stages of incubation, maturity, and decline; and like them reproducing itself in the blood of the affected person.

2d. As requiring a susceptibility to its influence in the individual exposed to it; and as varying in its operations according to the nature, seat, and extent of this susceptibility or predisposition.

3d. As generated within the living body "an animal miasm," or derived from without it by exposure to some forms of malaria—"civic miasm"—or arising from epidemic influence; each several source giving origin to a form of fever essentially distinct from all other forms of the disease.

4th. As presenting pathological phenomena, *primary or essential*, arising out of the affinity of the poison for certain tissues or organs, as well as out of the special predisposition of the organ: *secondary or accidental*, arising—(a) from some peculiar exciting cause; (b) from secondary blood contamination; or (c) from reactive irritation set up during the course of the fever.

5th. As having a more or less definite course and duration, and presenting certain phenomena at its resolution or crisis; and as exhibiting certain differences in the modes of its fatal termination, and its morbid anatomy, in different forms of the disease.

6th. As presenting data for a rational prognosis, or study of its tendencies towards a particular termination.

7th. As presenting rational, general indications of treatment and management common to all forms of fever, and special indications for different forms and complications.

We shall thus study the doctrinal or theoretical, as well as the clinical or practical bearings of the subject.

The question what is fever, or the theory of fever, is a subject which cannot be passed over in this study, and which yet we must approach with diffidence, and discuss with no little reserve. It has been well said that "such a theory presupposes and involves a knowledge of the intimate processes and relations of the living powers which has not yet been attained. It requires also a knowledge of the agency, or combination of agencies, which sets in motion that concatenation of disordered actions—that complex combination of morbid processes which constitutes fever. We must also know the seat and character of these morbid processes and modifications; their peculiarities; their tendencies; the differences which exist between them in the several forms of febrile disease. We must know their relations to each other—which are primary and essential, which secondary and accidental. Of these things, as of the causes of fever and their mode of action, we are, to a great extent, ignorant; they are known to us rather *analogically*, and by comparison with other morbid processes, than absolutely and positively." (Parkes.)

The morbid processes to which fever presents most striking analogies, are those set up in the body by the action of morbid poisons, or what are termed zymotic diseases, among which it is accordingly classed. The principal analogies to these may be thus stated:—

Fever, like other morbid poisons, has a latent period, or stage of incubation, during which none of the febrile phenomena are manifestly set up, but at the expiration of which its influence is felt by the entire organism.

Like other poisons, its influence seems directly proportioned to its intensity, and, like them, when this is such as to kill at once, there is the least post-mortem evidence of its action, the morbid appearances being found in such cases, not in the parenchyma of organs, but in the confluence, so to speak, between the blood and the tissues, and consisting of congestion or stagnation of blood in an altered condition; like other poisons, it displays peculiar elective affinities, these being modified, in some cases, and directed to certain tissues or organs, by their existing condition with regard to nutrition; it being well ascertained that parts undergoing active retrograde metamorphosis are peculiarly subject to the operations of morbid poisons. As with them its operations may be modified by temperament, or family and personal peculiarities, by epidemic influence, by the coexistence in the blood of other poisons, as well as by the contaminations originating in the body of the patient from defective elimination or retained secretions. In like manner it is reproduced in the blood as they are, while the susceptibility to its influence seems to become exhausted by the attack, the susceptibility to the influence of any other morbid poison remaining undiminished; and, lastly, like them it has a natural tendency to cure through the elimination of its products by critical evacuations in patients in whom there exists a healthy condition of the excreting organs.

All our observations tend to prove that the direct action of the poison of fever (and of all morbid poisons) is that of catalysis; in other words, that the poison has a force or power by which a compound body like the blood is decomposed by mere contact with it. If we look for the portion of the organism the most adapted by its nature, constitution, and functions, to become the subject of such catalytic change, we surely find it in the blood, in which the chemical and vital forces are in such perfect equilibrium that a change in it is effected by every disturbance, however trifling, or from whatever cause it may proceed. Moreover, we find that in patients killed by the direct action of an intense fever poison, the blood is fluid, and otherwise changed in character. Again, if we meet with cases of peculiar proneness to receive the infection of fever, or other zymotic diseases, we find the cause in a change in the blood, whether produced by over-crowding, or the breathing putrid emanations, or famine, or the decomposing action of water applied to the surface of the body, by which a breaking up and solution of the blood-corpuscles is effected; and if we examine the worst complications of fever, we find that they are the result of decompositions set up by the presence in the blood of the products of retrograde metamorphosis of tissue, such products giving rise to the worst forms of secondary inflammations which occur in fever.

Much labour has been expended in the controversy between the Solidists and Humoralists whether the *primary* action of the fever poison is exerted upon the nervous system or the blood. I have elsewhere examined the arguments for both theories, and expressed the conviction which I still entertain, that the phenomena of the latent period, or stage of incubation, and those of reproduction of the same poison, both point alike to the blood as the seat of the operations of the fever poison, while it is through the conjoined agency of the nervous system and the blood, that the phenomena of the formed disease are produced. I have just now mentioned, as one of the laws of morbid poisons, that these affect the entire organism, using the word in the comprehensive sense of the aggregate of the molecules of which the body consists. We have every reason to believe that this holds good of fever, and that the first step in the fever process is a modification of the function of nutrition, "of which function," says Mr. Paget, "the tissues and the blood are the principal factors in their mutual relations." I believe that it is in the alteration of this relation between blood and tissue that fever commences: here is the seat of its operations, not in the parenchyma of organs, but in the confluence between the blood and the tissues. If so, the first impression of the fever poison must be upon the blood into which it is received, in which it is rapidly augmented, and through which it is diffused. It is still a moot question whether this change in the attractions between the blood and tissues is immediate and direct, or is effected by the medium of the nervous system, more especially of the 8th pair, Virchow's definition of a theory of fever being—"Fever consists essentially in elevation of temperature, which must rise from an increased tissue change, and have its immediate cause in alterations of the nervous system." This is the theory which, put forward by Virchow, has been so clearly expounded by Dr. Parkes. I believe there is no good evidence of any such alteration during the stage of incubation, or up to the period of the initiatory rigor, which may be regarded as the first symptom of the moderating nervous centres becoming influenced by the fever poison. (*Appendix I.*)

But in all the phenomena which succeed the invasion of fever, the blood and nervous system are joint-factors. "As the blood," says Claude Bernard, "really describes a circle, the diastaltic nervous system describes a cycloid. As the blood diffuses its atoms into every minute space of the system, the diastaltic spinal system extends its influence over every one of these atoms. The blood undergoes its changes in the methæmatous or blood changing (or capillary) channels, placed between the alternate branches of the arteries and the incipient roots of the veins; to the same point the diastaltic spinal system extends its wondrous influence." In accordance with this, we find those functions deranged in fever which, like animal heat, are the product of the joint influence of these factors; and we find, moreover, that such derangements will vary in amount with the extent to which the nervous system, or the blood, or the molecular nutrition of an organ may be severally affected.

Of these derangements, usually the first to attract notice is diminution of nervo-muscular force. Even in the most insidious cases, this has been observed in the form of mental hebetude or dulness, and bodily languor, dizziness, feelings of fatigue, depressed spirits, or irritability of temper, increased sensibility to external impressions, as light, sound, and temperature (horripilatio). Even at this early period the respiration is disturbed, laboured breathing being a very early symptom, owing, probably, in part to the diminished tonicity of the right ventricle, but chiefly to the influence of the fever poison upon the eighth pair of nerves. Most, if not all of these symptoms, may be usually observed before the commencement of the stage of reaction; and in some rare cases, in which either the blood has been subjected to the action of an unusually intense poison, or the nervous system has been depressed by some severe shock or strong emotion, fever has begun and ended with these symptoms, and the stage of reaction has never succeeded at all; or if it has, the reaction has been imperfect, and the symptoms throughout have been those of depression of the vital energies, characterized especially by failure of mental and motor power, sinking of the temperature, congestion of the right side of the heart, and of the capillaries of all the viscera. Numerous cases are recorded of fever of this type, which terminated fatally at periods varying from six hours to three days.

This form of the disease has been admirably described by Armstrong in his chapter on congestive typhus. The following is one of the cases detailed by him: "A young man of slender make, had travelled several miles to see a relative sick of typhus, with whom he remained, and upon whom he attended several days. On returning home, he was suddenly attacked with vertigo, chilliness, sickness, and extreme weakness of the lower extremities, and when he reached his own house he appeared most strangely confused in his head and intellect, staggering and talking like one intoxicated. He was immediately put to bed, and did not complain much afterwards, but gradually fell into a profound coma, in which he lay without motion. At my first visit," says Dr. Armstrong, "the face was then pale, and somewhat livid; the breathing deep and impeded; the pulse small, frequent, and irregular; the tongue white, and covered with slimy saliva; the skin dingy, and partially damp; the heat of which felt nearly natural over the heart and belly, but the extremities were rather cold. The head was shaved and blistered without loss of time; mustard sinapisms were applied to the feet, and large cathartic injections repeatedly administered. In about twelve hours, when the bowels had been often and copiously moved, the patient gave some indications of returning sensibility; and in a few hours more looked up, and even recognized some intimate friends who were present, but spoke in a feeble and faltering accent. For several hours there was an appearance of improvement in some particulars; his pulse and breathing freer, his voice became more natural, and his skin of a warmer glow; but in contravention to these favourable symptoms, his hands were tremulous, his tongue fouler, and there were a few dark petechiæ scattered over the trunk and arms. Moderate portions of wine were now recommeded at short intervals, with a view to support his strength, but the debility increased under this plan, and he again sunk into a deep stupor, in which he expired slightly convulsed, about forty hours from the first attack." Several other similar cases are graphically narrated by Dr. Armstrong; the appearances after death being in each, a turgid condition of the veins of the pia mater and substance of the brain; bloody serum in the ventricles; the lungs, liver, and spleen congested with grumous blood; and the two latter viscera so softened in one case as to be ruptured by handling. Dr. Armstrong justly remarks that in these cases the stage of excitement never emerges at all without the interference of art, or does so very imperfectly; the energies of the system being either nearly extinguished, or so much oppressed as to be unable to create an universal excitement. He adds, "Sometimes, however, patients do not rapidly sink under the first shock of the attack; but that being passed over, they linger for many days in a state of stupor, or mental indifference, and die, at last, with a foul dark tongue, shrivelled cool skin, and deep sunken countenance. The open forms of fever, in which heat and arterial reaction are equably developed, will be found the least dangerous; and those of an obscure nature, in which neither heat nor arterial reaction are equably developed, the most perilous and unmanageable."

Fortunately, these cases are rare and exceptional. As a rule, reaction speedily follows upon the stage of depression, when, fever being established, you will observe the evidences of increased regressive metamorphosis of tissue in the marked rise of the temperature, the acceleration of the pulse and respiration, and in the notable increase of the excretion of urea—the product of this metamorphosis. Two effects of this change in the molecular nutrition follow, which have great influence upon the future progress of fever, more especially in its advanced stage; one is the alteration of the attraction between blood and tissue, upon which the capillary circulation depends, leading to derangements of this function to some extent general, but more especially in the organs which may be undergoing the most active disintegration at the access of the fever; the other is the disturbance of the balance between disintegration and excretion, leading to the accumulation in the blood of a secondary poison, the action of which is more or less that of a fever poison, or capable, apparently, of entering into combination with it. This secondary blood contamination you will find to be a fruitful source of the most serious nervous symptoms, and secondary inflammatory affections. It may present itself as simply the result of excess of disintegration over secretion, or as the retention or suppression of an excretion. The first form we meet in the over-fed, in the spirit drinker, and in those who have induced excessive disintegration of their nervous and muscular tissue by fatigue at, or subsequent to, the period of access of fever; the other, may occur from previous disease of the lung, liver, or kidney, or from congestion produced by cold, when this is the exciting cause of fever, or in the case of suppressed menstruation from a similar cause, or a mental shock occurring at the same period. Examples of both forms will occur to us hereafter.

Efforts at elimination of the morbid fever poison would seem to be made at different periods, and in different modes, in the various forms of the disease. Some would consider the initiatory rigor, and subsequent increase of heat, as such an effort. (?) There may be reason to view, in this light, some phenomena which we refer to the elective affinities of these poisons. Thus I should regard as eliminative efforts, not only the early vomiting of variola, scarlatina, and typhus, and the diarrhœa of typhoid, but also the coryza and catarrh of measles, the irritative cough of the early stages of typhus, and the eruption of all the exanthemata.

Crisis we of course regard as the great and final effort at elimination. Up to this period the products of the decompositions set up by the action of fever, go on increasing, the disturbance of the balance between metamorphosis of tissues and excretion increasing in like proportion, and with it, the danger of secondary blood contamination and its consequences. The recovery of the patient will depend upon his surviving these dangers, with the excreting organs still capable of eliminating the large amount "of partially metamorphosed substances, which," to use the words of Dr. Parkes, "have been retained in the body, and at length have been brought to that point of oxidation or change which permits their elimination by one or other organ.

We shall meet with abundant examples of the important part which retained effete material plays in the production of secondary complications in fever, as well as of their great influence upon our treatment during its course, and at its final stage.

Indeed, we have at present, in our wards, two most illustrative cases. In one of these, "Susan Williams," who came under my care on the ninth day of typhus, of a remarkably asthenic type, with fœtal sounds of the heart, pulse 152 in the minute, and with the surface thickly covered with dark maculæ. Symptoms of pulmonary congestion manifested themselves on the thirteenth day, and, on examination of the chest, pneumonia, engaging the lower lobe of the right lung, was found. Under the use of turpentine, given internally, and of turpentine epithems externally, aided by a liberal allowance of brandy, a favourable crisis occurred between the fourteenth and fifteenth days, and along with it rapid resolution of the pneumonia, the dulness on percussion and crepitus quickly disappearing.

We witnessed precisely similar occurrences in the case of the man "Hawkins," who suffered an attack of pneumonia, with the usual physical signs and well-marked rusty sputa, about the commencement of the third week of typhoid. In both these cases we believe that, owing to the altered attractions between the blood and pulmonary tissue, as well as to the diminished energy of the nerves of respiration, pulmonary congestion occurred. At an advanced period of the fever, there was set up in the seat of this stasis of depraved blood a low form of pneumonia, recognized in both cases by the ordinary physical signs of that affection. I need not dwell upon the importance of this condition, or its influence on the result if overlooked.

To recapitulate briefly: The order in which the operations of the fever poison occur are—1. Its primary toxic action on the blood, and, through this medium, upon each molecule of the body. 2. A form of inhibitory paresis of the moderating nervous centres, and more especially of those nerves connected with the heart, lungs, and stomach. 3. Increased disintegration of tissue, with increase of animal heat. 4. Derangements of capillary circulation. 5. Derangements of secretion. 6. Accumulation of the products of disintegration in the blood. 7. The phenomena of elimination and of crisis.

The clinical study of fever may be said to consist chiefly in the daily observation of these progressive changes, set up by the action, direct or indirect, of the fever poison; affecting the due performance of the several functions, and the physical condition of the organs themselves. Your success in this study will depend much upon your knowledge of the physiology of these functions, and of the manner in which they are influenced by the action of the fever poison; and by their deranged condition influence, in turn, the course and termination of the disease. You should daily make a careful examination of the several organs, and you will, in your investigation, bear in mind two facts: The essential nature of fever, and its true relation to the local lesion; and the influence of the existing state of nutrition of each organ in determining to it; in a greater or lesser degree, the special action of the fever poison.

This subject will come under our consideration in my next Lecture.

LECTURE II.

ETIOLOGY OF FEVER.

I HAVE to-day to draw your attention to the etiology of fever, or to those conditions within or without the body upon which its origin and diffusion depend.

The *essential* cause of fever is no doubt the presence of the poison in the blood, by the catalytic action of which the phenomena of the disease are set up.

The source from which this poison is derived is, properly speaking, the exciting cause of the fever, as inoculation may be said to be an exciting cause of variola; contagion, that of typhus, &c. But inasmuch as many persons exposed to the influence of these exciting causes escape the disease, while others, under a scarcely appreciable amount of exposure are attacked; we seek within the body for some reason for such a difference in degree of susceptibility, and we recognize a condition which we term predisposition; sometimes a personal or family peculiarity—an idiosyncrasy—but generally acquired, and due to certain causes, which we call predisposing causes. In a few instances these are found to act with such intensity as to generate a fever poison without the aid of any exciting or extraneous cause; but in the majority of cases, both the predisposing and exciting causes will be found to co-operate in the result. I shall first consider the nature of predisposition, and the mode of action of predisposing causes, and afterwards the exciting causes.

In my last lecture I endeavoured to illustrate, rather than attempted to explain, the nature and modus operandi of the poison of fever, by showing its analogy to other morbid poisons, and its conformity to their laws.

It appeared that, like them, it is introduced into the blood, upon which its primary action is produced, and in which, or in some material present therein, it finds its nidus and the substance capable of being converted into it, and out of which consequently it is reproduced.

Further, we found that it is carried, in this substance, to the surfaces destined by nature to separate these different materials from the blood, and there sets up certain specific irritations and gives rise to deposits, all of which are regarded by many as efforts at elimination depending upon the elective affinities or attractions of the fever poison.

Perhaps a more correct expression would be, that each form of fever has a special attraction for some specific material in the blood; and that the so called elective affinities of the poison, as of scarlatina for the tonsils, typhoid for the intestinal follicles, &c., are in reality examples of the attractions of this portion of the blood for these their normal excreting surfaces.

This view would seem to be supported by two considerations, to which I shall presently have more particularly to refer. The first, is the fact that susceptibility to the action of each morbid poison is, as a rule, exhausted by the attack, owing, we believe, to the material being exhausted upon which the action of the poison is exerted, while the susceptibility to other poisons remains the same as before. The second, is the fact that fever, in common with other morbid poisons, manifests a special attraction for whatever organ may happen to be undergoing retrograde metamorphosis at the time of seizure, as if it found in this organ its nidus and the material for its reproduction.

The fact that the aptitude is exhausted, as a rule, by an attack of the disease, is well known in regard to the exanthemata, and is generally believed to hold true of typhus. I have attended many persons in different forms of fever at different times, but I have never yet attended the same patient in two attacks of typhus; nor have I ever had satisfactory evidence of a typhus patient having suffered a previous attack of the disease. Of the disappearance from the blood of the material capable of conversion by the catalytic action of the poison, Mr. Simon remarks, "that the preparatory and permissive state is a peculiar chemical state dependent on the presence or the excess in the blood of a material *convertible into identity with the poison;* that the poison thus augmented endeavours to eliminate itself by surfaces, the choice of which is a distinctive mark of each poison respectively; that for a greater or less time after the fulfilment of this eliminative process, *the susceptibility of the disease is exhausted;* and finally, that the severity of the disease in each instance will depend, not on exterior circumstances, but on internal and personal conditions; on the patient's own possession, within the stream of his circulation, of a larger or less abundance of that specific material which constitutes on the one hand his susceptibility to infection, and, on the other, his power of expressing the disease. He must necessarily evolve symptoms in proportion to his richness in that which furnishes their material."—*Lectures on General Pathology.*

Not less remarkable is the second fact, that the aptitude or susceptibility is exhausted only in the case of the poison whose attack has been suffered, and that the susceptibility to all others remains unaffected.

This is beyond all question. Scarlatina affords no protection from typhus, or variola, or measles; nor does an attack of typhus protect from any other form of fever. It seems to me, that we cannot resist the conclusion that each of these poisons finds its own specific material in the blood; each distinct and different from the others.

What is the nature of this essential predisponent we know not, but it seems intimately associated with the presence in the blood of the products of forward and retrograde metamorphosis of tissue, particularly the latter; and it will appear that predisposing causes act by virtue of their power to increase this metamorphosis, either in the system generally, or in some special organ in particular; or by hindering the elimination from the blood of these products, and so retaining them in that state of imperfect oxidation which is favourable to their becoming the subjects of zymotic change. We may refer to fatigue and mental emotion as acting in the former mode; and to ochlesis, renal disease, and arrested menstruation, as acting in the latter.

Such is an outline of the theory of predisposition to the action of the fever poison; whether general, consisting in an increased susceptibility to its influence; or special, acting by determining the action of the poison to some one organ in particular, and so modifying its primary complications. Strictly speaking, this is, to use the words of Mr. Simon, "not theory, but the simple generalization of facts which lie within the sphere of our daily observation."

The student who wishes to thoroughly comprehend the phenomena of fever, will do well to study the nature and modus operandi of what are termed its predisposing causes; their influence upon the blood and nervous system, and their power of modifying or neutralizing the vital resistance to the action of the poison upon the organism.

I have mentioned that the condition of the organism which constitutes aptitude or susceptibility, is connected with the presence in it, from time to time, of the products of forward and retrograde metamorphosis of tissue; and accordingly, we meet with the greatest degree of aptitude, simply considered, in the young, in whom this metamorphosis is most active, and the balance between the forward and regressive changes most easily disturbed.

But although we say the young are susceptible of the action of the poison, we do not imply that they are predisposed, in the more strict sense of the term. Predisposition to fever supposes something which interferes with the elimination of the poison before it has set up its specific action; and it supposes a proneness in the blood, or some material in it, to become subject to this action. Let me illustrate my meaning by an example. A young man has been exposed to the emanations from a fever patient: after exposure he becomes intoxicated. The action of alcohol is to retard elimination, both from the lungs and from the kidneys, by checking the excretion of carbonic acid and of urea; susceptibility has thus increased to proneness; and this will be still further augmented, if to intoxication be added exposure to cold or moisture while intoxicated, or in the subsequent stage of depression; which acts as a further check to elimination. We know that under this combination of intoxication and exposure, the blood suffers decomposition, often to the extent of necræmia or death of a large portion of that fluid, as, for example, in gangrene of the lungs. (*Appendix II.*)

Take another example of this gradation in degree of susceptibility. The female, at the menstrual period, is known to be susceptible to zymotic diseases in a more than ordinary degree; but if menstruation be checked by cold or a mental shock, at the time of her exposure to infection, she becomes predisposed in a high degree; and in a still higher, if this exposure occur about the puerperal period. In the one case, that of arrested menstruation, a retained excretion furnishes the predisponent, illustrating an observation of Dr. Todd, "the uterus may be, and often is, a source of contamination of the blood;" while in the other—the puerperal state—the rapid disintegration of the uterus leads to the accumulation in the blood of effete matter in a state of progress to decomposition, while it determines the special action of the poison to the organ undergoing such disintegration.

This explanation of the peculiar proclivity of the puerperal female, we owe to Dr. Carpenter, who offers, also, a similar explanation of the action of two other predisposing causes, overcrowding and famine.

When I come to enumerate the exciting causes of fever, I shall adduce instances in proof of the power of ochlesis (or the poison generated, in the emanations arising from crowded collections of human beings) to cause fever, without the introduction of extraneous infection; and of a similar power to cause another form of fever in the exhalations from obstructed sewers, cesspools, &c. Inasmuch, however, as both act by producing in the blood an accumulation of matter in a state of progress to decomposition, they must be considered powerful predisposing agencies; and such, in fact, they are, not to fever only, but to all zymotic diseases; especially to epidemic cholera and scarlatina. Did our time permit, I could narrate many instances of the havoc produced by cholera in crowded barracks, schools, and prisons, in England and in India; and I could give you similar examples of the attraction which the same disease manifested, on each epidemic visitation, for those towns, or streets, or dwellings, in which the defective state of the sewerage, or the impurity of the water, furnished a predisposing agency.

Precisely the same may be said of scarlatina, which spreads with extraordinary rapidity and fatality in the overcrowded dormitories of schools, and also in the localities in which civic miasma abounds.

One of the proofs of this action of malaria as a predisponent to fever, is the long period of undefined ill health which often immediately precedes the invasion of fever in a person living in this atmosphere. I have known such a state to continue for six weeks before formed disease set in. During the entire time, the patient suffered from capricious appetite and occasional diarrhœa, and had daily copious deposits of lithates in the urine. I recently attended a case of typhoid in this city, caused, as there was every reason to believe, by defective sewerage, in which gastric derangement, headache, and languor preceded the attack of fever for at least three weeks; the fever itself running a course of four weeks' duration. A still more decisive proof is afforded by the observations of Dr. Potter, quoted by Dr. Tweedie. He found that the blood drawn from five persons, apparently healthy, who lived in an infectious district, in which malarious fever was prevalent, presented the same characters as that drawn from the fever patients (yellow serum and broken-up corpuscles); while the blood of five other persons, residing in a healthy part of the country, presented no such appearance. Again: he bled a young gentleman who had just returned from the country, and found the blood healthy; after fourteen days' residence in the infectious district he repeated the operation, and found the same appearances as in the fever patients; of the six persons whose blood had undergone the above change, four were shortly after seized with fever, the other two had headache, nausea, and other indications of disease, without formed fever.

I think it is impossible to resist the inference that the condition of the blood, constituting a predisponent to fever, existed in these cases.

On the influence of famine, Dr. Carpenter says: "We have not merely that general depression of the vital powers, which is a predisposing cause of almost any kind of malady, and pre-eminently so of zymotic diseases, but also the presence of a large amount of disintegrating matter in the blood and general system, which forms the most favourable nidus possible for the reception and multiplication of such poisons. And thus it happens that pestilential diseases most certainly follow in the wake of a famine, and carry off a far greater number than perish from actual starvation."

Fatigue is a powerful predisposing cause, and of its mode of action, by promoting increased disintegration of tissue, and consequent accumulation of its products in the blood, there can be no question. These effects frequently continue marked throughout the fever. Patients in whom muscular fatigue has been a predisponent, usually present the most formidable forms of secondary blood contamination, and the most remarkable lesions of the nervo-muscular functions, as violent tremor, jerking subsultus, spastic rigidity, and convulsions. Some of you may remember the case of the cattle dealer from Dundrum, who was admitted into this hospital about two years since, and who presented all these nervous symptoms in the highest possible degree, the jerking subsultus and rigidity being most remarkable. We found, on inquiry, that subsequently to the rigor which ushered in fever, this man had walked a distance of twelve miles to attend a fair, and having transacted his business, had walked several miles more.

Many similar instances of the connection between muscular exercise and a special form of nervous symptoms have occurred under our observation in these wards.

In Bright's disease, we have a typical example of the influence of certain blood disorders as predisponents; its modus operandi being the retention in the blood of nitrogenous matter in a state of progress to decomposition, which appears to act as a ferment capable of setting up in the blood phenomena, so closely resembling typhoid fever as to be not unfrequently mistaken for it.

Of the influence of prolonged exposure to cold and moisture in producing, in a lesser degree of intensity, increase of regressive metamorphosis of tissue; and in its higher degree, rapid decomposition of the blood itself, we have abundant proofs. Of the former, the experiments of Lehmann, Homolle, and others, furnish examples. Lehmann found that the effect of remaining in a cold sitz bath for fifteen minutes, while fasting, was to cause an appreciable loss of weight, and an increase in the amount of urine. The increment in the mere water of the urine amounted to 71 per cent., and that of the urea to 29 per cent.

In a higher degree, and more especially if aided by such causes as retard the elimination of the effete matters from the blood, such as alcohol, the consequences of immersion are not unfrequently fever, of an intensely septic type. Several instances of this have occurred under my observation. One patient, a young man, was employed by a gentleman, for several hours of a summer day, in wading along a river in search of young wild ducks. Fever followed, marked by oppressed breathing and extremely rapid pulse. In a few days his skin was covered with petechiæ and vibices; he had uncontrollable epistaxis and hæmorrhage from the gums, from the lungs, bowels, and bladder. In another fatal case, the patient, a young man of very intemperate habits, had passed the day in the water, often to his hips, fishing; after his return home he had a severe rigor, followed by intense heat of skin, stitches in the chest and pains in the abdomen; in a few days a purple typhus rash appeared over the body, followed by numerous boils on the thighs and arms, which rapidly burst, and discharged a dark tarry blood. When admitted into hospital, a week after seizure, his skin was thickly covered with black petechiæ, and he had hæmorrhage from the nose, stomach, bowels, and bladder. He died on the day following. In a very similar case, which was attended by a friend of mine, a man, of intemperate habits, had bathed in a river while partially intoxicated, and had afterwards walked a distance of eight miles to his home, where he arrived in a state of great exhaustion. In a few days fever set in, and on the second day of the fever petechiæ appeared, followed, on the third, by dark purple and black spots of a very unusual size; discharge of blood from the nose, gums, and bowels followed, and continued till his death, which occurred on the fifth day of the fever.

I think I have adduced sufficient examples to prove to you that causes which increase the amount of the products of regressive metamorphosis present in the blood from time to time, whether they act by increasing the activity of this metamorphosis, or by retarding the elimination of its products, act as predisposing causes of fever, and of its special complications.

Causes which act as predisponents, through the agency of the nervous system, are chiefly the depressing emotions of fear, grief, shame, anxiety, and the deadening influence of prolonged cold. The connection between them and the occurrence of fever in the individual subjected to them, has sometimes induced some to assign to them the power of generating fever, and to the ignoring its essential cause—a fever poison. For example, so acute and accurate an observer as Dr. Cheyne describes a class of cases having all the characters of typhus, the causes of which he states to have been "loss of property, of character, and wounded pride." "Of such patients," says Dr. C., "a great proportion die. The most remarkable part of the disease is, that it does not spread. I have no recollection of a second case of this kind of fever occurring in a family, and I have never been able to discover that the patient had been exposed to contagion. It would seem to arise solely from mental causes." In such cases I cannot but believe a source of contagion did exist, though not recognized; and I can regard the mental causes only in the light of most powerful predisponents, depressing vital energy, and neutralizing the power of resistance of the decomposing action of the fever poison, possessed by the healthy organism.

Perhaps the most remarkable instance of the action of fear on record is given by Dr. George Kennedy: "J. N., aged 26, of remarkably temperate habits, chimney-sweep, left his house in the morning, in perfect health, to assist in sweeping chimneys at Beggar's Bush Barracks. While engaged at his work he lost his balance, and was in the act of falling from the roof of a high building, when he saved himself by clinging to a spout, from which he remained suspended until his strength was exhausted before assistance arrived. On his return home, about two o'clock, he appeared quite well, detailed to his family, with perfect calmness, all the circumstances connected with the accident, but continued to recur frequently, during the course of the evening, to the providential escape he had from death. In the night, he awoke from sleep excited, and talking somewhat incoherently on the same subject. The following morning he was admitted, in high fever, with violent delirium. He lived twenty-eight days, having become completely paralyzed for some time before his death. The delirium continued from the time of his admission till a short time before he died, when consciousness returned for a little, and he recognized his wife."—*Report of Cork-street Fever Hospital*, 1839.

Here we have an instance of a predisponent, acting specially upon the great nervous centres, producing an unusual complication, paralysis.

The influence of predisposing causes and of the various forms of predisposition, does not terminate with the invasion of fever, but continues during its course, often determining the nature and seat of its complications and its mode of termination. Generally speaking, if the predisposing cause be mental or emotional, the most profound nervous symptoms will occur in the course of the fever; the seat of most active disintegration being the emotional centres of the nervous system, as instanced in the case I have just read. If it be an arrest of menstruation, the most prominent complication will be the hysterical, attended with marked alteration of the temperature. If Bright's disease, it will be uræmia.

The predisposing cause influences the result unfavourably, either by favouring decomposition of the blood, or by lessening the power of resistance to the action of the poison, or in some instances by both. In habitual spirit drinkers alcohol appears to act in both ways, by retarding elimination of effete materials and by depressing the nervous energy. Prolonged cold, more especially if conjoined with moisture, seems to have this double influence. Renal disease not only predisposes to the invasion of fever, but also to a complication in the form of secondary blood contamination, which we shall hereafter find to be of the most serious nature and usually fatal in its result.

Besides the ordinary forms of predisposition, we occasionally meet with influences which seem to act by impressing, as it were, an exceptional character upon the disease. Such are the cases of individual predisposition or idiosyncrasy, and of family forms of predisposition. Not only in some families is there a peculiar liability to fever, but to the same complications, however unusual these may be. Dr. Graves observes that in some families we find a very curious coincidence between the play of the various functions in disease as well as in health. He mentions the case of two brothers who had fever at the interval of a month, each of whom, on the ninth day of the fever and at the same hour was seized with formidable anomalous symptoms, quite out of the common course; the vital powers becoming suddenly depressed, the heart's action tumultuous and intermitting, and the surface cold. In the course of my life I have met several examples of this family idiosyncrasy. Some years ago I attended a gentleman in typhus whose nervous symptoms were unusually severe, and more especially those referable to the medulla oblongata. In particular, he had complete dysphagia, fluids being ejected from the pharynx with the force of a syringe. Several years afterwards his brother was attacked with fever, and presented precisely the same unusual group of symptoms, with complete dysphagia.

In another instance, a young gentleman suffered a very rare complication of typhus, pleuritis with effusion. Some years afterwards I attended his brother in typhoid fever, complicated also with pleuritis and effusion, the right side being attacked in both. Upon one occasion I was summoned, in great haste, to see a gentleman who had been under my care in fever, the patient and his friends being alarmed by an excess of shivering. I ventured to state that this might usher in a crisis of the fever (as, indeed, it did), when I was informed that fever had terminated in the same way in the patient's mother, and some members of his family.

Some years since, three brothers were admitted into this hospital in fever, each of whom had convulsions on the fourteenth day. In the account of the epidemic of 1847, Dr. Moss states that in one house, three cases of gangrene of the nose occurred. This, however, may have been owing to other causes than the one under consideration. The case, however, reported by Dr. G. Kennedy, of two sisters attacked at the same age with cancrum oris, is an instance in point; one occurring in 1826, the other in 1829.

These instances may serve to give you an idea of the important part which family types of constitution sometimes play in the development of fever and its complications.

I have said predisposition may be local or special as well as general; not only may the blood be in a condition peculiarly favourable to the action of the morbid poison, but certain portions of the organism may, in virtue of their existing condition, become the special seats of its action; or may manifest less power to resist this action. The overworked brain of the student is a well-known example of this condition; the nervo-muscular system of the patient who has undergone great fatigue at the period of exposure; and the uterus of the puerperal woman, afford similar examples. The same condition exists in these, viz., increased disintegration of tissue of the organ engaged. Perhaps the most conclusive of all is the patient of Dr. Budd, mentioned by Mr. Paget in his lectures on Surgical Pathology, who having fallen from the mast upon his nates, and having, soon after, been attacked by smallpox, had a thin crop of pustules everywhere except in the seat of former injury, where they were crowded as thickly as possible.

To recapitulate: We have seen that—

1. In order to produce its specific actions, or formed fever, it is necessary not only that the fever poison should be received into the blood, but that it should meet with a material in that fluid capable of becoming converted into a substance identical with itself.

2. That in a few individuals, this material would seem not to exist in the blood, in the case of one or other of the morbid poisons, to which these persons appear not to be susceptible under any amount of exposure; the poison being eliminated by the process of excretion.

3. That as a general rule, this susceptibility does exist, but would seem to be exhausted by an attack of the poison, either permanently or for a lengthened period.

4. That this susceptibility though exhausted with reference to this particular poison, remains undiminished as to others.

5. That the susceptibility is capable of being augmented in various degrees.

6. That the material in the blood on which it depends (what may be termed the essential predisponent), is probably different in each form of fever, and seems to exist in or connected with the

products of retrograde metamorphosis of tissue, increasing in amount as these are increased.

7. That the predisposing causes are of two kinds. Firstly, those which act immediately, by increasing the amount of the products of disintegration of tissue present in the blood, as ochlesis, malaria, famine, fatigue, cold, and moisture, or any other internal condition or external agency which may produce this result by checking elimination of effete matter, as by causing the retention of an excretion. Secondly, those which, by their depressing influence on the moderating nervous centres, produce, indirectly, the same effect, by lessening the vital resistance to molecular change.

8. That an organ may be said to be predisposed to become the seat of the special operations of fever in one sense, when, as in the case of an excreting surface, it is the normal outlet for the elimination of the matters upon which the poison acts; that in another sense, any organ is predisposed which is undergoing increased waste at the time of invasion of fever, and that the special or local predisposing causes act by producing this increased retrograde metamorphosis in a special organ.

9. That the nature and amount of predisposition has an important influence, not only on the genesis of fever, but also upon its course, its complications, and its result.

LECTURE III.

ETIOLOGY—EXCITING CAUSES.

At our last meeting I entered into the consideration of predisposition to the influence of the fever poison, and of what are termed predisposing causes. To-day we have to consider the exciting causes, or the sources from which the poison is derived. The study of these is important, not only from its bearing upon prophylactic or preventive measures, but also with reference to the classification of fevers. If it could be proved that there is but one source of continued fever, this fact would be a strong argument for the unity of species of fever. If, on the contrary, it should appear that the disease may arise from several causes, essentially different in their nature and effects upon the organism, we shall be led to regard these effects as constituting several distinct species of fever. These considerations are entirely distinct from, and irrespective of, the differences in symptomatology and pathology, which we observe in our clinical study of fever, and to which I am in the daily habit of inviting your attention; and I shall, accordingly, defer any argument on the vexed question of unity or diversity of species, until we have investigated that portion of the subject, and are prepared to give the due weight to considerations derived from clinical observation of the three forms of disease which we meet in our fever wards, from time to time; namely, infectious typhus, endemic enteric or typhoid fever, and the relapsing synocha, which has been epidemic for a season at different periods during the present century; the last being the famine-fever, so-called, of 1847-8-9.

In the meantime, as it seems to me that the history of fever, during a long series of years, furnishes conclusive evidence of the causal relation between all these forms of disease, and certain conditions of persons, places, and epidemic periods, I shall draw your attention to the facts bearing on the etiology of fever; first, with reference to the conditions under which it has occurred; and, secondly, to the forms of disease which have been observed from time to time under these conditions. In a review so extended, I must needs draw upon the testimony of many observers; but I shall advance nothing of which I have not myself witnessed examples in the course of a long experience, and tolerably close observation.

If, then, we institute inquiries as to the sources of fever in one of our great cities—in London or Edinburgh, Dublin or Glasgow—we find certain localities rarely, if ever, free from fever of a contagious type, as proved by its occasional deportation and diffusion elsewhere by an infected person. If we investigate the conditions under which this fever exists, we find, uniformly, the co-existence of close, ill-ventilated dwellings, crowded with human beings, want of cleanliness and of sufficient clothing, and frequently scarcity of food and of fuel; and in times of poverty, and consequent depression of strength and spirits, arising from want of employment, we see the fever, which was previously confined to a small number, spread with rapidity, and become more malignant in its type.

Again, in the same or similar localities, or in other situations where the conditions of over-crowding and wretchedness may be wanting, but in which sources of malaria exist in the forms of defective sewerage, open cesspools, half-dry sides of rivers, and such like foci of what is termed "civic miasm;" fever not unfrequently breaks out in the course of a dry summer and autumn as an endemic disease, prevailing in, but confined to the locality, seeming to be capable of deportation to a very limited extent, if at all, and manifesting a contagious property in a very feeble degree compared with the former.

Again, at different periods, coinciding usually, but not uniformly, with times of scarcity of food, fever will appear and spread with great rapidity among the poor in these crowded localities; but no longer confining itself to them, or depending on deportation from them for its diffusion elsewhere, will spread by atmospheric influence, as it would appear, rapidly over the country, breaking out almost at the same time in places far remote from each other, as, for example, in 1847, when it appeared at the same time in London, Galway, and Silesia.

It needs but a cursory review of these several conditions under which fever originates from time to time to satisfy us that they are essentially different and distinct in their nature and modes of operation.

With regard to the first of them, we shall find that the influential conditions appear to be the generation and retention of the effluvia emanating from a number of living beings; the circumstances which favour the generation of the poison being cold, want of fuel, and want of ventilation, and, in many instances, depression of mind. A number of histories of epidemics of fever thus originating in barracks, prisons, and other crowded dwellings, have been published by Palloni, Peebles, Ferriar, Murchison, and others, to which I must refer you; but one of the clearest and simplest narratives, as well as one of the earliest, given by Dr. Hunter, I shall shortly quote:—

"In the month of February, 1779," says Dr. Hunter, "I met with two examples of fever in the lodgings of some poor people whom I visited, that resembled in their symptoms the distemper which is called the jail or hospital fever.

"It appeared singular that this disease should show itself after three months of cold weather. Being, therefore, desirous of learning the circumstances upon which this depended, I neglected no opportunity of attending to similar cases. I soon found a sufficient number of them for the purposes of farther information.

"It appeared that the fever began in all in the same way, and originated from the same causes. A poor family were lodged in a small apartment, not exceeding twelve or fourteen feet in length, and as much in breadth. The support of these depended on the daily labour of the husband, who, with difficulty, could earn enough to purchase food necessary for their subsistence, without being able to provide sufficient clothing or fuel against the inclemencies of the season.

"In order, therefore, to defend themselves against the cold of the weather, their small apartment was closely shut up, and the air excluded by every possible means. They did not remain long in this situation, before the air became so vitiated as to affect their health, and produce a fever in some one of the miserable family. The fever was not violent at first, but generally crept on gradually; and the sickness of one of the family became an additional reason for still more effectually excluding the fresh air, and was also a means of keeping a greater proportion of the family in the apartment during the day. Soon after the first a second was seized with the fever, and in a few days the whole family, perhaps, were attacked, one after another, with the same distemper. The slow approach of the fever, the great loss of strength, the quickness of the pulse, with little hardness or fulness, the tremor of the hands, and the petechiæ or brown spots upon the skin, to which may be added the infectious nature of the distemper, left no doubt of its being the same with what is usually called the jail or hospital fever. It would appear there is no great power of infection in the body alone, provided the air be not confined. Cold is the cause of the air being confined, which gives rise to the poison; and thus directly opposite to the opinions usually received, there is more danger of producing this disease in a cold country, and in a cold season of the year, than in a warm one."

It has been long disputed whether the poison thus arising among crowded collections of living beings—ochlesis—is capable of generating typhus, or whether it is merely a powerful predisposing cause requiring extraneous infection to produce this result. That it acts in the latter mode cannot be questioned. That it is capable of generating, *de novo*, the typhus poison is, I think, also proved by a body of evidence which should carry conviction to any unprejudiced mind. Exactly as scarlatina breaks out spontaneously in the overcrowded dormitories of schools—an occurrence certain to follow from this cause—so in overcrowded lodgings, barracks, and gaols, will typhus appear, no evidence or suspicion existing of contagion from without. Numerous instances of this have been collected by Dr. Murchison, to whose work I must be content to refer you; I may, however, mention a fact of some importance, that in several well authenticated instances, persons brought out of these crowded collections of human beings have communicated fever to others, without having themselves been affected with the disease. (*Appendix II.*)

Fever thus generated by ochlesis, is rapidly diffused by contagion. The records of the disease show that those are the most certain to suffer who are most exposed to infection. Medical students frequently furnish examples of this property of typhus; so do physicians and nurses; all being brought into close contact with the sick. Those servants whose duty it is to carry away and wash the clothes of fever patients, suffer in an equal proportion. During one epidemic in Edinburgh, Dr. Christison states, that of thirty-eight nurses only two or three escaped, and of fifteen gentlemen who filled the office of resident clerk, only two escaped. In the London Fever Hospital, says Dr. Tweedie, the laundresses whose duty it is to wash the patients' clothes, are so invariably and frequently attacked with fever, that few women will undertake this duty. Dr. Reid and Dr. Cheyne both state that, in 1817, not a single person appointed to receive the clothes of the sick escaped the disease.

But you will meet, from time to time, men of curiously constituted minds, who are not satisfied with this kind of evidence of contagion, strong as it is. This is, however, a fact in the history of fever which should be conclusive. It is the spread of the disease by deportation. In other words, infection is proved, if the individual who communicates the disease goes from the place where he resides to the spot where the healthy person is, and there infects him.

It is this power of deportation which makes it so necessary to segregate the infected person; and which makes the separation of a fever patient so useful as a means of limiting the diffusion of the disease.

Many years since I had a very striking proof of the vast influence which deportation might exercise. In two years, 1835-6, there were admitted into a fever hospital, of which I had the charge, 363 cases of fever, of the abdominal type, none of which appeared to have arisen from contagion. From this time to the end of 1838, the number of such cases continued about the same, while the total number of admissions increased to 400 in 1837, and 600 in 1838, the difference being owing to an increasing number of cases of typhus, brought exclusively from two districts which were about fourteen miles apart, and about equidistant from the hospital. Being, at the time much interested in the subject of the origin of fever, I instituted careful inquiries as to the source of these local epidemics, and I was satisfied that they originated from two cases of imported infection. In the first, a young man had arrived in this country from America, after an unusually rapid voyage, during the whole, or nearly the whole, of which he had been indisposed. On landing, he was immediately removed to his father's house, and on his arrival there was seen by a medical man, who pronounced his disease to be fever. He died on the second day after his arrival. His father's house and neighbourhood was previously quite healthy, but in two days after his death the father sickened, and on the day following his sister. She communicated the disease to her husband, whose residence was half a-mile distant.

He was attended by his brother, who caught the disease, and was sent into hospital. The father was visited, before his death, by a brother residing nearly two miles distant; and on his return home he sickened, and communicated the disease to his son. A brother of the importer contracted the disease, apparently from his father, and died; as did all the above persons, with the exception of the one who was sent to hospital. In short, of this family eight out of nine persons were infected, and seven died. I learned that in the course of a short time several other families were completely exterminated. The fever spread with a rapidity and fatality perfectly unprecedented, and long prevailed in the town and neighbourhood to which it was thus carried; altogether its victims numbered more than 100 persons.

In the second case, the daughter of the man first admittted into hospital was a servant in Dublin, where she contracted typhus, then prevalent here, and died. Her brother visited her, and remained till her funeral took place; he sickened, returned home, and there died of what was described to me as a long spotted fever. After his death, seven individuals of the family sickened within a day or two of each other, and were sent to hospital, where the father died, and several of the others passed through very severe fever. Typhus spread from this house to the immediate neighbourhood, and subsequently to the surrounding country.

Many other instances of imported contagion might be adduced, but I have selected these on account of their wide influence, and the conclusive evidence they afford to the infectiousness of fever of this type—exanthematic typhus.

The characters of the disease thus generated by an animal miasm, whether spontaneously in jails, barracks, and over-crowded lodgings, or introduced from without and spreading by infection, present a striking uniformity in all times, places, and seasons. The typhus described by Huxham, Pringle, Hunter; by Burserius, Hoffman, Palloni, Hildenbrand, and Copland, is the same in every particular as that which Murchison and Jenner have so recently and so accurately described, and which prevails in London and Dublin at the present time. It appears to me that it has been rightly classed by Hoffman, Copland, and others among the exanthemata, to the laws of which it will be found to conform in most, if not all, particulars. The frequency and uniformity of the eruption is as marked as that of scarlatina; the alternation of the irritation of the air passages, with the exanthem, resembles that of measles; the eruption fades, if the bowels be much acted on, exactly as in the other exanthemata; like them, the disease, once commenced, runs its determinate course, and like them becomes infectious at its acme, and still more so after its crisis, it being a well-ascertained fact that a large proportion of cases of contagion occur during convalescence. Finally, like them, it confers immunity from future attacks, in the vast majority of instances. Its characters have been well summed up in the definition of Dr. Peebles: "This contagious febrile eruption is an exanthematous affection—the production of human effluvia—where society is placed in circumstances favourable to its development, and should be considered the effect of a poison, *sui generis*. It arises from a miasm which generates in the human body an eruptive fever, distinct from all others, as other exanthemata are distinct."

The second of the sources of a fever poison is most frequently met with in large towns, and hence has been termed by some "civic miasm;" but it is by no means confined to these; and some of the most remarkable outbreaks of endemic fever have occurred in country districts, and even in isolated houses. The essential condition was, however, the same in all the instances which I have met with or have read of, viz., either the diffusion in the locality— whether a room, a house, or a yard containing several houses—of the gases arising from fecal matter undergoing decomposition, in a sewer, a cesspool, or on the banks of a half-dry stream, or the drinking of water contaminated by sewage matter. In many of the recorded instances of this endemic fever, there was no overcrowding, or want of ventilation, or of personal cleanliness; and the disease affected rich and poor alike, and unlike infectious fever, did not usually commence in the winter, but rather in the summer and autumn, subsiding on the coming on of the rains of winter, and seldom if ever being carried out of the locality and diffused elsewhere. One of the most striking points of difference between this and the exciting cause we have before described is, that while the segregation of a typhus patient is found to arrest the spread of that disease, the same measure has no such influence upon the endemic fever arising from malaria. On the other hand, while I do not doubt that the fever poison originally received from this source may be reproduced in the blood of the patient, and diffused by contagion, I have reason to believe that the contagion is always feeble, in comparison with that of typhus and the other exanthemata; and that the spread of the disease in a locality is seldom, if ever, due to this cause. You are, perhaps, aware that Dr. W. Budd maintains that it is diffused by infection through the medium of the alvine discharges of the sick. Some of the instances he gives are apparently conclusive; but I must refer you to his papers in the *Lancet* for December, 1856, or to Dr. Murchison's chapter on the exciting causes of pythogenic fever, in which the question is fully and ably discussed. (*See Appendix.*) I may, however, mention, that if called to attend a case of enteric fever in an ill-drained house, you were to confine your preventive measures to the removal and disinfecting of the patient's excreta, as recommended by Dr. Budd, you would certainly fail to check the diffusion of a disease which has its primary origin in sources without the body—in the decomposing fecal matters of sewers, cesspools, &c.

I could detail a large number of instances of the outbreak of fever which have occurred from this miasm under my own observation, both in Dublin and in the country; but I would rather refer you to the decisive cases detailed by Dr. Murchison. The Croyden fever, the Windsor fever, the Bedford fever, the Birmingham fever; were all examples of pythogenic, typhoid, or enteric fever, as it has been variously termed by various writers.

On the characters of the fever produced by fecal miasm, it may be remarked, that the elective affinity of the poison seems to be for the blood-making organs; Peyer's glands, the mesenteric glands, the spleen, and the lungs, being each the seat, in turn, of its operations. When received into the blood in a concentrated dose, and of intense virulence, it seems to kill with a rapidity resembling that of the concentrated poison of typhus; producing an effect on the brain similar to that of a narcotic poison, but, unlike typhus, leaving, even at this early period, the traces of its specific action on the intestinal follicles and mesenteric glands.

Thus, for example, in the remarkable outbreak which followed the opening of a drain at the school at Clapham, one boy died comatose in 23 hours from the seizure; another died in 25 hours; but the post-mortem examination showed tumefaction of Peyer's patches and of the solitary glands, ulceration in one, and congestion and enlargement of the mesenteric glands.

Dr. Murchison relates a most decisive case of enteric fever originating in the emanations from a cesspool, which proved fatal in

47 hours, and in which the solitary glands and Peyer's patches were enlarged, and infiltrated with a yellowish deposit; and the mesenteric glands as large as hazelnuts, and congested.

It cannot be questioned that cases of fever are met with from time to time, though unfrequently, in which the phenomena of both forms of disease are present in the individual; and also, that of several members of the same family one has presented the distinctive characters of typhus, and another those of typhoid fever. Two explanations of these exceptional cases have been offered. By the one, it is assumed that both forms of disease arose from the same poison, derived from the same source—contagion—and were varieties of the one species. By the other it is assumed, that in the locality two sources of fever-poison were present—ochlesis— namely, and fecal miasm; that in one member of the family the former poison produced the fever; in another, the latter; while in a third, both set up their operations at the same time, or in rapid succession in the system; the characteristic eruptions being seen together, or one after the other.

In a paper on the co-existence of several morbid poisons, in the 24th volume of the *Medico-Chirurgical Review*, Dr. Murchison gives several most illustrative cases of this hybrid disease.

Sir D. Corrigan gives an admirable example in the 10th lecture of his work on fever.

Besides the two above sources of fever, originating in the same locality, and either continuing as the endemic disease of the locality, or diffused around it, or carried from it by contagion, I have mentioned a third mode of origin of fever, observed at periods of uncertain interval, in which bad harvests and want of food have been succeeded by the rapid spread of disease over a wide extent of country; or by its simultaneous outbreak in distant parts of the same or different countries. This last mentioned fact is sufficient to disprove the opinion that a condition affecting only one portion of a country, like the famine of 1847, is the invariable cause. We must, therefore, admit the existence of a third exciting cause of fever—viz., what has been termed atmospheric or epidemic influence.

We recognize this influence in the simultaneous outbreak or decline of fever in distant places, the fever being of uniform character in all; in the tendency to peculiar complications during the period;

and in the fact that concurrent diseases frequently present some of the characters of the fever of the period.

The epidemic of 1847–9 corresponded in time with the famine of that period, and so is often called famine fever; but a little consideration will show that this was rather a coincidence than a strictly causal relation. No doubt, famine is a powerful predisponent to this fever as well as to typhus, but relapsing synocha has, like typhus, prevailed as an epidemic when no famine existed.

Such was the case in Scotland and in parts of Ireland in 1844. The same form of fever which spread universally in 1848–9, prevailed as an epidemic in Belfast, Galway, and Meath in 1844–5; the fact has been recorded by Dr. Seaton Reid of Belfast, Dr. Lynch of Loughrea, and myself.[1] In both periods, the yellow tinge of skin was prevalent; the same congestions of the liver, stomach, and spleen; the same sudden and violent onset, early crisis, and frequent relapse; and as, during epidemics of cholera, a choleraic tendency has been observed in other abdominal affections, so, during this, I observed the same icteroid tinge in different blood diseases, several instances of scarlatina, for example, occurring; in which the yellow skin and scarlet rash were present together.

In a country, and in circumstances presenting the conditions for the generation of every form of fever, it could not be expected that any one form would prevail to the entire exclusion of the others, and accordingly we find that during the three great epidemic periods of 1817, 1826, and 1848, the several observers describe each form as that prevailing in different places. This was particularly the case in 1848, when in overcrowded prisons and workhouses, typhus constantly prevailed at the same time with, or subsequent to, the relapsing fever. Of the essential difference between the two there could be no better proof than that the one afforded no immunity from attacks of the other.

To sum up our observations upon the etiology of fever, we believe that two conditions must exist for its production. (1) A morbid poison derived either from animal miasm, or from fecal miasm, or from epidemic influence. (2) The presence in the blood of a variable proportion of matter in a state of retrograde metamorphosis—the true nidus of the fever poison—upon which its catalytic action is exerted, and in which it is augmented and reproduced.

[1] Report on Epidemic Fever, *Dublin Quarterly Journal*, vols. vii. and viii.

It would appear that the predisponent is necessary, inasmuch as we have proof that persons may be exposed during a lengthened period to the imbibition of the poison, more especially of civic miasm, and may eliminate it from day to day until the action of cold, or of spirits, or of a mental shock, or of an arrested excretion, may cause a sudden increase of decomposing matter in the blood, when fever at once follows. It would appear that some of the so-called exciting causes, such as cold, really act by suddenly increasing the existing predisposition, either by their directly increasing the amount of disintegrated material, or by their indirectly effecting this by their depressing action upon the nervous system.

With reference, then, to the relation between the above described conditions and the different forms of fever, it may be assumed to be proved that a person previously healthy, but susceptible of the influence of fever, will receive the infection of typhus if exposed to the emanations from a typhus patient, and, by the reproduction of the poison in his blood, will become a source of infection to others.

We think it has been proved that breathing the emanations from crowded collections of human beings in non-ventilated apartments is the most powerful predisponent to typhus; also, that it is capable of generating the poison *de novo*.

We think that a body of evidence, of the most conclusive nature, proves that the decomposing fecal matters of sewers and cesspools generate a poison capable of setting up fever of the abdominal type, under certain conditions.

It appears that this latter is strictly an endemic disease—the fever of the locality: however generally it may prevail, during hot and dry summers, it really is not diffused by infection, but has a focus in each locality; while the former, being carried from place to place, may be so rapidly diffused by infection, as to become a *quasi* epidemic; but the true epidemic fever of this country is relapsing synocha, called also famine fever. That this has sometimes appeared in a form pure and unmixed, so as to be readily recognized, as in Scotland, in 1843, but that, on most occasions, one or both of the two former fevers have also prevailed during a portion of the epidemic period, and so have been confounded by some observers with the true epidemic fever, while others have recognized and pointed out their separate coexistence.

There appears good reason to believe that the fevers produced are as distinct and different as the sources which give rise to them.

Exposure to the infection of typhus will only produce typhus; while exposure to malaria produces typhoid or enteric fever; neither does the one confer any immunity from the other. Only last summer you saw a female patient in this hospital who presented the rose-coloured patches, and other symptoms of typhoid, return to the same bed within a month after convalescence, with exquisitely-marked typhus. No fewer than three similar cases have occurred during the present spring. The converse happens occasionally, the patient usually manifesting the phenomena of typhoid during the latter stages, or immediately on the subsidence of typhus, as in the case I have referred to by Sir D. Corrigan.

During the different epidemic periods, abundant proofs have been afforded that relapsing synocha confers no immunity from typhus, or *vice versa*. In 1848, I had under my care twelve individuals, who, after recovery from the former, contracted typhus in the convalescent ward of the hospital, and were readmitted to the fever ward, either before or soon after their discharge from hospital.

In the same year, a professional friend, whom I attended through a most severe attack of relapsing fever, was, in less than three months afterwards, attacked by typhus, with profuse eruption. Some years after I was called to see this gentleman in well-marked typhoid.

At our next meeting I shall proceed to the clinical study of these different forms of fever; but as I do not intend to give you didactic descriptions of them, such as you will find in books, I shall not pursue the plan of treating of each form separately; but while endeavouring to build up in your minds a concrete idea of febrile phenomena as "a whole," I shall keep constantly in view the differences presented by diseases, of which one kills by its action upon the blood and nervous system, without producing appreciable local lesion; another is characterized by determinate and constant local lesions; while the third presents phenomena peculiar to itself and differing from both the former.

LECTURE IV.

PATHOLOGY AND SYMPTOMATOLOGY.

Having briefly stated the received doctrines regarding the nature of fever and its etiology, I have now to call your attention to the practical aspect of the subject; to the study of its pathology and symptomatology, and to the bearing of our conception of fever as a morbid poison on the explication of its phenomena as observed at the bedside and after death. While your clinical study of disease should in all cases be as comprehensive and methodical as possible, it should be especially so in fever; in which everything connected with the disease, with the patient, and with the surrounding conditions, has an influence upon the diagnosis, the treatment, and the result. It is, therefore, of the greatest importance that your notions upon all these subjects should be clear and defined, that the facts presented to your observation at the bedside, should not be allowed to flit, as it were, before your incognizant senses, but should be grasped, and reflected on by the intellect; and, so far as possible, be referred to principles, and studied in relation to the theory of the disease.

Your diagnosis of the species of fever you have to deal with will be materially aided by a knowledge of the etiology, mode of invasion, and early history of the case. Meanwhile the previous health, habits, occupation of the patient, and all the circumstances connected with the seizure, will, if conjoined with his existing condition, be data for a just estimate of his power of resisting disease, and enable you to foresee its course, and to judge of its tendencies to one or other event; thus I do not hesitate to say that in a fever distinctly traceable to communication with a typhus patient you may expect to meet with a case of typhus, or that, if the exciting cause be unquestionably civic miasm, enteric fever will follow. Another guide to diagnosis will be furnished by the mode of invasion, usually sudden and marked in typhus, slow and insidious in typhoid.

The contrast between the two forms in regard to the mode of invasion is most striking, the typhus patient being usually confined to bed by the fourth day; while the adult attacked with typhoid may go about his business for ten days, or more, and though languid and complaining of loss of appetite and disturbed sleep, will not seek advice, until the diarrhœa becoming obstinate, he consults a medical man upon that account. In his graphic sketch of typhoid, under the name of Epidemic Gastric Fever—(*Cyclopædia of Practical Medicine*)—Dr. Cheyne describes these cases as commonly called walking fevers. Many patients thus affected have pursued their usual avocations, until struck down by fatal peritonitis, from perforating intestinal ulcer.

The invasion of relapsing fever is, on the other hand, still more rapid than that of typhus, the patient passing in a few hours from a state of health into one of ardent fever, ushered in by remarkable shivering, with vomiting, often of a green fluid, and intense headache, &c.

In regard to prognosis, I shall hereafter have occasion to show you that the circumstances attending the invasion are not seldom such as render a fatal termination almost a matter of certainty, although that result may be deferred to the ordinary period of termination of the fever.

Having ascertained all that is requisite to know of his antecedents, you next enter on a careful examination into the existing condition of the patient, with regard to each of the functions in succession—an examination which will have to be repeated at daily or even shorter intervals, since the condition of the fever patient is one of progressive change, and this change is by no means uniform, but subject to perturbations arising from various influences. You will have to make daily note of these changes in the temperature; in the colour of the surface; in the various phenomena of the eruptions; in the nervo-muscular function; in the organs of respiration, their physical condition and function; in the varying conditions of the heart and circulation; in the gastro-intestinal, urinary, and cutaneous excretions; and in the functions of the cerebro-spinal system. You will, moreover, have to watch the influence upon these of your remedial measures, and of the hygienic conditions in which the patient is placed, and to continue, modify, or suspend your measures accordingly.

The pathological phenomena, set up by the action of the fever poison, are of two kinds—primary or essential—arising out of the action of the poison upon the blood and nervous system, and its affinity for certain tissues or organs, as well as out of the special predisposition of the organ, which, in certain states, may be said to attract the fever poison. Secondary or accidental—arising (*a*) from some peculiar exciting cause; (*b*) from secondary blood contamination; (*c*) from reactive irritation. Strictly speaking, the latter, or secondary phenomena, only, can be termed complications; those of the first class are either purely toxic, or arise from the tendency of the poison to be eliminated by certain surfaces or excreting organs.

To the toxic action of the poison we refer the majority of the nervous symptoms of typhus. To the eliminative effort of the system we ascribe the infiltration into Peyer's follicles and the mesenteric glands in typhoid, and that peculiar form of pneumonia to which Dr. Stokes has given the designation of "arrested or aborted typhus."

The immediate effect of the action of the poison upon the organs to which it is attracted, is to derange their capillary circulation by altering the attractions between the blood and the tissues. In the case of the brain, the derangement of the capillary circulation was long regarded as the cause of the delirium, and other nervous symptoms of fever; by some these were ascribed to inflammation, by others to congestion of the organ; but, guided by the analogy of the narcotic poisons, we now believe that, in many instances, they are due rather to the toxic action of the fever poison. The theory of this action is, that the *post-mortem* congestion, which was formerly believed to be the cause of the delirium and coma, is rather the effect of the injury done to the brain and to the blood by the poison; whereby the attraction of the materials from the blood, suited to the nutrition of the brain, is retarded, and ultimately stopped. But this force of the action between the blood and the tissues is a powerful agent in the maintenance of the capillary circulation; when, therefore, it is impaired, the blood moves slowly and feebly through the capillary system.

The examples I have already cited from Armstrong, sufficiently illustrate this toxic action of fever, in its most intense degree;

fluidity of blood, and general stasis in the capillaries, being the *post-mortem* appearances in these cases of congestive typhus.

In less deadly forms of disease, the congestion is most marked in the organs for which the particular poison has the greatest affinity: the brain, and lungs, and heart, in typhus; the intestinal follicles, mesenteric glands, and spleen, in typhoid; the liver, stomach, and spleen, in relapsing fever, manifest this elective affinity and its results, stasis of blood, disintegration and softening of tissue, and exudation of typhic deposits.

The advanced stages of fever furnish many examples of the secondary group, or true complications. Take the pulmonary complications of advanced typhus as an example. When we open the thorax of a patient who has suffered pulmonary congestion, and who has succumbed under this fatal complication, we find the lungs of a dark colour, heavy, and congested, with portions inclined to a brownish hue, softened and friable, from what is termed hypostatic pneumonia. Here we have the material result of a group of conditions which existed during life, acting and reacting upon one another. Thus, (*a*) the primary alteration of the attractions between blood and tissue, caused by the fever poison, and its immediate effect—congestion. (*b*) Change in the innervation, caused by the influence of the poison on the 8th pair of nerves. (*c*) Reactive inflammation, set up in the congested portions, with exudation of a low type. (*d*) Secondary blood contamination, arising partly from deficient aeration, and partly from resorption of depraved blood and exudative materials; with its consequence—secondary poisoning of the nervous centres; all have been in operation, and conjoined, have produced the fatal result.

Like every morbid process in fever, the complication, of which this is an outline, is capable of being rendered more complex by the coexistence of other blood contaminations.

It is to be observed that our knowledge of the laws which regulate the pathological phenomena of the different species, will enable us to distinguish such as are extraneous and accidental, from those which are the result of the normal and regular action of the fever poison.

Take, for example, diarrhœa, occurring in the course of typhoid or typhus. We know that while the attraction of the poison of typhoid is for the glands of Peyer, and diarrhœa is, consequently, a normal symptom of that affection, the poison of typhus observes no such law, and diarrhœa, if it then occurs, is either from catarrhal inflammation of the mucous membrane or is colliquative, or, in not a few instances, is critical.

Some years ago I witnessed two cases, which were running their course at the same time in our female fever ward, under the care of the late Dr. Lees, who kindly gave me his notes of the appearances after death. One of these was a case of well-marked typhoid in a young girl, the other of typhus, with profuse eruption, in an old woman. Both had diarrhœa, the latter more especially. On *post-mortem* examination, the usual appearances of typhoid were found in the intestinal glands of the younger patient, while no trace of such was to be seen in the other. (*Appendix III.*)

There are several methods according to which these pathological phenomena might be presented for your clinical study. We might, for example, take them in the above order of primary and secondary; or as they occur in each form of fever studied in succession—the plan adopted by Bartlett, Jenner, Tweedie, and Murchison, and other writers on fever; but I think the best method is to adhere to the simple arrangement pursued in our clinical investigation of disease in general, and to study the derangements of function, with reference to the organs or systems affected, taking these in succession.

Following the rule I have often impressed upon you of proceeding in your examination from without inwards, noting external sensible phenomena before you study internal symptoms; you will, on each occasion, first carefully observe the expression of face, manner, decubitus, and colour of the patient, the condition of the surface with regard to heat and moisture, and the presence and characters of the eruption. These will all be found not only to differ from the appearance of health, but also to present striking differences in the different species of fever, as well as at different periods of the same fever. The derangement of temperature, for example, will be found to have a constant relation to the form of fever, to the period of the disease, and to diurnal periods, as you will see in the charts before you, which are copied from those in Dr. Aitken's work.

You will observe that in typhus the temperature attains its maximum height on the fourth day, varying not more than a degree from this to the ninth day; that until the ninth day it is higher in the morning than in the evening, but afterwards the reverse is the case; that between the eleventh and twelfth days the difference between evening and morning amounts to two degrees of temperature; and that between the fifteenth and sixteenth there is a difference of five degrees; that between the sixteenth and seventeenth the evening temperature has fallen three degrees; the following morning (seventeenth) being three degrees lower still, or but one degree above the standard of health.

In typhoid, you will observe that, in the first place, the maximun temperature is never so high as in typhus; next that it is attained much more slowly, ascending gradually till the fourteenth day, when the morning range corresponds with the evening of typhus at the same period; that up to the fifth day the evening temperature is two degrees higher than the morning, the morning being the higher from this to the fourteenth; that from the fourteenth day the evening range again becomes the higher, continuing steady, (at 104 degrees), until the commencement of the fourth week, when the difference between evening and morning becomes the greatest amounting to five or six degrees—in fact, to a marked remission of fever at this period. You will observe that from the fourth to the fourteenth day, the difference between morning and evening does not exceed one degree; and Dr. Aitken considers that this is characteristic of typhoid fever. He also remarks that typhoid may be excluded from our diagnosis if the temperature, on the first or second day, rises to 104 degrees, or if, from the eighth to the eleventh day, it falls below 103 degrees. (*Appendix III.*)

I am not aware of any accurate observations upon the range of temperature in relapsing fever. It is known, however, to be on an average, higher than in either typhoid or typhus, amounting, in some cases, to 107 degrees. This is in accordance with the greater increase in the rate of the pulse, and more rapid disintegration of the tissues, in this form of fever. (*Appendix III.*)

It has been suggested that the following short rules should be observed in using the thermometer: 1. To make morning and evening observations; 8 A. M., and the same hour P. M., have been recommended. 2. To place the patient in a medium or diagonal position between the back and side; neither fully on the back or side. 3. To insert the thermometer in the axilla of the side on which the patient has been lying, as it has been protected from external cold. 4. To be careful that it is placed in contact with the skin, no clothing intervening, and to place the arm in contact with the side. 5. To allow it to remain in the axilla for at least five minutes.

I could give you no better illustration of the value of daily thermometric observations in fever, than the case of Darcy which is now under your observation. In this man, on the morning of the thirteenth day of maculated typhus, the temperature, which had been slowly and steadily falling for several days, was found to have suddenly risen about four degrees; the only other appreciable symptom being increased rapidity of breathing, which was loud and somewhat nasal in character. There was no cough or expectoration, and the most careful examination of his chest failed to detect anything save the slightest possible trace of dulness on percussion, under the left clavicle, with slightly roughened breathing. I can say that, taken alone, these signs would not have justified the diagnosis of pneumonia, which was made chiefly from the marked and sudden rise in the temperature, the character of the respiration being such as might have been ascribed to the condition of the nervous centres. You are aware how soon the breathing was relieved by taking away a few ounces of blood by cupping, and how the ordinary signs of pneumonia, and the rusty expectoration soon followed.

The colour and expression of the patient will vary according to the type of fever, according to its stage and to its internal complications. The muddy, dusky, congested look of the face and eye of typhus are familiar to you; the expression of the eye has been aptly compared by Jenner to that of a person awaking from a drunken slumber. However distinct the maculæ may appear, the intervening skin is not, as in typhoid, pale or of its natural colour, but is darker than natural, and has a dusky mottled appearance, well depicted in Dr. Murchison's plate, now placed before you.

The changes in the appearance of the surface in the progress of a case of typhus, furnish valuable indications of the tendency to one or other termination, and should be attentively studied from day to day, as also should the energy or languor of the capillary circulation, as indicated by the effects of pressure. In a favourable and uncomplicated case, the rash which resembles measles—without, however, the distinctly crescentic form of that eruption—appears about the fifth or sixth day, and declines about the fourteenth to the sixteenth, not disappearing altogether until crisis has occurred. I believe it to be a permanent rash, and that the spots, unlike those of typhoid, which appear in successive crops, remain unchanged until they either disappear or are converted into ecchymotic spots —true petechiæ—a conversion which takes place, in most cases, to some extent, towards the termination of fever, and which is one of the phenomena it is most important you should observe, inasmuch as you may often measure the gravity of the case by the extent to which this conversion, and the appearance of petechiæ at the latter periods, occurs. In many cases the eruption is never distinct, a dull marbling, or mottling, of the surface only being present until the petechiæ appear. In such cases the application of a cupping-glass causes the spots to become distinctly visible in a few seconds.

The colour and expression of the face in typhoid usually present a striking contrast to typhus. Instead of the dusky face, injected eye with contracted pupil, and heavy, sleepy expression; you will observe a clear bright eye, with somewhat dilated pupil, a face perhaps pale, perhaps slightly flushed, and in cases in which the intestinal affection is severe, presenting a vivid circumscribed blush, totally unlike the diffused flush which we sometimes see in typhus, with cerebral complications.

The surface of the body corresponds in appearance, being pale after the first days, when a general blush often pervades the entire surface. About the ninth or tenth day, on an average, you will find a few rose-coloured spots, of a round or elliptical form, upon the surface of the chest and abdomen, or upon the back. Pass your finger over these, and you will find them slightly papular; press them, and they will completely disappear, returning in a few seconds. Watching these from day to day, you will observe that they become pale, of a rather brown tint, and vanish—never being converted into petechiæ—and are succeeded by another crop, which, after a few days, is, in turn, followed by another; and so on until the termination of the fever. Unlike typhus, in which the changes in the eruption furnish the most certain indications of the tendency to a fatal or a favourable termination, I do not think a fatal result is ever indicated by the rose spots of typhoid, which seem alike in the mildest and in the gravest cases.

At rare intervals we meet with pale blue patches upon the surface of the body. I have most frequently seen them upon the sides of the chest. They occurred in several instances, in this hospital two years ago. Dr. Murchison has accurately figured them in his work on fever.

Another appearance on the skin, frequently observed in typhoid, is what are termed sudamina. These are minute colourless vesicles, appearing on the neck and front of the chest, or less frequently on the abdomen. Although not peculiar to typhoid, they are certainly more frequent in it than in other forms of fever. They resemble the ice plant or the frost upon a pane of glass, more nearly than any other object with which I am acquainted.

You will occasionally find it by no means easy to distinguish the one form of eruption from the other. An imperfectly developed, or discrete typhus eruption, in its earlier stages, may be readily mistaken for a typhoid, if this is more than usually abundant; the frequent uncertainty of the date of access of typhoid favouring this error. Such cases require the exercise of caution and farther investigation, and you will best arrive at an accurate diagnosis by a careful inquiry into the origin and previous history, as well as the accompanying symptoms of each case. Not unfrequently you will have reason to believe that the two poisons coexist in the case, the eruptions being commingled, or the one succeeding the other, as in the patient Morrit, who was so recently under your observation.

This young man was admitted, labouring under a very mild form of typhoid, not more than three or four rose-coloured spots being visible. He appeared to be convalescent, and was removed to the convalescent ward, where he contracted typhus, with profuse eruption. Immediately after crisis, he again manifested the symptoms of typhoid, with a much better marked eruption of rose spots than on his first attack. (*Appendix III.*)

The colour of a patient in epidemic relapsing fever is peculiar It is usually a straw colour, or pale yellow—deepening at times, and in certain complications into jaundice—but not, as a rule, depending on the presence of bile in the blood, at least in uncomplicated cases. Occasionally in the worst forms a bronzed tint has been observed at the commencement, ushering in a highly congestive type of the disease. In the primary attack there is no form of eruption; but in many cases which came under my observation, from 1845 to 1849, I met with the typhoid spots in second or third attacks, owing, I believe, to the coexistence of the two species of fever in the patient, the one continuing after the subsidence of the other.

True, or ecchymotic petechiæ, appear not to be uncommon, and, according to many observers, are not confined to the advanced stage, but seem to depend upon the previous state of the patient as to destitution; or the existence of other causes of a depraved condition of the blood. According to Dr. Murchison they differ from the petechiæ of typhus in not being developed in the centre of the exanthem, but occurring primarily.

The phenomena referable to the surface which appear occasionally, or accidentally, in the several forms of fever indifferently, are —herpes labialis, sometimes observed in the early stage of typhus, more especially in that form in which the fever ushers in pneumonia; purpling or lividity of the nose, which seems more common in relapsing than in other forms; vibices, which, as indicative of decomposition of the blood, are met with in typhus and relapsing fevers; bullæ, which are not peculiar to any form, but also indicate a depraved condition of blood: bedsores; which properly belong to the sequelæ of fever.

You must bear in mind that changes in the colour of a fever patient are frequently due, not to the fever so much as to some visceral complication, or some form of secondary blood contamination. Thus, as I shall hereafter point out to you, the cerebral, pulmonic, bronchial, hepatic, and renal complications, have each their own modification of the colour of the face or general surface. To this I shall, thereafter, direct your attention, as affording guides to the diagnosis of these complications.

The cutaneous transpiration is modified differently in different forms of fever. As a rule the skin is dry in typhus, until the period of crisis, and then the perspiration is usually by no means abundant, except in fatal cases, in which it is often cold and clammy, sometimes warm, and generally profuse.

In typhoid, perspiration not unfrequently occurs at the close of the febrile paroxysm, in the advanced period; when, exacerbations and remissions of fever become marked; the tendency to nocturnal or early-morning sweats usually continuing for some time after crisis.

In the relapsing synocha, intense heat of skin is often accompanied by sweating throughout; and the copious perspiration following a rigor, which marks the crisis, is one of the most striking characteristics of that form of disease.

The odour from the surface is remarkable and peculiar in typhus —not so marked in the other forms. A cadaveric odour is frequently present in all forms for a day or two before death, and a similar odour sometimes pervades the room in cases of typhoid, with extensive ulceration of the bowels, being most marked after the evacuations by stool.

Besides these that I have mentioned, there are other external phenomena, such as those of the muscular motions, which are of value, as signs of serious cerebro-spinal lesions. Indeed, it may be said, that the sensible phenomena which you observe at the first glance, will often give you a valuable clue to the diagnosis of serious complications in the course of fever. At your daily visit, therefore all changes in the countenance, temperature, decubitus, muscular movements, &c., should at once attract your attention, and lead to a more careful examination of your patient.

Thus a cerebral or cerebro-spinal complication may reveal itself by the turgid, purple, congested face, flushed brow, suffused eye, and contracted pupil; by the retracted head and stiffened neck or arms; by the rolling of the head on the pillow; by increase of the cutaneous sensibility; or by loud nasal breathing, and spasmodic sighing respiration. Similarly, pneumonia will be recognized by the suffused face, hurried painful breathing, and suppressed cough. Peritonitis from perforation has sometimes been detected by the change in the patient's countenance, expressing the collapse, which was its only symptom; and uremic poisoning produces a change in the countenance and manner, which cannot fail to strike those who have once observed it. These symptoms or signs will come under our consideration in their bearing upon the different complications; but I now allude to them as illustrating the mode in which your daily observations should be conducted; eye, ear, and touch being trained to receive the different impressions made upon them with each change in the case; while the judgment must be informed, in order that it may rightly interpret each impression, and assign its due value to each sensible phenomenon. Always remember that our great observers have all been men of highly cultivated minds, of varied attainments, great thinkers. Of one of these it has been well said—"Gooch was no mere chronicler of so-called facts, but an historian, who could see in the very germ the laws by which the subject was evolved; a very few cases set his penetrating intellect on the true track. He worked by insight rather than sight—intensively rather than extensively. 'The period of my life,' he says, 'when I improved most rapidly, was when I gained clear and orderly notions of the objects of examination. The faculty of observation requires rather to be guided than to be sharpened; the finger soon gains the power of feeling when the mind has acquired the knowledge of what to feel for.'"—*Ferguson's Prefatory Essay to Gooch on Diseases of Women.*

LECTURE V.

CARDIAC AND PULMONARY LESIONS.

WE have to-day to consider the study of those deranged conditions of the functions of circulation and respiration which occur in fever, and the complications arising from the secondary affections of the thoracic viscera.

In this study are comprised the cardiac phenomena, the radial pulse, the capillary circulation, the varied modes of disturbed respiration, and the physical signs referable to the heart and lungs, observed during life, as well as the anatomical appearances presented by these organs on examination after death.

I present the circulating and respiratory systems for your study conjointly, rather than separately, on account of their close anatomical connection, and the mutual interdependence of their functions. While the condition of the left ventricle of the heart must be studied in connection with the radial pulse, and the systemic capillary circulation; that of the right ventricle, is associated with the pulmonary capillaries. Situated midway between these and the systemic veins, a weakened right ventricle influences both; and the loaded condition of the right cavities, which may be owing to obstructed pulmonary circulation, is recognized by its effect in causing remora in the great veins which form the descending cava.

The study of the functions of circulation and respiration furnishes us with symptoms or signs which are not limited in their applicancy to the organs within the chest. We derive the most important and general indications, in regard to both diagnosis and treatment, from the radial pulse; the cardiac sounds and impulse, furnish the most unerring sign of cerebral irritation; and the same may be said of the mode of respiration which has been termed cerebral breathing.

When, therefore, you are called upon to examine a patient in fever, you should carefully investigate the phenomena of these symptoms in succession; with constant reference to each other, as well as to the conditions from which they arise or with which they may be associated.

Your study of the systemic circulation will comprise—(a) the heart, or, more particularly, the left ventricle; (b) the arterial pulse; (c) the capillary circulation.

The phenomena of the respiratory system which it is most important for you to note are—the mode of breathing, its rate and ratio to the circulation; the physical signs referable to the lungs, and the rational symptoms of thoracic complications, if any.

Lastly, you will observe the signs or symptoms of a weakened, dilated, and congested condition of the right ventricle and auricle, sometimes associated with the signs of pulmonary congestion.

The condition of the left ventricle in fever may be either—(a) healthy throughout its course; (b) softened, and becoming gradually weaker from an early period to the end of fever; (c) strong and excited throughout the fever; (d) weak during the earlier period, but becoming excited as disease advances, and presenting the combination of an apparently strong heart, with small and feeble radial pulse.

I need not tell you that we are indebted to Dr. Stokes for a full and complete investigation into the weakened condition of the left ventricle in typhus; and for determining the signs of this condition, and their value, as indications for the exhibition of wine and stimulants. I shall give you Dr. Stokes's description of this change:—

"The condition now before us may be thus described: the heart is little, if at all, altered in volume. It is generally of a livid hue, but this it may have, in common with other internal organs, as is often seen in fever. It feels extremely soft, especially in its left portions, and the left ventricle frequently pits on pressure. Nothing remarkable is to be found as to the pericardium or endocardium, and the valves are unaffected. The principal change is found in the muscular structure, which is often infiltrated with adhesive, as it were, gummy secretion. The left ventricle exhibits a singular appearance, for the traces of muscular fibre are lost, and the external layer, to the depth of the eighth of an inch, converted into homogeneous structure, in which no fibre can be found. The colour of this altered portion is generally dark, and it resembles the cortical structure of the kidney. In some cases this change occurs in patches, varying in depth, and from a quarter to three-quarters of an inch in breadth. The change may affect the septum cordis, and—but to a much less degree—the right ventricle. The internal network of fleshy bundles appear less engaged, and though these may be pale, their firmness seems but little altered. The right ventricle is almost always more firm and hard than the left, which may be so softened as to break down under a slight pressure. In one case both ventricles seemed almost equally engaged, and so great was the softening of the organ, that when the heart was grasped by the great vessels, and held with its apex pointing upwards, it fell down over the hand, covering it like the cap of a large mushroom. Yet, even in this case, the left ventricle was more softened than the right."

The observations forming the substance of Dr. Stokes's Memoir, were made during an epidemic of typhus in 1837 and 1838. He mentions that he did not meet with the same condition during the epidemic of relapsing fever of 1847, and he justly remarks that—"The phenomena of typhus softening of the heart appear to be under the same influences as those of other secondary affections, and that their frequency varies with the epidemic constitution."

I may mention that having had many opportunities of examining the condition of the heart after death, during the epidemic period referred to by Dr. Stokes, I not only met with the appearances he has described, but occasionally also with a form of softening in which there was none of the concrete exudation, causing an appearance resembling the cortical substance of the kidney, but a loosened and macerated appearance of the muscular fibres, which seemed to be infiltrated with a glairy sticky fluid.

In some cases the softening would appear to be due to rapid degeneration of the muscular fibres into fat, as observed by Dr. Murchison.

Dr. Stokes thus describes the signs of this softened condition as derived from the modification of the heart's impulse and sounds:

"We observe that the impulse becomes less and less distinct, the change being generally gradual, but in some cases more sudden. The loss of impulse is first perceived at the apex and to the left side; and we frequently find that while it has ceased in this situation, it can be discovered under the ensiform cartilage. In the most extreme cases of loss of impulse, we cannot discover it even by careful examination; but this is rare, for in many instances in which the impulse appears to have completely subsided, we may find it by examining when the patient is turned on his left side, or by pressing with the fingers in the intercostal spaces at the end of expiration, when we discover it like a feeble vermicular sensation.

"In most cases the diminution of impulse is attended with corresponding loss of sound: but it must not be forgotten that impulse and sound are not always proportionate or corresponding, either in the invasion or retrocession of the disease.

"The re-establishment of the impulse generally takes place before that of the sounds, and in cases where all impulse has subsided, we observe its return first in the inferior sternal region, and next under the left mamma.

"In most instances, however, the diminution and return of the first sound are accompanied with diminution and return of the impulse; but in some the sounds may become distinct before the impulse returns, while in others the impulse reappears long before the first sound has been fully restored. In one case we found that on the eighth day the sounds were not in proportion to the impulse; on the tenth the impulse continued, but the first sound was totally absent; and on the eleventh day no impulse could be felt, and yet the first sound was distinctly audible. In another case, on the twelfth day the impulse was less perceptible than on the day previous, but the first sound had more strength.

"It is hardly necessary to observe that the value of want of impulse of the heart in typhus fever, as an indication of the change in question, depends entirely on the circumstance that in the earlier periods of the case a distinct impulse had been observed. This will assure us that the want of impulse, under the circumstances in question, is not the natural condition.

"*Modifications of the Sounds.*—We shall arrange these characters in the order of their frequency. The first, and unquestionably the most important, is the diminution of the systolic sound, which may go on to its complete extinction, or to such a degree as to give a singular predominance to the second sound. The progress of this lessening or extinction of the first sound follows the same laws as the diminution of the impulse—that is to say, we observe it first as affecting the left side of the heart, and next as spreading to the right; and the diminution or extinction of the systolic sound, as of the impulse when observed at both sides of the heart, is indicative of an extreme degree of weakening.

"Hence, as might be expected, we meet with the following combination of circumstances:—

"1. Feebleness of the systolic sound, causing predominance of the second sound, evident at the left side of the heart only; under sternum no loss of proportion between the sounds can be discovered.

"2. Cessation of the first sound over the left ventricle. We have then in this situation the heart acting with a single sound, and that sound the second, while beneath the sternum the double sound continues.

"3. Cessation of the first sound at both sides of the heart, so that, no matter where we examine, we find the heart acting with a single sound, which is the second.

"4. In one or two extreme cases we observe the complete extinction of all sounds of the heart, and yet the pulse at the wrist could be perceived, and life was continued for more than thirty-six hours. One of these cases presented the greatest amount of softening of the heart that I have met with.

"During the period of restoration of sounds the phenomena follow an inverse course, the first sound re-appearing under the sternum, soon to be followed by its restoration at the left side. Under these circumstances, too, we may occasionally observe that the progress of restoration of the systolic sound is from the base towards the apex. Thus, in a patient who presented well-marked signs of softening of the heart, we found that on the twelfth day, and when the pulse had fallen to 80, although the sounds were feeble at the base of the heart they were quite proportionate, yet at the apex and to the left of the ensiform cartilage, the first sound still predominated.

"In the next series of cases the signs, though indicative of great weakness, are very different in character. Like the former, they are attended with diminution or loss of impulse; but the disease seems to act more on the entire heart, and we neither observe the predominance of morbid signs on the left as compared with the right side, nor that of the second over the first sound. In these cases there is no extinction of either sound, but both are diminished in loudness, and become of a nearly similar character. To this condition I have given the name of the fœtal character, from the close resemblance of the phenomena to those of the heart of the fœtus in utero—a resemblance which, especially when the pulse is acting at the rate of from 125 to 140, is almost complete. We have not yet been able to point out any anatomical difference between this condition and the more ordinary instance of diminution of the first sound, but it may be observed that although we meet cases in which this character exists throughout, yet that in others both sets of phenomena occur at different periods, and, as it were, run one into the other. Coupling this with the fact, that the two classes of cases are met with under the same epidemic constitution, and without any ascertained difference in the general symptoms, and finally, that the treatment by stimulants is equally called for in either case; it may be concluded that there is a close relationship between the conditions in question.

"Finally, we have observed, in a very small number of instances, a diminution of the second sound, while the first was but slightly, or not at all, affected. It is not easy to explain this condition; possibly it is connected with a want of resiliency of the aorta itself, for in many cases the pulse is found to be remarkably compressible."

The important practical result of these investigations will be best stated in the words of Dr. Stokes himself. "It is not too much to assert that our researches on the condition of the heart in fever have given a great facility in judging as to the necessity of stimulants in any given case of that disease; and they not only furnish a rule by which the junior and unexperienced man may be guided, but give to every practitioner a greater degree of confidence in himself, when he has to determine the use, the increase, the diminution of stimulants in fever. And independent of all this they have furnished us with new prognostics of great value. With very few exceptions it may be laid down, that the return of the impulse and of the first sound under the use of stimulants, justifies a good prognosis, more especially if the rate of the pulse is falling; and conversely we find that the existence of an excited state of the heart with a strong impulse and clear and proportionate sounds indicates danger, and the more so if the pulse be weak and rapid, and that its rate increases rather than diminishes under the use of stimulants." (*On Diseases of the Heart and Aorta*, p. 378.)

When we come to consider the study of the cerebro-spinal lesions in fever, you will find the excited condition of the heart referred to by Dr. Stokes, has frequently its origin in the great nervous centres, being mostly observed in combination with other symptoms referable to the brain or medulla oblongata.

I think it might safely be said, that in any case in which the heart's impulse becomes strong and jerking, and its sounds exaggerated during the latter period of fever, the great nervous centres will be found to be the seat of irritation. This is a fact of which you have had several remarkable illustrations in the course of the present year, when cerebro-spinal arachnitis has been unusually prevalent. The study of the cardiac phenomena thus becomes of increased importance from their bearing upon the diagnosis of an otherwise obscure complication.

But, again, the condition of the left ventricle has a close connection with the entire systemic capillary circulation. It appears to be a rule of general applicancy, that if, in any blood disease, the capillary circulation becomes retarded—as, for example, in Bright's disease—the left ventricle is called into increased action to aid the circulation in the capillaries, which is rendered difficult by the altered relation between the blood and the tissues. We may, therefore, readily imagine that in fever the existing general congestion in the capillaries will be notably increased by any diminution in the power of the left ventricle, and such is no doubt the fact. Hence the effect of wine and brandy, when these act beneficially, is not only to increase the power of the ventricle, but through it to arouse the stagnant circulation in the capillaries. The worst cases which we meet with are those in which the one effect is produced without the other—in which, under the influence of wine, the heart's action becomes strong, while the radial pulse becomes gradually imperceptible, the surface cold, and the maculæ darker in colour; while, on the other hand, in those in which wine exerts its beneficial influence, the heat returns to the extremities, the colour of the surface improves, and the maculæ change from dark purple to a brighter tint.

You should make the radial pulse, as well as that of the larger arteries, an object of your attentive daily study. You should learn to estimate its apparent haste, as well as its frequency, its force, its volume, its hardness or softness, its distinctness or its dicrotous character, its regularity or irregularity, its equability or its tendency to falter and become uneven.

Each of these characters you will find to have relation either to the condition of the heart itself, or of the pulmonary, or cerebral, or capillary circulation, or the type of fever, or the approach of a favourable or unfavourable crisis. In all cases you should examine it in connection with the heart, and with the respiration, inasmuch as signs of value are deduced from its want of correspondence with the former, and of due proportion in frequency to the latter.

The function of respiration is deranged in several modes and from several conditions in fever: thus, at the commencement of fever a semi-voluntary form of breathing is observed, differing from the dyspnœa of thoracic disease, a *besoin de respirer*, as it has been termed, which, to the experienced eye, presents one of the most characteristic signs of the initiatory period. This modification may exist without any physical signs referable to the lungs; sometimes an engorged and dilated right ventricle co-exists, but it seems to be less due to this condition than to the change produced by the poison in the vital attraction between the blood and the pulmonary tissue, to which must be added the altered innervation caused by the direct toxic action of the poison upon the medulla oblongata and eighth pair of nerves, an alteration which we shall find to exercise an influence throughout the course of fever, producing the most serious complications in its later periods. It is most important that you should distinguish the derangement of the respiration thus caused from that arising from the physical conditions of the lungs, which I have just now to notice. This cerebral respiration is graphically described by Dr. Graves, by whom it was first accurately discriminated. "When in typhus you find the patient's breathing permanently irregular, and interrupted by frequent sighing; when it goes on for one or two minutes at one rate, and then for a quarter or half a minute at another rate, you may rely upon it that sooner or later an affection of the brain will make its appearance. . . . . . I speak here of breathing independent of any pectoral affection. I am in the habit of calling this kind of breathing cerebral respiration, because my experience has told me that it is almost invariably connected with congestion or oppression of the brain."

It has been justly remarked by other observers that the diagnosis of cerebral breathing is assisted by the presence of a nasal stertor, and by the absence of the congested livid face which is present in severe pulmonary complication; to which is, of course, to be added, the absence of the physical signs of such complication. Of the two forms, the cerebral and the pulmonary complication, the former is that which most easily attracts observation; unless we add to the disordered respiration, the congested livid face, the dark colour of the eruption, the character of the cough, and the physical signs in the latter. Both may be attended with that sudden rise in the temperature which we lately observed in the case of Darcy.

The anatomical lesions referable to the lungs in fatal cases of fever, are those of bronchitis, pulmonic congestion, and pneumonia. The first is common to each form of fever, the second is almost constant in typhus, the third is more frequent in typhoid in the shape of lobular and hypostatic pneumonia, and occurs in that form of disease which Dr. Stokes has described as aborted or arrested typhus. Bronchitis may occur at any period of typhus, more especially in the epidemics of winter and spring. You are not to suppose that the frequent and irritative cough of the first few days is invariably, or even usually, due to this condition; it may be merely a cough of irritation resembling that of the early period of measles, and like it may subside on the appearance of the eruption; but on the other hand, it may continue, and on examination of the patient's chest you may find stethoscopic signs of bronchitis, namely, sonorous, sibilous, and mucous rales. This may become more intense in degree and more suffocative in its character as the fever progresses, and may greatly and fatally influence the result, more especially in advanced age, when attended with great prostration and debility; or, if associated with pulmonic congestion.

When this combination occurs, the broncho-pneumonia may be recognized by the progressive frequency of respiration, increasing difficulty of expectoration, and increased viscidity of the sputa—which may assume a brick red or prune juice tinge—and by the supervention of dulness on percussion, with ill-defined crepitus or tubular breathing in the depending portions of the chest.

In typhoid, the bronchitic complication belongs rather to the advanced stage, seldom occurring before the third or fourth week, when it frequently becomes a source of great danger to the patient, and of difficulty and embarrassment to the physician, the differential diagnosis between it and acute phthisis being both difficult and uncertain. This we shall have to notice when considering the sequelæ of fever.

The second lesion, pulmonic congestion, is almost constant in typhus, much less frequent in typhoid; it was found by various observers in 140 out of 146 fatal cases of the former. It may be so marked as to cause the lung to sink in water, and to become softened; but it is still to be distinguished from pneumonia by the absence of the concrete exudation which gives a granular appearance to the section of a hepatized lung. The colour also differs, being purple, resembling the spleen, and both lungs are usually affected, "stagnation of, for the most part, diseased blood being found equally diffused along the posterior surface of the lung; while the anterior half of the lung is generally bloodless, dry,—but perfectly sound otherwise."—(Hasse.) One or both, however, of these congestions, or a portion of them, may become the seat of reactive inflammation—hypostatic pneumonia—an occurrence which is not peculiar to either form of fever, but is met with in the advanced periods of both; more commonly, however, in the latter stage of typhoid. This condition is defined by Hasse to be due to inflammatory irritation, resulting from genuine hypostasis of blood in the lungs. He states that it frequently occurs in only one lung, the other exhibiting simple stagnation.—(*Appendix VI.*)

The signs of this congestive pneumonia correspond very nearly with those of the ordinary sthenic form of inflammation, the patient's countenance presenting the lurid flush and sallow tinge characteristic of this disease; dulness on percussion, with crepitating rale, and tubular breathing being usually met with on a stethoscopic examination of the posterior portion of the chest. But these signs are not unfrequently latent to some extent; there may be no crepitus or tubular breathing, and the condition of splenization may be revealed only by comparative dulness on percussion, and a muffled and suppressed kind of inspiratory murmur. In many cases the rational symptoms are still more latent; there is no complaint of pain or oppression, the respiration is little more disturbed than in other cases of fever; there may be little or no cough, and no expectoration. In this condition I have frequently drawn your attention to a sign which should at once lead you to make a careful examination of the patient's chest. You have seen, in many instances, that when on percussing the anterior portion of the chest of a patient who perhaps presented no symptoms of pulmonary affection, a sound was elicited on one side, differing in kind from the normal resonance, higher in pitch, and having a tympanitic character, we found dulness on percussion on examining the posterior portion of the same side, with other signs of congestion or pneumonia. At each daily visit, then, you should at least examine the anterior portion of the chest, and if you detect this change in the percussion sound, especially if conjoined with puerile respiration, accelerated breathing, and elevation of temperature, you should carefully examine the depending portions of the chest, no matter how marked may be the adynamic condition of the patient.

This is not the time to enter into any detailed explanation of the phenomenon of heightened clearness, were such an explanation possible. Two causes may, however, be mentioned as *probably* concurring to produce it. One is the slight increase of tension of the parietes corresponding to the unaffected portion of lung; the other is the shortening of the vibrations set up by the percussion stroke, by the condensed lung, which may be said to act as a kind of sounding-board. (*Appendix IV.*)

LECTURE VI.

CARDIAC AND PULMONARY LESIONS.

BESIDES the pulmonary complications noticed in my last Lecture, I have to direct your attention to the peculiar form of pneumonia which occasionally seems to replace, as it were, the symptoms of fever, and to which, accordingly, Dr. Stokes has given the name of aborted or arrested typhus.

This condition has been regarded by others as, strictly speaking, less a complication of typhus than the localization by elective affinity, of a blood poison in the lung; the general febrile condition terminating thus—by a form of crisis, as it were.

The close relation between fever or pyrexia, and pneumonia, and the frequent replacement of the former by the latter, has long been observed by pathologists; and it has been supposed by Wunderlich, as quoted by Dr. Parkes, that the exudation in the lungs coincides with the end of the pyrexia—that is to say, that the defervescence commences when the lungs become hepatized. "If it could be satisfactorily made out," says Dr. Parkes, "it would certainly imply that the exudation into the air-cells relieved or cured the fever; in other words, that the lung disease is not a primary but a secondary condition, and that it succeeds to, and brings to an end, by purifying the blood, a condition of general pyrexia, arising from blood disease. Without believing that this relation is quite determined (if it were determined the case would be settled), there is no doubt that the fever ends spontaneously, or very greatly lessens, at the time when the inflammation of the lungs is very great."

Upon the hypothesis that the pneumonia is but the localization by elective affinity of a blood disease, by a process analogous to that of gout, is explained, says Dr. Parkes, "the previous malaise; the sudden outbursts of fever, when the diseased blood implicates, at last, the nervous system; the singular and rapid termination of the pyrexia, at a time when the lung lesion is yet intense; and the enormous elimination of urea during the very first days, before the lung exudation has softened down."

You have had frequent opportunities of witnessing this rapid subsidence of the symptoms of fever on the supervention of pneumonia, and you are aware that Dr. Stokes is of opinion that specific typhus itself is not unfrequently thus arrested.

Dr. Stokes thus states his views of this affection: "The patient is attacked with the usual symptoms of typhus fever, and he comes into hospital after two or three days' illness. There is nothing about him to make one think that his disease will not run the usual course of the epidemic of the day, and we are prepared to expect a fever of at least a fortnight's duration. On admission he may have no symptom which would call attention to his chest; but as early in some cases as the beginning of the fourth, and in others, of the fifth day, it is discovered that the upper lobe of one lung is solid, or nearly so. The clavicle is quite dull on percussion, so is the scapular spine, and the dulness extends to the line of the mamma. This discovery has been so often made accidentally, that I am sure many of such cases have passed unnoticed, at least where the attendant is not well informed as to the insidious nature of typhus local diseases, and does not make it a practice and a duty to examine daily, as far as he can, the condition of every organ."

"But the most remarkable circumstance in these cases, is the constitutional disease seems to be cut short. The expression of fever leaves the countenance; the peculiar colour or hue of typhus disappears; the eye bright and intelligent; the tongue cleans; and the pulse comes down to a natural state. And thus we have seen patients so altered in the course of twenty-four hours, that one had some difficulty in recognizing them; all the symptoms of typhus were gone, and nothing remained but the consolidation of the lung. And this, too, is not attended with any notable suffering. There may be a little cough, some dull pain, or an inability to lie on one side; but that is all. The respiration is scarcely, if at all, accelerated. In fact, it would seem that there was no irritation or excitement of the organ; and the case is another proof how much less the sufferings in disease are connected with the mechanical than with the vital conditions of organs."

"How," asks Dr. Stokes, "are we to look at such cases? That they are no examples of inflammation of the lung it is plain; and it appears probable that if this local disease had not occurred, the patient would have gone through the course of the fever of the day. Does it not seem as if the constitutional disease exhausted itself, as it were, on the production of the local affection, just as in certain cases of simple variola we see the fever to subside on the appearance of the pustule? I do not know whether such cases have been observed elsewhere; but of their existence we have had abundant proofs. It is worthy of remark, too, that when we compare these cases with the ordinary forms of typhus attended with secondary disease of the lung, the local affection is here developed at an unusually early period; and it may be that in the more protracted cases of fever, the nature of which is to develop local affections, the periods of this development, and of the cessation of the fever, may also be coincident. We do not, however, find that this is so common as to establish a rule."[1]

Before I pass on from this part of our subject, I would wish to mention an occurrence which I have not seen noticed by any writer. In several instances, in which fever was attended with cerebral symptoms, particularly with great emotional excitement, and with marked cerebral breathing; these symptoms have been succeeded, after some days, by inflammatory congestion of the upper lobe of one or other lung. This complication differs from the affection above described by Dr. Stokes: first, in the seriod in which it has occurred; secondly, in not being attended with resolution of the fever; and, lastly, in having been—in my experience, at least—always preceded by highly-marked excitement of the emotional nervous centres, as if originating in the nerves of the respiratory organs.

I am unable to offer you any satisfactory explanation of this occurrence; of which, however, I have seen too many examples to have any doubt.

The condition of the right ventricle should be studied, in connection with the respiratory system, upon which it exercises great influence, and in relation to which it certainly deserves more attention than it has usually received.

Viewed as a simply febrile phenomenon, diminished contractility with passive congestion of the right ventricle is of frequent occurrence in the earlier periods of typhus and relapsing fever. It manifests itself in a feeling of anxiety and embarrassment of the

---

[1] "Lectures on Fever:" *Medical Times and Gazette.* 1855.

præcordia, frequent sighing, moaning, and jactitation; there is a sense of weight, with tenderness of the epigastrium, and, on examination by percussion, the præcordial dulness will be found to extend across the lower third, and sometimes to the right of, the sternum. Along with these signs I have sometimes observed turgescence of the jugular veins, and some degree of lividity of the lips, with purplish flush of the cheeks. I have often seen immediate relief of these symptoms follow the application of a few leeches over the lower end of the sternum, or the abstraction of a few ounces of blood by cupping. I have even known relief follow dry cupping. In other cases I have known this condition continue, with the distressing symptoms due to it, throughout the entire course of the fever; the signs above mentioned disappearing suddenly on the occurrence of crisis.

This fortunate termination is not, however, constant. Occasionally prolonged stasis of blood leads to the formation of fibrinous clots in the right cavities; the signs of this occurrence being marked increase of dyspnœa, with irregular intermitting action of the heart and failure of the general circulation. It may be reasonably presumed that this congestion of the right cavities is due to several co-existing causes, in various proportions. Thus it may be owing to the direct action of the fever poison upon the nutrition of the organ diminishing its tonicity; to the influence of deranged innervation; occasionally to pre-existing disease, such as contracted mitral opening, or to a weakened condition of the right ventricle caused by fatty degeneration; or to long-existing chronic bronchitis and emphysema; or, finally, to the constant co-existence of pulmonary congestion, set up during fever, opposing a resistance to the entrance of blood into the lung by the pulmonary artery, which the weakened ventricle is not able to overcome. When the first and second conditions alone exist, the cardiac congestion may be removed by the conjoined employment of derivatives and stimulants; but in the more complex cases, although the condition of the ventricle may seem to improve for a time under the use of stimulants, it returns at the more advanced periods of fever, and powerfully conduces to a fatal result. Thus, in a case which I saw a few years since, under the care of Dr. G. Kennedy: A middle-aged woman was admitted into Cork-street Hospital on the eleventh day of typhus, much maculated. She had urgent dyspnœa—the respirations being 60 in the minute—and cough with frothy mucous expectoration. The lower portions of both lungs were dull on percussion, with muffled feeble respiration mixed with bronchitic rales; the heart's action heaving, with a small, compressible, radial pulse of 128. The præcordial region was dull on percussion from the left nipple to the right of the sternum, and downwards to the border of the ribs. The impulse and sounds much stronger on the right than on the left side of the heart. Under the administration of wine and stimulants, aided by a large blister to the front of the chest, the patient rallied for a time; the maculæ became bright in colour, the countenance more clear, the respiration easy, and the præcordial dulness diminished in a marked degree. This improvement was maintained for about a week, when gradually the former symptoms returned; and on the day before her death, which occurred on the twenty-first after seizure, she presented a dusky face, with dark petechiæ, and a low temperature of the surface, which was covered with cold greasy sweat; the pulse was faltering; respiration hurried and sighing; heart's action heaving and irregular; and there was great increase of præcordial dulness to the right side.

You had lately an opportunity of studying this condition of the heart in the case of Jane Fisher, in our female fever ward. In this case you observed the signs I have enumerated, more especially the sighing respiration, the lividity of countenance, and the marked increase of præcordial dulness. The signs disappeared, in a great measure, under the influence of repeated dry cupping, conjoined with the free administration of stimulants; but returned, from time to time, in a less degree, until crisis occurred.

In this case the loaded condition of the right ventricle was associated with pulmonary congestion; and it may be questioned if it ever occurs singly, unless at the invasion of fever, inasmuch as pulmonary congestion, to a greater or lesser extent, may be considered a constant condition in typhus. If, as I have before remarked, the influence of the fever poison produces a change in the systemic capillary circulation, the same cause is certain to produce the same effect in that of the lungs; in which we have reason to believe that the vital attractions between blood and tissue are even more readily altered, and in which we have a powerful disturbing influence in the deranged condition of the nerves of respiration. Moreover, the diminished power of the right ventricle must have an influence on the pulmonic capillaries, similar to that observed in the systemic circulation in cases of weakened left ventricle. When this derangement of the pulmo-cardiac circulation becomes extreme, we have the most urgent dyspnœa and præcordial anxiety, with deranged action of the heart, feeble and irregular pulse, and sinking of the temperature. After death we find, along with bronchial and pulmonic congestion, a loaded condition of the right cavities of the heart, and frequently fibrinous coagula of firm consistence extending into the pulmonary artery. (*Appendix IV.*)

No doubt prolonged stasis of blood in these cavities will alone be sufficient to produce the effect; and it becomes therefore most desirable that this condition should be removed, and the pulmo-cardiac circulation relieved with as little delay as possible. Various measures may be resorted to for this purpose, with more or less advantage—brandy, turpentine, and ammonia, dry cupping, revulsives, local bloodletting by leeches or cupping. But in extreme cases of pulmo-cardiac congestion, I have conjoined with stimulants and revulsive remedies, the abstraction of a few ounces of blood from the arm by venesection with advantage. I will read you an abstract of my notes of the two of the most urgent cases of pulmo-cardiac complication I have ever seen, which I treated in this way:—

Margaret Colman, aged 18, strong, and previously healthy, was brought four miles to hospital on a cold day in the month of January, 1840, in the sixth day of typhus, and on arrival was much chilled.

On the 8th day she was reported to have raved during the night; pulse, 100; eyes suffused; surface of arms, chest, and back thickly covered with dark-coloured maculæ; urine pale.

On the 10th day she was reported to have breathed rapidly, and with difficulty, and to have coughed incessantly during the night; the extremities having become cold and livid.

The resident apothecary had applied a blister between the shoulders, warmth to the extremities, and had administered decoction of senega, with camphor and carbonate of ammonia.

Notwithstanding these measures, the respiration was gradually becoming shorter, more hurried, and laborious, accompanied by moaning and jactitation; and she complained of weight at the heart. The lividity of the face and arms had much increased during the hour preceding my visit, and the extremities were cold and covered with dark maculæ. The physical signs were those of bronchitis, with congestion of the lower lobes of both lungs, and of congested right cavities of the heart, the action of which organ was heaving and tumultuous, with weak, rapid, and compressible pulse. Her intelligence was perfect, but she manifested little sensibility to external impressions; such as the heat of mustard cataplasms and turpentine epithems, which were freely employed.

Considering that the indications were obviously to relieve the loaded right auricle and ventricle, and at the same time to arouse the heart's action by stimulation internal and external; I took six ounces of blood from the arm, and ordered the senega and ammonia to be continued, with the addition of 20 minims of oil of turpentine to each dose; a wineglassful of warm punch being given at intervals.

A large sinapism was also applied to the chest—followed by turpentine epithems—and sinapisms to the legs.

On visiting her in the evening I learned that the incessant cough had been at once relieved by the bleeding, that she had slept quietly, and had passed some high-coloured urine. The countenance was visibly improved, being much less livid; skin warm and dry; respiration easier, with less moaning; and the pulse fuller and stronger than in the morning.

11th and 12th. Little change. Ordered small flying blisters above and below the clavicles, and to have four ounces of wine.

13th. Was reported to have passed a sleepless night; lividity has returned with laboured respiration; has low muttering delirium, with involuntary discharge of urine. Four ounces of blood were taken from the chest by cupping; Senega mixture with turpentine continued, and a musk and camphor bolus given every fourth hour.

14th. Has had no sleep; cough and expectoration improved; no other change. Blister under each clavicle; eight ounces of wine.

15th. Had a few hours' sleep. During this day she had occasional tendency to faintness, which was averted by the free administration of stimulants. During the following night she slept soundly, and awoke on the 16th day with a pulse of 100 soft and full, respiration easy, skin warm, and, for the first time, slightly moist. Henceforward her recovery was steady and rapid.

The other case to which I have alluded was in many respects the most remarkable, and the effect of treatment the most striking, I have ever witnessed. The young lady, the subject of it, was a relative of the late Dr. Graves, who published, in the second edition of his *Clinical Medicine,* a short account, which I wrote to him at the time.

Miss ——, æt. 14, was attacked by typhoid fever attended with severe diarrhœa, causing much prostration at the time of my first visit on the 11th day. From this time till the end of the third week the tendency to diarrhœa continued, yielding finally to port wine, decoction of bark, with aromatic confection, and enemata of acetate of lead and opium. Towards the close of the third week the bronchitis attending the later period of typhoid fever set in; and during the night following the twenty-first day, became suddenly and alarmingly aggravated; the breathing becoming loud and laborious, the pulse fluttering, and at times imperceptible at the wrist, the surface cold and hot by turns, and covered with profuse greasy perspiration.

On my visit in the following forenoon, I found her apparently moribund. She was breathing hurriedly and with stertor, the cheeks puffing slightly with each expiration; she had also convulsive twitchings of the mouth; sight, hearing, and consciousness appeared to be totally lost; the eyes were fixed and glassy, the pupils contracted to a point; the face had become bloated and livid, the lips purple, and the jugular veins distended; the respiration was accompanied by loud râles over the front of the chest; the heart's action was tumultuous and jerking, but the radial pulse was scarcely to be felt; finally, the abdomen was enormously distended and tympanitic. It appeared to me that death was impending from sudden congestion of the lungs and right side of the heart, and I proposed to try the desperate chance of a small bleeding with a view of relieving the loaded right ventricle, and so gaining time for further measures. I accordingly took about four ounces of blood from the arm, and followed this operation immediately by the application of sinapisms to the spine and to the soles of the feet, and relays of hot flannel sprinkled with oil of turpentine to the abdomen; giving, at short intervals, a few drops of oil of turpentine in brandy punch. The turgescence and livid colour left the face after the bleeding, and did not return; the breathing also became easier, but no other signs of improvement appeared; and at the end of three hours it was still impossible to arouse her to consciousness. I then left her, giving directions that the external and internal stimulants should be continued; brandy and water being given every half-hour, while the power of swallowing remained.

During the night following there was little change in her condition; but in the course of the morning of the twenty-third day she began to manifest some return of consciousness, and in the afternoon I found her evidently able to recognize those around her. The pulse was still fluttering and irregular, and the breathing high and laborious, accompanied by loud bronchial râles, but without the stertor of the day before. I continued the turpentine and brandy, and I applied small flying-blisters over the course of the 8th pair of nerves—a practice frequently recommended by Dr. Graves. A gradual change for the better took place during the night, and on the twenty-fourth day the pulse had become full and steady, though still over 100; the breathing less hurried and laborious; and the meteorism had entirely subsided; the profuse greasy perspiration had also disappeared, and the skin was moderately warm and soft. I now desired brandy and water to be given at longer intervals, and withdrew the turpentine, giving decoction of senega and carbonate of ammonia for a few days, while the cough and copious expectoration continued. The convalescence was rapid and complete.

Dr. Graves remarks that no case could teach more decidedly than this the necessity for a cautious prognosis in fever, or of never relaxing our treatment in despair of recovery. When we consider the subject of treatment, I shall have to impress upon you the supreme importance of the measures you adopt in the last paroxysm of fever, upon which the fate of the patient so often depends; and the nature of which measures, will necessarily vary with the indications requiring to be fulfilled. In the other two cases I have narrated, I have not a doubt that life was saved by the abstraction of blood. The explanation of its salutary effect we find in the experiments of Coleman, as quoted by Wilson in his lectures on the vascular system, and John Reid, as given in his physiological and pathological researches. Both of these experimenters found that in animals asphyxiated by hanging, or by the operation of narcotic poisons, the right cavities of the heart became turgid with blood, and ceased to contract on the application of a stimulus; but that their contractions were at once excited upon the distension being lessened by the abstraction of a small quantity of blood from the overloaded auricle. (*Appendix IV.*)

Of course I should never advise the abstraction of blood without the most energetic measures being at the same time adopted to sustain the heart's action, and arouse the failing nervous energy. Warm punch at short intervals, turpentine, ammonia, and the repeated application of flying blisters, sinapisms, and turpentine epithems, are the best means of fulfilling these indications.

To recapitulate briefly—The rules you will observe in your study of the circulating and respiratory systems in fever are:—

1. To study them conjointly, bearing in mind the intimate connection of the two functions, and the close anatomical connection between the organs.

2. Never to be content with the observation of one portion of the systemic circulation; but, in all cases, to note the condition of the heart, arterial system, and capillaries.

3. Similarly to conjoin your observation of the congested condition of the right cavities of the heart, with that of the larger systemic veins on the one hand, and the pulmonary circulation on the other.

4. To bear in mind that the heart, though weakened by the action of the typhus poison, may become excited under the influence of cerebral or cerebro-spinal irritation, when the phenomena of jerking heart and small radial pulse will be observed.

5. To connect the observation of the rate of the respiration with that of the pulse, and carefully to note any deviation from the normal ratio of the former to the latter; especially when this is conjoined with sudden elevation of temperature.

6. To note any alteration in the percussion sound of the anterior portion of the chest; and no matter how latent may be the rational symptoms, to make a careful examination of the posterior portion of the chest, when any such alteration as a heightened clearness is observed.

7. To be careful to discriminate disturbed respiration of cerebral origin, from that arising from pulmonary congestions; taking, as your guide, the associated symptoms such as the excited action of the heart in the former, and the more marked rise of temperature, with the physical signs, in the latter.

8. In all cases, when called to attend fever during its early period, to make a daily and careful examination of the chest, in order that the supervention of pneumonia may not occur unobserved.

LECTURE VII.

ABDOMINAL LESIONS.

PROCEEDING in our study of the pathology and symptomatology of fever, we have next to consider the derangements of the digestive system, and the anatomical lesions found in the abdominal organs after death. These you will find to differ in the various forms of fever, their importance being of the greatest in typhoid, and comparatively little in typhus.

1. The symptoms which you have to study are briefly derived from—(*a*) The tongue and fauces; (*b*) The desire for food or drink; (*c*) The quiescence or otherwise of the stomach; (*d*) The alvine excretions, their frequency and character; (*e*) Pain in the abdomen; the signs revealed by exploration of this region—meteorism, retracted condition, gargouillement, tenderness, &c.;—(*f*) The symptoms of peritonitis from perforation.

The tongue of the earlier stage of typhus—soft, blanched, tremulous, notched by the teeth, and sometimes shrivelled,—is highly characteristic, more especially when the disease is of the fatal congestive type. In these cases the character of the tongue often approaches that of cholera. The tongue of the typhoid patient usually differs at this period, being preternaturally red, more especially at the tip and sides, and with a creamy fur sometimes coated on the dorsum, and with a triangular red patch on the tip. As the disease advances, the upper surface becomes dry, the tip continuing red; and in typhoid especially it is fissured sometimes so deeply as to bleed. In other cases of typhoid the dorsum of the tongue is glazed, and occasionally presents aphthous spots, or a curdy coating, with sometimes one or more ulcers. In the more advanced stage of typhus it is protruded with difficulty, tremulously, and uncertainly; frequently it is not protruded at all. In the worst cases it seems to be retracted, drawn back as it were into the pharynx. Changes in the mode of protrusion and appearance of the tongue usually follow other signs of improvement. In many cases of typhus, however, a line of moisture at the sides of the tongue is one of the earliest indications of the approach of a favourable crisis.

Anorexia is almost constant in fever from its invasion—with the exception of relapsing synocha, in which a morbid appetite has been frequently observed—but patients often appear to relish liquid nourishment, such as beef tea, throughout the disease. Thirst is present as a rule. It usually becomes less marked as the fever advances, and its total absence constitutes one of the most serious symptoms, more especially when conjoined with other anæsthesiæ in the later periods. By some, absence of thirst has been attributed (erroneously in my opinion) to arrest of the function of nutrition. By the older writers it was correctly ascribed to diminished sensibility—an explanation which is supported by the fact, that a patient who can no longer ask for drink often swallows it with avidity when presented to him.

Nausea and vomiting are normal phenomena, when occurring in the early periods of typhus, typhoid, or relapsing fever, especially in the two last. In the more advanced stages vomiting is a symptom of a formidable character, arising usually from enteric, hepatic, renal, or cerebral complication.

The early vomiting of typhus or typhoid may be ascribed, in all probability, to the action of the fever poison upon the 8th pair of nerves; at a later period, in typhoid, irritation in the mucous membrane of the stomach, or of the ileum, may be the cause, it being associated with tenderness on pressure in these situations. With reference to this, Dr. Bright remarks "the most alarming symptom is the irritable state of the stomach, accompanied by frequent vomiting, when a quantity of green fluid is usually thrown up." In relapsing synocha the very frequent vomiting of green fluid—sometimes of black—seems to depend on congestion of the liver, stomach, and spleen. Other forms of vomiting depend upon complications to be hereafter noticed. The most frequent of these is, probably, the cerebral; the next the renal; and lastly, there is a form of vomiting which, in my experience, has been uniformly the precursor of a fatal termination, due, apparently, to paralysis of the 8th pair of nerves, and associated with other symptoms of this lesion. In this case, the stomach seems, for a time, inert, all fluids swallowed accumulating in it; until at length—generally during an attempt to swallow—it suddenly and violently ejects a copious quantity. I may remark that this form of vomiting is very frequently associated with another serious symptom—obstinate hiccup—and also with yawning.

You will arrive at the differential diagnosis of the conditions causing vomiting, by careful study of the coexisting symptoms; as well as by the species of fever in which it occurs. Thus, there will be little difficulty in recognizing the nature of the bilious vomiting, of relapsing fever, which occurs in about seven-tenths of the persons affected;—not unfrequently passing into black vomit,—and is occasionally, though rarely, followed by hæmatemesis. Again, in typhoid, when it occurs associated with tenderness over the ileocæcal region, meteorism and ochrey diarrhœa, and with constant nausea and frequent ineffectual retching; you will seldom err in ascribing it to follicular enteritis. Renal vomiting will be recognized by the presence of albumen, or of blood corpuscles in the urine, or by the suppression of this excretion; also, by the marked urinary tremors, facial spasms, &c., of the renal complication. The cerebral vomiting is associated with headache and intolerance of light and sound, with excited jerking action of the heart, soreness of the surface, and other symptoms of arachnitis. It is sudden, and the fluid vomited is almost invariably of a bright green colour. Its frequency led Dr. Graves to state as an axiom, that "in all feverish complaints, where, during the course of the disease, the stomach becomes irritable without any obvious cause; and where vomiting occurs without any epigastric tenderness, you may expect congestion or incipient inflammation of the brain, or its membranes."

Crisis, by bilious vomiting, has been observed by some writers.[1]

True Jaundice as distinguished from the characteristic sallow tinge of skin, occurs occasionally in relapsing synocha, and has been noticed by different authors. Cheyne and O'Brien mention, that in the epidemics of 1818, and 1826, it was met with in those who, from previous intemperance, had suffered disorders of the biliary system. It appears from the observations of Graves and Cormack to be unconnected with hepatitis. In a fatal case reported by Dr. Law,[2] the liver and spleen were found much enlarged, congested, and softened, and I have reported a case[3] which terminated

---

[1] See Dr. Gordon's Lectures on the Complications of Fever, Dublin Hospital Gazette, 1854.

[2] Dublin Quarterly Journal, vol. 8.   [3] Ibid.

favourably, in which there was discharge of blood from the bowels, with deeply jaundiced skin, there being enlargement of the liver, with tenderness on pressure. Jaundice is a very rare complication in typhus, and is said, by some, never to occur in typhoid. This, however, is not strictly true, as a few cases have been recorded by Andral and others.

The frequency and character of the alvine discharges in fever are of much importance, and should be daily inquired into, as they have a direct bearing on diagnosis, prognosis, and treatment. No fact in medicine is better established than that ochrey diarrhœa— the recognized sign of ulceration of Peyer's glands—is the all but invariable rule in typhoid, while it is the rare exception in other forms of fever. It becomes, therefore, a valuable diagnostic sign of this disease: while in typhus, if diarrhœa occurs at all, it is either the effect of hypercatharsis; or a catarrhal diarrhœa; or, coming on at the close of the case, it may be critical; or, lastly, the profuse liquid discharges, in some fatal instances, may be properly termed colliquative.

As in typhus, so in relapsing fever, constipation is the rule; at least in the primary attack. At the time of crisis, however, diarrhœa sometimes sets in along with sweating; and in the relapse, occurs (as well as other phenomena of typhoid) not unfrequently.

It is, then, as one of the most constant phenomena of typhoid, depending on one of its most essential pathological conditions, that diarrhœa claims your attention.

It may be the first, or one of the first, symptoms, or it may not come on till the end of a week or ten days, or, in more rare instances, it may be postponed until a still later period. It may also continue after the termination of the fever, or, having subsided previously, may return during the period of convalescence. In each of these cases diarrhœa will have relation to a different pathological condition, and will possess a different diagnostic and prognostic value. Thus occurring as an initiatory symptom at the commencement of typhoid, I should regard it as an eliminative discharge from the liver and intestinal glands. (*Appendix V.*) In the second stage it is a sign of ulceration of Peyer's glands, and of inflammatory irritation of the surrounding mucous membrane. Continuing after fever, it is a certain indication of the persistence of ulceration. Its return during convalescence is usually caused by reactive irritation, or, as too frequently happens, by a new attack of enteritis and ulceration, arising from indulgence in improper food, or some other mismanagement. Finally, it has occasionally appeared to me that diarrhœa originated from suppression of the cutaneous transpiration after fever:—To this condition I shall have to draw your attention when speaking of the treatment of the convalescent.

The characters of the stools in typhoid fever are strikingly uniform. Usually they resemble pea-soup in colour and consistence, and, on standing, separate into a supernatant ochrey fluid, and gritty flaky deposit. According to Murchison, the fluid contains about 4 per cent. of albumen and salts, chiefly chloride of sodium; and the deposit consists of particles of undigested food, disintegrated intestinal epithelium, blood corpuscles, and shreds of sloughs, which have separated from the intestinal ulcers, together with multitudes of crystals of triple phosphate. Not unfrequently I have observed, instead of the characteristic stools, evacuations consisting entirely of a green fluid, evidently containing the colouring matter of the blood which had undergone a change. Occasionally, too, the stools are mucous and mixed with blood, resembling those of dysentery. Lastly, pure blood, in greater or less quantity, is sometimes discharged by stool. This latter occurrence is one which should immediately arouse your attention, as, although it may sometimes be salutary—a form of crisis, in fact—it is often a source of great danger.

You must not, however, forget, that although diarrhœa is the rule in typhoid, this rule is not without exceptions. Ulceration of Peyer's glands has been found after death, in cases in which no diarrhœa had existed during life. In the first volume of the third series of "Guy's Hospital Reports," Dr. Wilks mentions a case of typhoid in which constipation had existed, and in which the intestine was full of firm scybala, under each of which was an ulcer.

Intestinal hemorrhage in typhoid fever seems to follow upon one or other of three conditions: 1. Ulceration, involving a considerable vessel; 2. Congestion of the venous capillaries, with or without extravasation into the submucous tissue, this extravasation sometimes occupying a large portion of the intestine; 3. A septic condition of the blood, sometimes primary, but no doubt often caused by reabsorption of matters from the surface of typhoid ulcers, more especially in protracted cases. I have seen several examples of the latter, and I may refer you to cases reported by Dr. Todd and by Dr. M'Dowell. (*Appendix V.*)

In relapsing fever, hemorrhage from the bowels is by no means unfrequent. It seemed to me, during the epidemic of 1848, to be often connected with a highly congested condition of the spleen in that form of fever. I may here mention that epistaxis, which was also of frequent occurrence in the epidemic relapsing fever, seemed to be also connected with the same condition of the spleen, the physical sign of which, is dulness on percussion of the lateral and posterior portion of the left side; this dulness being frequently accompanied by pain in the epigastric and left hypochondriac regions; both, no doubt, produced by congestion and enlargement of this organ. During the epidemic of relapsing fever of 1848-9, I so frequently observed this condition of the spleen existing after the crisis, and followed by relapse, that I was led to connect the latter with it, and to regard it as probably due to the reabsorption of depraved blood which had lain by, as it were, in this organ, and so had not shared in the depuration of the mass during crisis. A somewhat similar observation is made by Dr. Henderson: "When the enlargement of the spleen took place during a paroxysm of the epidemic fever," says Dr. H., "it did not subside when the latter terminated, so that the treatment of the splenic disease was, for the most part, the business of the intervals between the paroxysms. None of the cases terminated fatally, and I cannot, therefore, give an account of the changes of structure or aspect which the spleen may have undergone; yet that the disease was truly inflammatory, the great degree of pain and tenderness which accompanied it, and the occasional development of a symptomatic fever, distinct from the epidemic fever, and unlike it in several important particulars, appear sufficiently to attest."—*Edinburgh Medical and Surgical Journal*, v. 61.

Pain in the abdomen is of frequent occurrence in the early periods of typhoid; indeed, it is often the first thing complained of in association with diarrhœa, when it is of a griping character, resembling that of ordinary bilious diarrhœa.

Careful and daily repeated examination of the abdomen is necessary in all cases of fever, more especially in typhoid. This reveals, in general, more or less meteorism, which sometimes, in the latter stages of typhus or typhoid, amounts to very considerable tympanites. In typhus, this symptom is indicative of a depressed condition of the nervous system; and is usually associated with other symptoms of this condition; it acts injuriously by embarrassing the respiration, and moreover is of very unfavourable augury. In typhoid it sometimes appears to indicate extensive disease of the intestinal follicles, being usually associated with obstinate diarrhœa. Occasionally you will remark a localized meteorism in cases of typhoid, the cæcum being distended with air, and presenting an evident globular tumour. The opposite condition—or that in which the abdomen is retracted—is sometimes met with at an advanced stage of typhoid, or during an imperfect convalescence. It is usually associated with uncontrollable diarrhœa, and is a sign of the worst possible augury.

Gargouillement is a sign of much value in typhoid fever, indicating, as it does, a condition of the ileo-cœcal valve, which allows of the reflux of the fluid contents of the cæcum into the small intestine. In cases in which diarrhœa has not yet occurred, you may expect it, if, on making steady pressure over the seat of the valve, you feel a crepitation or gurgling. This is not to be confounded with the gurgling produced by displacing the air in any other part of the intestinal canal. It is a totally different phenomenon, giving the idea of the passage of fluid from a larger into a smaller space, and being strictly localized, which the other is not.

It has been observed that diarrhœa occurring in typhus is not attended with gargouillement. Thus, Dr. Da Costa remarked, that in an epidemic of typhus in Philadelphia, attended with diarrhœa, there was in all the cases a total absence of gargouillement, as well as of affection of Peyer's patches in the cases which terminated fatally. My experience agrees with this.

In many cases of fever you find general tenderness of the surface, or cutaneous hyperæsthesia, in which the surface of the abdomen shares. This, however, is a sign of no value, *quoad* the abdominal organs; the tenderness of dothinenteritis is localized, confined to the ileo-cœcal region, and even to a portion only of this region; and if it is produced by pressure elsewhere, is still referred to this part, or, as in some cases, to the umbilicus. You should carefully distinguish these forms of tenderness, as they are both signs of much value, but of conditions each wholly different from the other.

You should always attach much importance to tenderness on pressure over the ileo-cœcal region; but at the same time you should not infer too much from its absence, since extensive disease may exist in the intestinal follicles without its being felt. Instances of fatal peritonitis have occurred in which no complaint of pain or tenderness has been made, up to the occurrence of perforation of the bowels. These cases, however, are the exception, not the rule; and usually, steady pressure will elicit a complaint of pain or uneasiness, provided the patient's consciousness be not to much impaired.

The symptoms which indicate that peritonitis has supervened during the course of typhoid fever, or during its convalescence, are —rigor; vomiting of green fluid; pain, commencing in the ileo-cæcal region and extending over the abdomen; most frequently tympanitic distension of the abdomen, but sometimes a tight, retracted, and rigid condition; arrest of diarrhœa, if it previously existed, is usual, though not constant; and in some cases I have observed a frequent desire to pass urine. To these succeed all the signs of collapse, rapid thready pulse, cold skin with greasy perspiration, hiccup, hippocratic face, and death usually within forty-eight hours —often in a much shorter period.

These symptoms, or most of them, may be present, and yet no perforation be found after death. The fact has been stated by several observers, and occurred to me in two instances during the epidemic of 1847-8, in which I found extensive ulceration but no perforation. But in the vast majority, the peritonitis is the direct result of perforation of the intestine, or rupture of some other organ —as of the spleen, the gall-bladder, or a tumefied and softened mesenteric gland. In an immense majority of instances the first-mentioned lesion is the cause; and this may take place at any period from the ninth or tenth day to an advanced period of convalescence. In reading the history of these cases you cannot fail to be struck with the fact, that in a large number the symptoms of peritonitis were so obscure, owing to the defective consciousness of the patient at the time, that neither it, nor the perforation which caused it, was even suspected. In many cases collapse was the only symptom observed.

The anatomical lesions found in the abdomen after death in the different forms of fever, vary in each. Thus in typhus, nothing is met beyond a congested condition of the capillaries and of the spleen; ·and (while diarrhœa is not unfrequent either as the effect of purgative medicines, in the early stage, or as a critical evacuation at the close, or as a sign of great secondary blood-contamination in the advanced stage of the disease) in no case, as already observed, is the characteristic ulceration of the intestinal follicles of typhoid present. Full proof of this interesting fact is to be found in Dr. Wilks' Report on Fever, *Guy's Hospital Reports*, Third Series, vol. 1.

The vomiting, which occurs in seven-tenths of the cases of relapsing fever, and the jaundice, which was observed in nearly one-fifth of the cases in Scotland in 1843, are accounted for by the existence of various degrees of congestion of the mucous membrane of the stomach—amounting sometimes to submucous extravasation,—and to more or less engorgement and softening of the liver and spleen. In the remarkable case reported by Dr. Law—in which the patient passed in the course of less than two days from a slightly jaundiced hue into deep jaundice, with delirium, lethargy, coma, and death—the liver and spleen were found much enlarged and congested, and so softened in structure that they seemed as if they had been soaked in blood.

In a case under the care of my friend, Dr. O'Reilly, of Trim, during the epidemic of 1847-8, this condition of the spleen led to fatal rupture. Dr. Murchison states that in one case Dr. Jenner found the spleen to weigh 38 ounces. I have recorded a case of relapsing fever in which it weighed 46 ounces. There can be no doubt that congestion of this organ, is much more marked in relapsing synocha than in typhus or even in typhoid fever, while the affection of the intestinal glands, so constant in typhoid, is altogether absent in the primary attacks of relapsing fever, though frequently present in the typhoid form of relapse.

Pathologists have remarked, that the liver is not prone to become the seat of any peculiar affection in either form of fever, beyond congestion and some degree of softening. A distinction has been drawn between the frequency of occurrence of abscess of the liver, in connection with ulcerated colon in dysentery, and the extreme rarity of the same occurrence in ulceration of the glands of the small intestine. Some years ago, Dr. Hare presented to the Pathological Society of London (*Transactions*, Vol. III.), an example of acute abscess following enteritis, the peculiarity of the case being that the small intestine was the seat of the inflammation and ulceration; but, even in this case, neither the symptoms nor morbid appearances were those of typhoid fever. Dr. Gordon has however published an interesting case of acute hepatic abscess complicating typhoid fever, as well as one of abscess complicating typhus. Frerichs and Gordon relate cases of acute atrophy of the liver occurring in fever with jaundice. (*Appendix V.*)

The constant and characteristic anatomical lesion of typhoid is the affection of the glands of Peyer (which is called by some dothinenterite; by others, acute follicular enteritis) with which is constantly associated a congested, enlarged, and softened condition of the mesenteric glands, and of the spleen.

The morbid process in these organs has been so accurately described by Rokitansky, that I shall refer you to his account of the successive changes, which you may compare with the plates by Bright and Cruveilhier, now placed before you.

In a few words, the order of the morbid process in the agminated and solitary glands, seems, according to the most recent and accurate observations to be—first, infiltration of the gland by the deposit; next, escape of the exudative material into the submucous tissue covering the gland; then inflammation and sloughing. The infiltration appears to occur in the early stages of the fever; ulceration usually from the ninth to the eleventh day; and sloughing with perforation, sometimes as early as the fourteenth or fifteenth. If this direct and rapid sloughing does not occur, the ulcers gradually cicatrize. Dr. Murchison states that in a case which terminated fatally from other complications, about the fortieth day, he found all the ulcers cicatrized. Dr. Peacock, however, gives a case, fatal on the sixtieth day, in which healing was still incomplete. Much will depend on circumstances independent of fever; but you cannot be too guarded in your prognosis with regard to this question, since numerous cases have occurred of fatal peritonitis from perforation long after diarrhœa had ceased, and even after convalescence had appeared to be established. This may occur in examples of the uncicatrized chronic, or what Rokitansky has termed the atonic ulcer; or it may arise from inflammation being set up in and around the newly-healed ulcers. It is obvious that the possibility of such an occurrence should suggest the greatest caution in the management of the convalescence from typhoid, more especially with regard to diet.

Along with the changes in the intestinal follicles, we meet with corresponding changes in the mesenteric glands situated opposite to the ulcers of the intestine. From the first, they become enlarged and congested; this state increases, and they occasionally suppurate. As the ulcers heal, they gradually shrink, and return to their natural size and condition.

The spleen is constantly enlarged and softened in typhoid, and, according to Rokitansky, the congestion and softening of this organ extend to the mucous membrane of the larger curvature of the stomach, the vessels of which become turgid, and the mucous membrane softened and friable.

The morbid appearances in fatal peritonitis from perforating ulcer, present nothing peculiar, with the exception, perhpas, of a greater tendency to be circumscribed than is the case in idiopathic peritonitis. Circumscribed peritoneal abscess has been met with by several observers. (*Appendix V.*)

LECTURE VIII.

DERANGEMENTS OF THE URINARY SYSTEM.

WE have to-day to consider certain disordered states of the function of secretion, and more especially of the secretion of urine, as one which is immediately connected with the molecular nutrition of the body, and the derangement of which constitutes one of the most serious complications of fever.

You note in the dryness of the tongue and mouth of a fever patient, and in his total loss of appetite, the signs of arrest of molecular nutrition or assimilation; a similar arrest of these changes would seem to take place in the lungs, since, according to the observations of Dr. Malcolm, the excretion of carbonic acid is notably diminished in typhus. On the other hand, Liebig's statement, that the exhalations from a fever patient are highly charged with ammonia, has been confirmed by many observers. "It is well-known," says Dr. Murchison, "that in severe cases of typhus, the breath has often an ammoniacal odour, and that thick white fumes are produced on holding a glass rod, previously dipped in hydrochloric acid, close to the mouth of the patient. Dr. Richardson found the breath in one case so ammoniacal, that it coated a glass slide moistened with hydrochloric acid with crystals of chloride of ammonium, and restored the blue colour to reddened litmus paper."

We may presume, that the secretion of gastric juice is arrested, or nearly so, in cases in which little or no solid food is taken by the patient. I am not able to refer you to any observations of importance regarding the secretions of the stomach, or of those of the pancreas or liver.

I have long entertained the belief, that in most cases of typhoid, a deranged condition of the biliary excretion is the first step in the morbid process. That, in fact an effort is made to eliminate the poison through this channel, as in the case with regard to other putrid substances received into the circulation; and that the diarrhœa of the early stage of typhoid (often its first symptom) is due to this cause.

But of the state of the biliary excretion and the part it plays in fever, we may be said to know nothing of a satisfactory kind. Neither do we derive any information of value from the intestinal excretion, unless in the case of the ochrey diarrhœa, attending ulceration of Peyer's glands in typhoid; of which affection it may be said to be a pathognomonic sign.

It is far otherwise with the secretion of urine, the importance of which, in relation to fever, has been recognized by physicians of all ages. This importance arises from the facts, that healthy urine is an accurate indication of healthy assimilation; and that retention of the excreted urine in the bladder, or suppression of its secretion by the kidneys, or faulty or imperfect secretion, are, each and all, attended with a form of secondary blood contamination, constituting one of the most serious complications of fever.

Your investigation of this subject will be much aided by your previous knowledge of (a) the constituents of healthy urine, and the relation of this excretion to the waste of the tissues; (b) the changes of its constituents produced by disease; and (c) the symptoms set up by the retention or suppression of the excretion, or of one of its constituents (urea).

In my Lecture on the Theory of Fever, I dwelt strongly upon the importance of the preservation of the balance between disintegration of tissue and elimination of effete material; and also on the serious blood contamination resulting from an arrest of the latter in a disease like fever, in which regressive metamorphosis is notably increased. Some portion of typhoid or adynamic state undoubtedly is, as Parkes observes, "an urinæmic condition, due, not to diminished formation, but to diminished excretion of urea." And this condition is most formidable in those patients in whom the previous mode of living has produced the accumulation in the blood of the greatest amount of materials in a state of metamorphosis, and so has incurred the greatest necessity for the active exercise of the excreting function of the kidney. We shall hereafter find that some of the most serious of the cerebro-spinal complications of fever may be ascribed to the poisonous action of these nitrogenous matters—urea and its cognate compounds—upon the nervous centres. (*Appendix VI.*)

If, then, from a faulty action of the kidney, due to pre-existing disease of these organs or any other cause—or even from simple retention—the excretion of these substances is checked, the complications alluded to may be expected to arise. Your daily attention is therefore required in order to ascertain that the patient is able to void urine by his own unaided efforts; a daily examination of the urine, in regard to its quantity and sensible qualities, is also necessary; and to this should be added, in many cases, the occasional determination of the specific gravity, and of the presence or absence of albumen or of blood.

Urine may be discharged involuntarily, or retained, or suppressed, or altered in quality, in various modes, as above mentioned.

The first occurrence indicates diminished sensibility, or paralysis of the sphincter vesicæ; the second, paralysis of the muscular coat of the bladder; in the combined conditions we have retention, with dribbling, often mistaken for involuntary discharge. Thus you can understand why it is so necessary to examine carefully the hypogastric region in all cases in which the urine is observed to dribble.

You should never rely on the statement of the nurse, but ascertain for yourselves, by percussion over the pubes, if the bladder is emptied; or if, on the contrary, the urine discharged into the bed is but the overflow of the distended viscus.

The general characters of febrile urine, according to Dr. Parkes, are—deficiency of water and chloride of sodium; increase of other solids, more particularly of urea, uric, hippuric, sulphuric, and phosphoric acids, and the pigments. The deep colour of the febrile urine, according to Dr. Parkes, is only partially due to its concentration. If febrile urine is diluted to the usual amount of healthy urine, it is still darker than usual. Professor Vogel has shown that the colouring matter is sometimes increased fourfold.

"The importance of this pigment," says Dr. Parkes, "consists in the fact that it may, apparently, be considered as a measure of the metamorphosis of the blood globules; and from its amount, we are led to conclude that in some cases of fever the disintegration of the blood particles may be four times as rapid as in health."

Observers of all ages have regarded any deviation in the sensible qualities of febrile urine from the above type, as indicative of danger.[1] Thus, the frequent coincidence of pale, watery urine, with convulsions and other serious nervous lesions, has been noticed by Willis and other old English writers on fever. I have myself most frequently observed it in cases in which the nervous derangement was of an emotional kind; some examples of this will come under review when we are considering the lesions of the cerebro-spinal system.

Albumen has been found in the urine of typhus and typhoid patients, as an occasional occurrence independent of any lesion of the kidney, by Edwards, Warburton Begbie, Sidey, Parkes, and other observers. It would seem to occur more frequently in some epidemics than others, and it has appeared to me to be occasionally caused by the exhibition of saline purgatives; a very common practice among our hospital patients. Upon one occasion two brothers were admitted into this hospital, one of whom had been severely purged, by salts, before admission. In him, the urine was albuminous for several days; while his brother presented no trace of it, the course of the disease in both being similar in all other respects. Taken alone, the presence of albumen would usually seem to indicate the participation of the kidneys in the general congestion of typhus. This congestion is, however, sometimes scarcely appreciable after death, even though symptoms of uræmic poisoning may have existed during life. Of this fact I have witnessed several examples in this hospital; and a case published by Mr. M. Taylor is quoted by Dr. Murchison: "A man, aged fifty-three, died on the twelfth day of an attack of typhus; the eruption was well marked.

---

[1] Recent observers have added but little to the descriptions of the sensible qualities of the febrile urine by the older writers. Take, for example, the following description of the appearances of the urine of exanthematous typhus, by Burserius (Institutes, vol. iii.):—"At first it is at one time thin and watery; at another time natural, and exhibits a globular, unequal palish cloud floating in it. Sometimes, also, at the beginning it is whitish, but copious; shortly after it grows confused, (?) like pomegranate wine, or yellowish, thick, turbid, and deposits a sediment. It likewise sometimes grows black, as if it were mixed with soot, or turns red, being slightly tinged with blood. Sometimes, during the increase, and at the height of the complaint, it is nearly suppressed, which must be considered as a fatal symptom, unless it quickly comes off thick, and deposits a sediment. Trollius, in his patients, always found it proper in quantity, seldom thin and pellucid, but generally free of sediment; sometimes of a dusky red, but never concocted or having a proper sediment. Pinaroli, however, found the urine, on the first days of the complaint, pale, clear, and scanty; during its increase, somewhat red and confused; at its decline, turbid, and thick, but not uniformly so.

Death had been preceded for four days by stupor and muttering delirium. Some hours before death three pints of urine were drawn off by the catheter. After death, the kidneys were found perfectly healthy—not even congested—and urea was discovered in considerable quantity in the blood removed from the heart and large veins."

We sometimes meet with blood in the urine; (a) It may occur from the nature of the exciting cause; cold applied to the loins or prolonged immersion in water. In the former case, there is usually, with hæmaturia, suppression of secretion of urine from acute nephritis. In the other, there are the symptoms of septicæmia, as in cases already detailed.

Some years ago I attended two young gentlemen, in each of whom fever followed upon cold applied to the lumbar region while perspiring. In both, hæmaturia came on in an early stage of the fever, and was rapidly followed by complete suppression of the secretion, by strabismus, convulsions, and coma, ending in death.

(b) It may be an indication of the highly congestive type of fever, as noticed by Sir D. Corrigan. "The urine in such cases," says Sir D. C., "is dark-coloured, muddy and without sediment, properly so called, but occasionally showing a small quantity of a dark-coloured, pasty deposit, such as is seen in the dark-coloured urine of hæmaturia after scarlatina. It is of low specific gravity, is small in quantity, and coagulates when heated, from the serum it contains."

(c) It is one of the most frequent signs of the septic type of fever; of that condition in which the blood is decomposed by the action upon it of putrid substances received into it, either through the stomach or by respiration. Such cases are invariably fatal.

In a patient, æt. 35, previously healthy, with the exception of a constipated condition of the bowels, fever came on after violent sickness caused apparently by eating mushrooms. He was profusely maculated when I first saw him, on the ninth day. His consciousness was perfect, but his manner listless and apathetic; his face and brow flushed; the eye injected, and the pupil contracted; distressing vomiting still continued, and during the ensuing thirty-six hours he had frequent returns of facial spasms, and constant tremor of the muscles generally. The attacks of spasm were accompanied by a peculiar tremulous moan, characteristic of uræmic poisoning.

The urine was a smoky-looking fluid, coagulating on the application of heat; and containing a large quantity of blood corpuscles. The facial spasms gradually increased in frequency, and ended in a fatal attack of convulsions on the eleventh day. Apparently the cause of the hæmaturia in this case was septicæmia.

Uræmic poisoning of the blood and nervous centres is the consequence of the different morbid conditions, of which the above-mentioned states of the urinary excretion are the signs. This poisoning may arise from—1st, retention; which may be either simple or complicated with diseased prostate or kidney.

2d. Rapid disintegration and consequent accumulation in the blood of the products of regressive metamorphosis.

3d. Nephritis, either acute or chronic.

I. Cases have been recorded, which prove that uræmic poisoning may occur in fever, simply from retention. Dr. Todd's observations also prove that these symptoms may be produced, in nonfebrile cases, by enlarged prostate. He records the case of a gentleman who suffered a succession of attacks of loss of consciousness, with partial paralysis of sensation, which, however, ceased altogether upon the difficulty of micturition having been relieved by mechanical means. I have met with similar cases, and I have sometimes seen the symptoms of uræmia come on during fever in old men who were the subjects of enlarged prostate, apparently from resorption of the retained urine. I have several times known general muscular tremor subside immediately after the introduction of the catheter in these cases; and I may refer you to examples published by Dr. Brinton and Sir D. Corrigan, of uræmic poisoning, evidently due to retention, and relieved by the same operation. (*Appendix V.*)

But less complete or prolonged retention will be followed by the same result, when this is associated with either chronic nephritis or acute congestion of the kidney. "In most cases of typhoid fever," Rayer observes, "in which there was affection of the kidneys, the patients were prostrated, after one or more days, by stupor; the bladder distended, formed a tumour in the hypogastric region, and the urine was retained, or dribbled from an overflow. The condition of the patients did not suffer them to make a spontaneous complaint of pain in the region of the kidneys; and the nephritis was diagnosed, during life, by the alkaline state of the urine, the presence of mucous or sometimes of blood globules, and that of albumen, in this liquid. . . . . The practitioner should be aware of the possibility and of the gravity of this complication, and of the danger which in this respect attends retention of urine, and of the necessity there exists not to allow urine to accumulate in the bladder; because nephritis, once developed, determines or hastens the death of the patient."[1]

II. Uræmic poisoning, caused by the accumulation in the blood of the products of rapid disintegration, was of frequent occurrence in the epidemic relapsing fever of 1843 in Scotland. I believe this fact was first noticed by Dr. Henderson, in a paper published in the sixty-first volume of the *Edinburgh Medical and Surgical Journal*.

It was long since observed, that persons of full habit, and who have been habitually luxurious in their mode of living, are peculiarly liable to suffer convulsions in fever; and as modern observations tend more and more to show that uræmia is the pathological condition upon which these convulsions depend, and inasmuch as numerous dissections have revealed a state of the kidneys either perfectly healthy, or at most but slightly congested, we are led to the conclusion, that in such cases the uræmia is due to excessive formation of urea from disintegration of tissue on the one hand, and to defective elimination on the other.

III. Lastly, uræmia in fever often depends upon acute, or chronic nephritis. I have already alluded to cases in which albuminuria, hæmaturia, and suppression of urine, occurred in consequence of the application of cold to the lumbar region. The cause of these symptoms in such instances, is acute congestion or inflammation of the kidney. The result, we may say, is usually fatal. Dr. Parkes mentions that in two cases of typhoid fever, the patients left the hospital well of the fever, albuminuria still existing. "In one of these two," says Dr. Parkes, "it existed when the urine was first examined; in the other the urine, at first non-albuminous, contained afterwards blood in small quantity, casts and albumen in large quantity. Although I had no opportunity of knowing what course the kidney disease afterwards took in this case, there was, evidently, a very profound kidney lesion which did not pre-exist, but was immediately excited by the fever."—*On Urine*, p. 250. Equally fatal is the pre-existence of chronic disease of the kidney; and of its frequency we may judge from the statement of Dr.

[1] *Maladies des Reins*, tom. ii. p. 204.

Christison, that "every instance of death from sudden convulsions in fever, which has of late occurred in the Edinburgh Infirmary, and underwent proper investigation, has proved to have been connected with an albuminous state of the urine, and with organic disease of the kidneys." Experience proves, that patients suffering from Bright's disease may go on for months enjoying tolerable health, if the blood is from time to time relieved by some evacuation, by means of which urea is eliminated; but such eliminative actions, being checked by fever, symptoms of blood-poisoning manifest themselves, and so commingled with those of fever, as often to render difficult an accurate diagnosis of the exact share of each in the production of the group. Such are the cases described by Fenger, under the name of masked forms of Bright's disease.

The extent to which this uræmic poisoning of the blood will affect the nervous centres in fever, depends much upon other causes, influencing both the nutrition of the brain, and the condition of the circulating fluid. In this respect the fever patient resembles the puerperal female, in whose blood, urea may exist in large quantity without producing convulsions, provided this fluid be otherwise healthy, and the brain free from congestion. (*Appendix V.*)

I reserve a detailed description of the nervous symptoms characteristic of uræmia, merely mentioning that the aspect is so peculiar as not to be mistaken by those who have once witnessed it. Some present a brick-red suffusion of face and brow, with a glazy blood-shot eye and contracted pupil; others, in whom the disease of the kidney is more advanced, being pale, waxy, and bloated-looking: the facial spasms also being perfectly pathognomonic of this condition. Dr. Christison has recorded a most characteristic case in his work on diseases of the kidney, which I shall read to you.

" A night-watchman, of regular habits, aged 32, about four years previous to his attack of fever, had suffered some form of urinary complaint, and had since been liable to frequent micturition, and at times retention, and also to hæmaturia. His fever commenced with nausea, vomiting, general pains—especially in his loins—headache, giddiness, and singing in the ears. On the fifth day he had frequent calls, with much inability to pass urine, and on the seventh complete retention. On the evening of the seventh day sixteen ounces of urine were withdrawn by the catheter, and on the following evening twelve ounces. On the ninth he complained of great weakness, general soreness, frequent greenish vomiting; the face was brownish, flushed, oppressed, and characteristically febrile. The pulse 100, and feeble; skin warm, and covered with pale, irregular, diffuse petechiæ; tongue dry and furred; bowels constipated. His answers were clear and intelligent, but he had muttering delirium and subsultus tendium. He was ordered a laxative solution, and a little wine, cautiously, with a morphia draught at night; eight ounces of urine were withdrawn by the catheter, but not preserved for examination. He passed a restless night, and on the morning of the tenth day he was incoherent, torpid, less easily roused, and affected with frequent twitchings of the face. The pulse was 120, and weaker; the tongue covered with sordes; deglutition difficult; the face flushed, and the eyes injected. Soon after the visit he was suddenly attacked with violent convulsions of the whole body, attended with deep, imperturbable coma, squinting of the eyes, a full, strongly jarring pulse, and powerful action of the heart. Three ounces of urine were withdrawn by the catheter; specific gravity, 1014, strongly coagulable by heat. The coma continued profound, with occasional spasmodic tremor and contractions of the limbs; and death took place only two hours after the appearance of convulsions.

"On examination, the only organs presenting any abnormal appearances were the kidneys. Both were very large, elongated, one-third at least beyond their average length, and very flabby. The cortical portion presented much greater breadth than usual, its fibrous structure was very obscure everywhere, and in most parts obliterated—generally darker than natural—exuding a good deal of blood from a fresh cut surface, and presenting a remarkable mottled appearance, composed of the ordinary brown cortical, but checkered uniformly with small, brighter, brownish-red, or blood-red points. Two ounces of fluid blood were cautiously removed from the heart, in which, upon examination, an appreciable quantity of urea was discovered."

"I have several times found," says Dr. Christison, "the blood loaded to a much greater degree with urea than in the present case, without any tendency to coma being manifested; and a greater amount of diminution of the urine not unfrequently occurs without the functions of the brain seeming, for some time, to suffer.

But in this instance the brain and nervous system had also to encounter the oppression arising from the typhoid form of continued fever."

Besides its poisonous action upon the nervous centres, urea occasionally manifests its presence in the blood during fever, by the occurrence of bullæ, always considered symptomatic of danger by physicians of experience. These bullæ are ascribed by Henderson and others to the conversion of the retained urea into carbonate of ammonia, which has the power of dissolving the fibrine of the blood.

To recapitulate. In your clinical study of fever you will always remember that—

I. The excretion of urea bears in health a constant proportion to the quantity of nitrogenous food ingested, and to the amount of disintegration of tissue by bodily exercise, mental labour, anxiety, or other cause.

II. That, therefore, the balance between the work to be done by the kidney, and its due performance—in other words, between the products of disintegration of issue present in the blood, and the excretion of urea—may be disturbed in either of two modes, viz., by excessive accumulation of the one, or by deficient excretion of the other.

III. That the consequence of this disturbance is in each case the same; uræmic poisoning being set up by the one mode in those who, by luxurious living, have caused the accumulation in the blood of a large amount of nitrogenous matters, which the kidney is unable perfectly to eliminate; by the other, in those who suffer from congestion of the kidney, produced by cold or by septicæmia; or from acute or chronic nephritis, or from simple retention, and consequent resorption of the urine.

IV. That your daily attention to the state of the bladder is imperatively necessary; and that the catheter should be at once employed in cases of retention, with or without dribbling.

V. That the quantity and sensible qualities of the urine being daily observed, any marked deviation from the normal colour or amount should lead you to note the specific gravity, and apply the tests for albumen, or blood.

VI. You should bear in mind that the poisonous action of urea upon the nervous centres, acts, like all poisons, with greater intensity, in cases in which there exists, also, congestion of these centres.

VII. You should also bear in mind, that what may appear, at first view, to be simply the adynamia of fever, is often, in a great measure, due to secondary blood contamination—in fact, to uræmia.

VIII. The occurrence of bullæ during the course of fever should lead you to pay more than ordinary attention to the state of the urinary secretion. (*Appendix V.*)

LECTURE IX.

CEREBRO-SPINAL LESIONS.

I have now to direct your attention to the study of a group of derangements of the highest importance in themselves, and of which it may be said more truly, perhaps, than of any other, that while they are not seldom obscure, latent, and difficult of diagnosis the safety of the patient will often depend upon the sagacity with which they may be recognized, or even anticipated; and the promptitude with which they may be met by suitable and energetic treatment.

The difficulty of a diagnosis of the nature of the various nervous derangements occurring in fever, and of their relation to this disease, arises from several sources; for example, we sometimes find it not easy to determine whether cerebral or meningeal inflammation is primary or secondary, the cause or effect of fever; in other cases, whether cerebral symptoms are due simply to the fever poison, or to its complication with inflammation; or whether, with this last condition, there is also one of the many forms of secondary blood contamination. Occasionally, also, the period of crisis is marked by various nervous derangements peculiar to itself, but yet not easily discriminated from those of a more serious nature. In conditions so complex, perfect accuracy of diagnosis is not to be attained, but you will be assisted in an approximation to it, by a careful consideration of the following points: The influences acting on the function of innervation in fever; the various modes in which these functions are deranged in this disease, as well as the physiological relation of each mode of derangement to different portions of the nervous centres;. the results of the investigations of morbid anatomy; and lastly, the effects of certain methods of treatment, of which we shall meet with illustrations hereafter.

The functions of the cerebro-spinal system, the derangements of which become objects of your study in fever, are those of ideation, or the intellect, consciousness, the voluntary or sensori motor, involuntary or excito-motor functions; the special senses of sight, hearing, taste, and touch; and the cutaneous sensibility. Your attentive consideration will also be demanded by the derangement of animal heat, and, most important of all, of sleep.

The modes in which these functions may become deranged will vary; (*a*) according to the type of fever—those set up by the true action of the fever poison being more frequent in typhus, while those due to reactive irritation are rather met with in the advanced stages of typhoid; (*b*) according to pre-existing states of nutrition of the cerebro-spinal centres—you will find that these states of nutrition have a powerful influence on the production of nervous complications in fever, in conformity with the law I have often referred to, that the elective affinity of a morbid poison is for the organ which may be undergoing the most active retrograde metamorphosis; (*c*) again, different modes of deranged cerebral capillary circulation exercise a marked influence upon these derangements—this holds true equally of active hyperæmia, anæmia, or passive congestion. Lastly, different forms of secondary blood contamination predispose to corresponding forms of nervous derangement in fever, or modify those already existing.

In your analysis of each case of fever, you will find the function of innervation is affected variously, according to the influence of one or other of the above conditions; I shall presently adduce some instances illustrating this influence.

Of the cerebro-spinal functions the intellect is the one most frequently deranged in fever, more especially in typhus, and in a mode characteristic of the intoxicating action of this poison. The aspect of the typhus patient has been aptly compared to that of a man awaking from a drunken slumber; his apprehension is, in a similar manner, confused and defective; impressions made upon his senses are imperfectly perceived by the sensorium, producing the phenomena of dreaming without sleep; he talks incoherently about external objects, confounding them with internal impressions, or with the faintly remembered objects of past sensations; occasionally he seems to be for days together under the influence of some painful dominant idea, the product, as it were, of confused perception of the present, and imperfect recollection of the past. This distressing state is sometimes produced by injudiciously excluding the light. Cullen mentions in his Physiological Lectures that in his own fever the admission of light, by rectifying the perception, and supplying the different parts of the complex idea to his mind, removed much of his delirium. The memory is so deficient, that the patient, who seems while in fever to recognize his attendants, calling them by name, becomes immediately afterwards unconscious of their previous presence; while the judgment occasionally is but little impaired, the patient being sometimes capable of sustaining conversation, and giving judicious directions: there is, however, one exception to this remark, namely, that he seldom judges rightly of his own condition. The character of the delirium varies, according to the age, mental activity, temperament, previous habits, &c., of the patient, and according to the state of the cerebral circulation, and the existence of secondary blood contaminations, &c. It may be observed, with truth, that nothing can be more varied than the forms of delirium met with in various pathological states, and at the same time nothing can be more unscientific or unsafe, than to attempt to diagnose these states by the kind of delirium. Thus the delirium of exhaustion, and that pneumonia complicating in typhus, are sometimes equally noisy and impulsive; while depletion in the one, would be a scarcely less fatal mode of treatment than a full opiate in the other. Any such diagnosis you will best arrive at, by studying the group of other nervous symptoms with which the delirium is associated. Yet some recent writers have proposed to deduce from the form or character of the delirium a diagnosis of the portion of the brain affected with inflammatory irritation in fever.

The normal state of a fever patient with regard to his consciousness, may be described as one of blunted sensibility. He sleeps from time to time, but nevertheless complains of want of sleep, and generally manifests a dulness of perception, and a drowsy apathy. *Ceteris paribus*, such is, no doubt, the most favourable condition with regard to consciousness; but this function may be variously modified by the intoxicating action of the fever poison, its heightened state being marked by watchfulness, often resisting all our efforts to procure sleep; the opposite condition, by coma, or stupor. In some cerebral complications we witness the alarming combination of watchfulness and stupor, but usually associated with the former there is a morbid quickness of apprehension; the patient talks rapidly, and answers instantly: his senses of hearing, taste, smell, and touch, are morbidly acute; and, while his skin is hot, he is much alive to the impressions of cold upon the surface of the body. This form of derangement is chiefly met with in youth, or early manhood; in the over-worked brain of the student, anxious professional man, or merchant; and in the spirit-drinker if young.

The opposite affection of the consciousness manifests itself, when, early in the fever, the patient shows more than usual somnolence and stupor, with sluggishness of perception; from which condition, aroused with difficulty, immediately he lapses into slumber. In this state, he protrudes the tongue slowly and hesitatingly, and does not withdraw it promptly; he refuses to drink, or does so listlessly and imperfectly, allowing the fluid to run from his mouth or coughing after each attempt. When such symptoms are associated with other serious nervous lesions, we may fear that the stupor will merge in fatal coma.

The most unfavourable cases are, indubitably, those which present the combination of watchfulness and delirium with stupor; symptoms of heightened consciousness, associated with those belonging to the opposite state.

The motor functions are differently affected in different stages of fever and in different complications. The fever poison, *per se*, depresses nervous energy, and the lesions of nervo-muscular force caused by its direct and uncomplicated action, seldom amount to more than various degrees of debility, or, in other words, diminution of muscular power; or to diminished control of volition over the muscular movements. The former is first manifested in the voluntary muscles; the limbs and the tongue tremble, and their movements are performed languidly and feebly; but as fever advances, the automatic and involuntary muscles suffer; swallowing sometimes becomes difficult, respiration laborious, and the heart's action faltering and irregular. There is usually, in the advanced stage, more or less subsultus, which, in graver cases, passes into carphology, or involuntary picking of the bedclothes. The want of consensus between the volition and the muscular movements—or the power of co-ordination in these cases, is shown in the vain attempt to protrude the tongue, and in the difficulty of swallowing. Meantime, as the control of the will is withdrawn, involuntary excito-motor actions occur, upon such impressions being made on the nervous centres as should normally excite voluntary movements of another kind; thus we occasionally observe violent facial twitchings to follow the putting a spoon to the lips; and Dr. Laycock narrates a curious instance in a glass-blower who, on attempting to drink, blew involuntarily into the cup every time its edge touched his lips—the will being powerless against the incident excitor stimulation, which had been habitually applied in the exercise of his daily occupation.[1]

Derangements of sensation, like those of the motor function, are among the early and normal effects of the fever poison, but become signs of cerebral irritation when they occur out of their order as to time—such as early deafness, or the headache of a later period—or when they occur with undue severity. Towards the termination, indeed, of simple and uncomplicated fever, if severe, profound lesions of the sensational consciousness are sometimes exhibited—such as anæsthesia of sight and hearing, insensibility to cold or to thirst, and to the impressions by which the active contraction of the sphincters are regulated. It has been maintained by some writers, that want of thirst arises from a lesion of the function of nutrition. I think the old writers were correct, who, like Willis and others, taught that it is a lesion of sensation—a nervous symptom forming one of a group of the most serious character.[2] Willis's expression regarding those in whom the symptom was associated with low temperature is: "Quippe ad hunc modum affecti circa morbi statum, plerumque in deliria, motus convulsivos, et non raro in maniam incidunt; e quibus brevi in mortem precipitantur."

With regard to the special senses, I may observe, that hearing is defective in a large proportion of cases of fever; and that in typhus the sense of taste is usually perverted, so that, in many instances, cold water alone is relished.

The cutaneous sensibility is augmented in some forms of cerebral complication; but in severe, uncomplicated typhus, is usually lost, or nearly so, in the advanced stages, the patient manifesting complete insensibility to cold.

Sleep, as a rule, is more or less deranged in all forms of fever. In favourable cases of typhus, the patient sleeps from time to time, often without being conscious that he does so. In typhoid, the sleep is restless and unquiet from an early stage, the wakeful period usually corresponding with the febrile exacerbation. In both we meet with obstinate sleeplessness, associated with the worst forms of delirium.

It cannot be said, that there is any casual relation between the

---

[1] Clinical Lecture on Fever, *London Medical Gazette*, vol. xxxix.

[2] "Sometimes there is no thirst present, even when the fauces and mouth are parched, which must certainly be ascribed to impaired sensibility."—*Burserius*.

nervous derangements of uncomplicated fever, and the appearances discovered after death. The appearances most usually observed, are, venous congestion of the membranes and substance of the brain, and subarachnoid effusion. With regard to the first, it is no doubt due to the altered attractions between the blood and tissues, whereby the nutrition of the organ becomes modified; and perhaps in some cases it may be partly caused by the position of the body, the state of the pulmonary circulation, or the mode of death. With regard to the second, Dr. John Reid has shown it to be the effect of deranged nutrition, and more frequently connected with the age of the patient, probably with his previous habits of intemperance, and want of sleep, than with either fever or inflammation. (*Appendix VII.*)

In cases of true inflammatory complication, and more especially during certain epidemic periods at which the complication prevails, you will meet with its products in the form of minute arterial injection of the surface of the convolutions; opacity and thickening of the arachnoid membrane; exudation of lymph, and of turbid or gelatiniform serum under the arachnoid, and in the ventricles; great augmentation in the number of bloody points on section of the brain, and increased firmness or perhaps acute softening of its substance. At different periods recently, the appearances found after death, and the symptoms during life, have been those of the peculiar inflammatory affection termed cerebro-spinal arachnitis, which has, from time to time, occurred as an epidemic in this country, and on the continent.

Of the symptoms of acute congestion of the brain, I may observe, that in such patients their behaviour is at first quick, excited, and restless, but in some it becomes at once heavy, stupid, and somnolent; the face and forehead flush, and revert again to pale; there is a constant knitting of the brow, increased on the admission of light; one or both eyes become suffused, and the sclerotic tinged from the injection of the minute vessels, the pupil being usually contracted; there is generally intolerance of sound, and tinnitus, sometimes early deafness. While the normal headache of fever declines about the period that delirium commences, in this complication it increases, or perhaps comes on for the first time at an advanced period; the patient, in general, is loud in his complaint of the pain, which is of a heavy, throbbing character: at first, all motion of the head is instinctively avoided, but at a later stage it is rolled upon the pillow incessantly, sometimes tremulously; the respiration (cerebral) is hurried, or sometimes fast and at others slow, interrupted by frequent sighing, which has at times, a catching, spasmodic character. The circulation is variously affected: the pulse, at first rapidly accelerated, usually soon falls below the natural standard, or becomes still more hurried and irregular on the slightest movement, being small and constrained at the wrist, while the carotids throb, and the heart's impulse is strong and jerking, its first sound being augmented. The muscles of the neck, and sometimes of the limbs, become rigid, and there is frequently muscular, as well as cutaneous dysæsthesia, the patient screaming on being attempted to be moved. There is green vomiting, persisting for several days, in a large proportion of cases. Frequently the pasty fur on the tongue presents a bright green colour. The urine is usually scanty, and of high specific gravity.

When this group of symptoms is presented at an early stage of the fever, your diagnosis will be comparatively easy; but it is not so in the case of a patient seen at a more advanced period, in which the symptoms of acute congestion may have become masked by those of the typhous state, and the patient's condition may, on a superficial view, present nothing differing from the simple toxic action of the fever poison. On this subject, Dr. Graves justly observes: "It is in many instances extremely difficult to distinguish the cerebral symptoms produced by the poisonous influence of fever on the brain from those which depend on true inflammation. The one gives rise to delirium and fatal coma as well as the other; and in the advanced stage of fever, when the manifestations of nervous energy are feeble and imperfect, and when the circulating and respiratory organs act with diminished power, the distinction between irritation and actual inflammation becomes a matter of great difficulty." Daily experience proves the truth of this observation, and yet no practical distinction can be more necessary. Nor do I believe that accuracy of diagnosis can always be attained even by the most experienced, or that we have any single, positive, and certain sign by which the two forms of disease can be in all cases distinguished. There are, however, several worthy of your attention, more especially, if associated together. The first is the state of the consciousness, which, in cerebral congestion at this period of fever presents the double derangement of watchfulness and stupor, usually conjoined with delirium of an impulsive character. If unable to rise, the patient is yet restless, rolling his head on the pillow, and not unfrequently rolling out of bed; in a word, he does not present the supineness of the intoxicated fever patient. Neither is the expression of his countenance the same: there is usually some degree of knitting of the brow, with contracted pupil and suffused conjunctiva; his muscles, whether flexed or extended, are rigid: his teeth firmly set; if he lies on his back, with his knees drawn up, it will be found that his heels are firmly pressed against the buttocks; at other times the extended limbs will present the combination of rigidity with tremor. The frequency of respiration is disproportionate to the rate of the pulse, it is noisy, and there is a loud and forced inspiratory murmur heard over the chest. The heart's sounds present that kind of augmentation caused by running or other over-exertion; but if the heart's tonicity is impaired by the fever poison, its muscular structure having become softened, the first sound may be feeble, and its impulse imperceptible. Cerebral irritation may be, even in this condition, detected by the character of the sounds; which, though feeble, are rapid, and give the idea of a jerking impatient contraction of the organ. You also will have a valuable sign in the greatly increased frequency or irregularity of the pulse on the slightest movement or excitement, or on the exhibition of wine or stimulants; while, at the same time, under their use, the head becomes hotter, and the extremities colder, the respiration more hurried and irregular, and the reflex actions, rigidity of muscles, restlessness, and jactitation more marked.[1] Whenever these symptoms follow the exhibition of wine, there can be little doubt of the existence of active cerebral congestion, no matter how advanced the period of fever, or how apparently debilitated the patient; and accordingly, here, treatment becomes a valuable aid to diagnosis. From my notes of many similar examples, I select short abstracts of a few cases which came under my own observation, and in which the symptoms during life, and the appearances after death, illustrate the observations I have just now made.

A girl, aged 16, was admitted into hospital on the twelfth day of fever. She was not maculated: nothing unusual or unfavourable

---

[1] Stoll thus sums up the signs—Constans delirium, vero et ferox cum urinâ paucâ pallidâ, sine contentis; .... tendinum subsultu, artuum tremore, inflexione eâ que violentâ, si eos extendere conetur medicus, maxillarum contractione ad ingerenda; depositionem seri ad ventriculos cerebri, infra tentorium cerebelli, atque ad thecam vertebrarum factam significat.

occurred after admission until the evening of the nineteenth day, when she became suddenly delirious, having been found by the nurse walking about the ward.

On the twentieth day the delirium continued, and she frequently attempted to leave her bed. The pulse had risen from 90 (on the preceding day) to a 100, but was small and constrained, while the heart's impulse was strong and widely felt.

On the twenty-first day there was less active delirium, but a tendency to stupor; the pulse rose in the evening to 120; heart as before.

Twenty-second day. She is reported to have slept for a very short time, for the first time since the nineteenth; less stupor, but great irritability; the arm is strongly retracted involuntarily, on attempting to feel the pulse; the heart's action strong and jerking.

Twenty-third day. Sleeplessness, restlessness, and general soreness of the surface, coldness of extremities; thready pulse, 120; with jerking action of the heart.

Twenty-fourth day. Stupor, passing into coma. Death.

On examination, the surface of the brain was found minutely injected, the convolutions slightly flattened, the arachnoid thickened and opaque in patches on the surface of both hemispheres. The substance of the brain was unusually firm and extremely vascular, presenting very numerous red points on being cut across; there was considerable effusion of serum in the ventricles and at the base, and the large arteries of the base were in a most congested condition. The heart was empty, contracted, firm in structure, and quite healthy in appearance in every respect.

In the above instance the cerebral congestion, no doubt previously latent, did not manifest itself until an advanced period of the fever; in the following it set in early.

The patient, a middle-aged man, who had contracted fever by exposure to contagion, was admitted into hospital on the sixth day. He was maculated, his extremities were cold, there were tendency to stupor and low delirium, an inability to protrude the tongue, and involuntary evacuation of urine and feces.

On the day following (the seventh), prostration was more marked; there was low delirium, with stupor and sleeplessness; pulse 112, and weak; heart's impulse weak, and first sound indistinct. He was ordered four ounces of wine.

On the eighth it was found that sleeplessness had been continuous, added to which, while stupor had increased, dysphagia and difficulty of protruding the tongue had come on. The pulse was 120, and weak; heart's impulse stronger, and first sound more distinct.

On the ninth day he had convulsions, which continued for nearly three hours without intermission, but ceased on the apothecary taking five ounces of blood from the temporal artery, leaving distorted mouth, and ptosis of one eyelid. He had one slight return of convulsion before his death on the following day, and gradually sank into coma, the pulse becoming imperceptible, but the heart's action jerking and excited to the last.

On dissection the appearances were similar to those of the preceeding case; the convolutions of the brain being flattened, their surface red, and minutely injected, and there was very copious effusion of serum in the ventricles. The heart was firmly contracted, its cavities empty, and its structure apparently healthy.

Dr. Anderson thus narrates a similar case:—

"Elizabeth Riddell, aged 15, became suddenly stupid and very restless on the twelfth day of typhus, having previously complained of headache and oppression; stare vacant; death in a few hours. *Inspection* after 26 hours. Upper surface of arachnoid and convolutions rosy; much effusion in the pia mater, though but slight congestion. The cerebral substance was very soft, the cortical of a pink colour; the ventricles contained bloody serum; purpura spots on the pericardium."

There can be no doubt that active congestion of the nervous centres manifests itself in different modes at different times, and that it is much under the influence of what is termed the epidemic constitution. You have a striking example of this in the occasional epidemical occurrence of cerebro-spinal arachnitis, which was very prevalent in this city from ten to twelve years since, and of which several cases have occurred in this and other hospitals during the past winter. It appears moreover to have prevailed in some places during the epidemic period of 1847–9, and is noticed by several observers, as by Dr. Cullinan of Ennis, and Professor Law of this city. Dr. Graves describes an outbreak of cerebro-spinal arachnitis, in a family at Rathmines, at a still earlier period. Recently, this complication has been admirably illustrated by Dr. Fritz of Paris.[1] (*Appendix VII.*)

The group of symptoms usually present in the instances which

---

[1] "Etude clinique sur divers symptomes spinaux, observés dans la fièvre typhoide." Paris, 1864.

have come under my own observation, all of which were in cases of typhoid or enteric fever, were :—

1. Headache, with intolerance of light.
2. Congested purplish flush of the face.
3. High temperature of the surface.
4. Marked soreness of the surface; cutaneous hyperæsthesia; and tenderness of the cervical vertebræ.
5. Pain and stiffness of the muscles of the neck, with partial opisthotonos—the patient usually lying with the head drawn back.
6. Sighing, irregular and panting respiration, or noisy respiration.
7. Jerking, sometimes rasping, action of the heart, much moderated occasionally by bloodletting.
8. Pasty thick fur on the tongue, usually of a light green colour.
9. Vomiting of a green fluid.
10. In a few cases, partial paralysis, or tonic contraction, with pain in the limb so affected.
11. In several instances, the patient has complained of acute pain with tinnitus in one ear.

A fact noticed by several observers, is, the great resemblance these cases sometimes bear to cholera, or to peritonitis ; vomiting, diarrhœa, and abdominal tenderness being so marked as to mask for a time the cerebral characters of the disease. Another occasional source of difficulty is their great similitude to hysteria. Such was the case in the woman Donnelly, who was recently under your observation; accurate notes of whose case were taken by Mr. Todhunter, our clinical clerk. (*Appendix VII.*)

I have met with several examples of this complication in private practice at different times. Two of these (both cases of typhoid) occurred in a brother and sister residing near Portobello, in close vicinity to an offensive open sewer.

In the sister, aged 20, the purple congested face, extreme cutaneous hyperæsthesia and green vomiting were well marked ; and partial paralysis of both hands continued after the termination of the fever, until the mouth was slightly touched by mercury.

In the brother, aged 18, the most prominent symptoms were rigors ; sickness and vomiting brought on by the slightest motion ; notable acceleration of pulse on change of position; severe headache, and rigidity with spasmodic jerking of the muscles of the neck. This condition also eventually yielded to the action of mercury pushed to ptyalism.

While these patients were under my care, I was in attendance upon a young lady in whom typhoid was apparently brought on by the same cause, an offensive drain in the house in which she resided. Towards the close of the fever, the cerebro-spinal complication appeared in the form of pain and numbness of the left arm, continuing for several weeks, and yielding only to the same measures, viz., repeated leeching, blistering the neck, and mercury, pushed to slight ptyalism.

Two of the cases which have fallen under my observation, were in medical students, in whom the disease was watched by myself and others with more than ordinary care.

In the first of these the cerebral symptoms set in on the invasion of fever, masking it completely for several days. This gentleman was seized with shivering, followed by headache and remarkable soreness of the surface, green vomiting, and strong jerking action of the heart. To these symptoms were soon added painful rigidity and retraction of the neck; the left arm, and, subsequently the leg, becoming also rigid, numb, and painful; the fingers being strongly and painfully flexed on the palm. Leeches were applied in relays, first of eight, and afterwards of six, behind the ears; each time with partial relief of the painful state of the neck and limbs, and, on the third application, with complete relief. Rose-coloured spots appeared on the ninth day, and the fever ran through its usual course. At the period of crisis there was a partial return of the rigidity of the neck, with pain in the arm and slight contraction of the fingers, but these symptoms yielded on the application of a blister to the neck.

A few days after crisis this gentleman was attacked by pleuritis, with signs of copious fibrinous exudation, from which he slowly recovered.

In the second patient the cerebral symptoms set in at a more advanced period of the fever, the access of which was very gradual. When I saw him, on the fourteenth day from the first feeling of indisposition, he presented the usual phenomena of typhoid fever. He had distinct rose-coloured papulæ on the trunk and arms, meteorism, gurgling over the cæcum, great increase of the limit of dulness over the spleen. There was but slight diarrhœa, but he had had repeated returns of vomiting and epistaxis during the preceding week; the urine was high-coloured, and in small quantity.

His face at this time was clear, with a bright flush; pupils rather dilated; he had very slight headache; the heart's action was natural, rather weak than otherwise; pulse, 100; respiration rather high and accelerated.

On the eighteenth day, headache had become very severe, attended with slight delirium; tremor of the hands and arms was also first observed; the pulse rose to 128; respiration loud and nasal, 36 in the minute. In the course of the next twenty-four hours, a marked change in the face was observed; the right eye became minutely injected, while the left remained clear; the flush became deeper and constant, and spread over the forehead; the brow was knitted. At intervals of ten or fifteen minutes a slight spasm seemed to pass over the face, the eyelids closing, and the eyes rolling under them; at the same time he moaned faintly, and thrust out the right arm, clutching the bedclothes with the hand. I happened to be engaged in sponging his hands at the moment when one of these paroxysms occurred, and on the instant the fingers closed on the sponge like a vice. He had slept for four hours on the night between the eighteenth and nineteeth days, but his sleep was noisy and distressing, his respiration appearing more laborious than when awake, and accompanied by loud moaning. As he had now had retention of urine for twelve hours, it was drawn off, and was found to be albuminous, and containing blood-corpuscles.

On the twentieth day there was little change: he had involuntary discharge of urine, with partial retention. The cerebral symptoms continued as before, with injection of both conjunctivæ, and a dark flush of the face and forehead; the heart's action was remarkably strong and jerking.

Twenty-first day. He had a good night; consciousness much improved; had passed urine voluntarily; pulse 108; rose spots fading fast.

Twenty-second day. Still improving: urine loaded with pale lithates, and not albuminous.

Twenty-third day. All the signs of reactive irritation of the brain had set in. The sleep had become more noisy and laborious than ever; the eyes much injected; the face deeply flushed and of a purple hue; skin hot; great thirst; the heart's action was jerking and impulsive; pulse 132; respiration high, laborious, and irregular.

In consultation it was agreed to apply four leeches behind the ear, a bladder half filled with ice to the head, and to give three grains each of gray and antimonial powder, three times a day.

Early on the morning of the twenty-fourth day a gradual change took place: he breathed more easily, slept more quietly, the face became less flushed, and in the course of the day the pulse came down gradually to 92, at which point it remained for several days, while he steadily improved in all respects.

On the twenty-seventh day, however, another change occurred. After a sudden and violent fit of coughing, he was found on our visit with a flushed face, hot, dry skin, rapid pulse, and hurried breathing. He had no headache or nervous symptoms, and his intelligence was perfect; his only complaint beyond the cough, being of the intolerable heat of his skin, and he constantly demanding a warm bath. On examination, the right side of the chest was found to be dull on percussion, both anteriorly and posteriorly in its superior half, in which situation the respiratory murmur was muffled, almost inaudible, and mixed with slight sibilant rales in this region. The expectoration was scanty and extremely viscid, and during the following three days was occasionally tinged with blood; examined under the microscope, it was found to contain a few pus-corpuscles.

This pulmonary congestion was, apparently, resolved by a critical effort. On the twenty-ninth and thirtieth days, copious sweating came on, accompanied by the deposit of lithates in the urine; and from this date the signs of solidification gradually disappeared, and his convalescence proceeded.

LECTURE X.

CEREBRO-SPINAL LESIONS.

I HAVE mentioned that the occurrence of nervous lesions in fever, is much influenced by various pre-existing states of nutrition of the nervous centres. Overwork, anxiety, and excitement seem to predispose to active congestion of the brain, as was evident in the following case.

A lady, aged 35, of nervous and excitable temperament, who had recently been subjected to unusual exertion and mental anxiety, was attacked by typhoid fever, which, for a fortnight, presented no peculiar symptoms. I was first called to see her upon the sixteenth day, when there was much prostration, with a dusky, muddy hue of the face; dark, thinly-scattered eruption; tympanitic distension of the abdomen, with tenderness and gurgling over the ileum, and gritty, ochrey diarrhœa. The urine was natural in appearance and quantity, the breathing was high and frequent, without any appreciable bronchitis. The heart's action feeble, and the pulse weak, ranging from 120 to 130; the intelligence was dull, but otherwise undisturbed. With a view to moderate the diarrhœa and support the strength, a small blister was applied over the ileum, and she was ordered decoction of bark and muriatic acid, with port wine. A gradual improvement followed these measures, and between the eighteenth and nineteenth days a favourable crisis seemed to take place; slight perspiration occurring, with copious deposit of pale lithates in the urine. On the following day, however, my attention was aroused by the turgescence and purple colour of the patient's face. Her breathing had become hurried, 52 in the minute, with laboured, jerking inspirations, while careful examination of the chest discovered no pulmonary lesion. The most marked sensibility and soreness of the surface had supervened; and she had, for the first time during her illness, involuntary discharge of urine. Her manner was hysterical and excited; and lastly, the heart's action, for days previous so feeble, had become jerking and forcible; pulse 112. The diagnosis of cerebral congestion was confirmed in the course of the day, by the gradual supervention of paralysis of the right arm and hand. The head was shaved, and relays of leeches were applied to the mastoid process; a large blister was placed over the occiput, and gray powder and James's powder given three times a day, mercurial inunction being at the same time employed. No improvement appeared until the twenty-third day, when the gums became slightly touched. The mercurial action was immediately followed by a change in the appearance of the face and manner, by the respiration becoming easy, and decreasing to 24 in the minute; by the pulse falling at once to 84; and the heart's action becoming tranquil; and by the loss of soreness of the surface. The paralysis of the arm and hand, however, remained, and only gradually disappeared, so that some months elapsed before she could resume her professional avocations as a teacher of music.

In this patient, I remarked that, for a considerable period after recovery, attacks of cerebral congestion, indicated by headache, flushing, brightness of the eyes, &c., were induced by any mental exertion or emotion.

This case also illustrates a fact we have often observed in this hospital, viz: the latency of cerebral lesions during the same progress of fever. They seem often to be masked by the febrile state, to appear at or subsequent to crisis. A remarkable example of this occurred here, about three years since, in a man named Cromwell, who was admitted in the first stage of typhus, having been severely beaten over the head two or three days previously. No symptoms whatever of active congestion of the brain appeared during the fever of fourteen days; on the contrary, those of asthenia were so marked, as to require the exhibition of 20 ounces of wine, and 12 ounces of brandy, daily, for a short time before crisis. Immediately after crisis, however, the most complete intolerance of all stimulants was manifested; and shortly acute headache came on, attended with constant vomiting of green fluid, marked cerebral breathing, jerking impulse, and rapid action of the heart, together with all the other symptoms of arachnitis. This state continued highly marked during a week, and yielded eventually to the daily application of leeches, and the administration of mercury pushed to ptyalism.

It sometimes happens that exposure to a noonday sun, more especially if combined with severe muscular or mental exertion, will produce active congestion of the membranes of the brain, with fever of a gastric type; the latter more or less masking the former. Some years ago, I attended a gentleman, aged 25, who had for several weeks undergone much mental excitement and bodily fatigue, with prolonged exposure to the sun during a hot season. For several days before fever set in, he complained of pain, tightness and weight in the occipital region, with intolerance of light, and more especially of sound; he also complained much of the distress occasioned by the forcible and jerking action of the heart, which could only be moderated by the repeated application of leeches to the mastoid process. After the termination of the fever, the recurrence of the same cerebral symptoms required repeated leeching, and they eventually yielded only to the full influence of mercury. Although his recovery was complete it was extremely slow, and for several months, any fatigue or excitement, or a single glass of wine were each invariably followed by headache, flushing of the face and brow, sleeplessness, and irritability of temper.

A less fortunate case came under my observation more recently. A gentleman imprudently sat for a considerable time reading, with the head bent forward under the direct rays of a noonday sun in summer. Headache soon followed, attended with vomiting of green fluid. To these succeeded fever of a gastric type, which was treated by a medical man without peculiar reference to the head symptoms. After convalescence from the fever, a constant headache remained, and when, after an interval of some months I first saw him, this gentleman appeared to be suffering from chronic arachnitis, attended with partial hemiplegia and epileptiform convulsions; he had also partial amaurosis. Shortly afterwards he had an attack of apoplexy, which ended fatally an hour after seizure.

In a few instances, I have observed the cerebral lesions due to previous disease or injury, set in with the commencement of fever, continue during its course, and undergo, first exacerbation, and then resolution, at the period of crisis. The following is an example of this occurrence.

A young woman was admitted into hospital under my care on the seventh day of maculated fever, having been for many months previously subject to frequent attacks of headache, ascribed to a severe injury from the fall of a heavy piece of timber upon the head.

The fever was ushered in with rigors and headache, followed by repeated vomiting; and on admission the face was flushed, the eyes dull and suffused, skin hot and dry, pulse 90 and full, heart's action violent, the tongue coated with a white adhesive fur; there was also intense thirst, and ardor urinæ. Soon after admission she vomited about eight ounces of green fluid, and vomiting or retching recurred at intervals throughout the fever. On the thirteenth day, after more than usually severe retching, she complained of great increase of headache, with intolerance of light, the eyes being much suffused. Six leeches were now placed behind the ears, and the same number twice afterwards. Subsequently she was also ordered to take a grain of calomel every third hour. On the fifteenth day the report states she had not been relieved by the leeching; she had constant rolling of the head on the pillow; the head being hot, the skin dry, tongue brown, urine scanty and high-coloured. A blister was applied to the nucha, and on the seventeenth day, there being no amendment, another to the occiput. On the evening of this day she had a convulsive seizure of a mixed character (hysterical epilepsy) and, after an interval of an hour, a second. On the following day the cerebral symptoms were relieved, and she appeared free from fever, but had mercurial griping and slight ptyalism; her convalescence was rapid and complete.

One of the most remarkable instances I have ever witnessed, of the influence of pre-existing disease in determining the early occurrence of nervous lesions in typhoid fever, as well as their peculiar mode of manifestation, occurred in a valued professional friend, attended by myself and others; one of whom (who was in constant attendance) has kindly given me some memoranda of the case.

Our patient was a gentleman of remarkable ability, and distinguished for scholarly attainments, but was, unhappily, of a strumous constitution, manifested in various forms. Shortly before his last illness, he had presented a group of symptoms which the late Dr. Hutton was of opinion arose from strumous disease of the base of the brain. The nervous derangements which set in at an early period of his fever (typhoid) were most peculiar, but were distinctly to be referred to the same portion of the nervous centres. The first was excruciating facial neuralgia, which was not relieved by the topical application of sedatives, but yielded at once to blistering the nape; a practice suggested by our belief in its centric origin. To this succeeded the most remarkable paroxysms of hysteria, more especially of hysteric dyspnœa, similar to the paroxysms of spasmodic dyspnœa, which we occasionally, though rarely, observe in epileptics, and which appear to replace the epileptic fit; also resembling somewhat a form of paroxysmal dyspnœa, I have occasionally seen in aneurism of the arch of the aorta. In these paroxysms, he would start up in bed, gasping for breath, the mouth open, the eyeballs staring, crying out for "brandy, give me brandy." This condition lasted for a few moments, and gradually subsided. To it succeeded coma, convulsions, partial paralysis of the left arm and hand, low delirium, failure of the circulation and death. (*Appendix VII.*)

In your clinical study of the cerebro-spinal derangements of fever, besides the difficulty of diagnosis to which I have referred in my last lecture, you will sometimes experience much difficulty in distinguishing the symptoms of anæmia from those of hyperæmia of the brain, those of inflammatory excitement from those of exhaustion; since opposite states of the cerebral circulation in fever, may produce similar derangements of innervation. In like manner, as we know, that in cerebritis, after the inflammation has been subdued a set of symptoms termed pseudo-inflammatory, often present themselves, arising from exhaustion, and demanding the exhibition of opium and wine.

The following case, which was under my care many years ago, and of which I took daily notes at the bedside, so well illustrates this fact with regard to cerebritis, that I am induced to offer it, not as a guide to your treatment of cerebral complications, since it occurred at a period when bleeding and calomel were more freely employed than in the present day; but as showing both the tendency to cerebral symptoms of the inflammatory form in those fever patients who have previously suffered arachnitis from injury of the head, disease of the internal ear, &c. &c., and also the fact, that sometimes the symptoms of inflammation are so commingled with those of adynamia and exhaustion, as to make the treatment of the combination neither simple nor easy.

M. Suttle, a shoemaker, aged 20, of intemperate habits, was admitted into the Navan Fever Hospital, on the 10th of January, 1840, complaining of violent throbbing pain in the head, especially of the forehead, intolerance of light, deafness, with discharge of reddish fluid from the right ear. He states, that 12 months since, he had a similar attack in the head and ear, for which he was 6 weeks under treatment in a Dublin Hospital. During the interval, there has been no discharge from the ear, but it has remained perfectly deaf.

On the 26th of December (the 16th day before admission), and for three following days, he was constantly intoxicated. While in this condition, he fell into water, and slept in his wet clothes. On the 31st (11 days since), he had a violent shivering, which continued with little intermission for two days, and has since frequently recurred. His health was not, he says, otherwise much affected, till the 7th instant, when acute headache set in, accompanied with flushing, alternating with shivering. On the 9th, the discharge from the ear recommenced without any relief of the headache.

The pain is sometimes very violent, coming on during sleep, and awaking him: he is drowsy, yawns frequently, and says he feels heavy and oppressed. His face and head are flushed and hot, temporal arteries throbbing, eyes suffused and heavy, pupils sluggish, pulse 88, full and vibrating, tongue soft and clean, bowels costive, urine natural.

He was bled to sixteen ounces, and ordered a purgative enema, and to take a scruple of calomel in the course of 24 hours.

January 11th. The enema produced three copious light-coloured evacuations: he fainted after the bleeding, and complains that he passed a bad night, feeling very cold, and shivering frequently. At noon he was sleeping, the trunk and limbs being agitated with tremor and twitching, his feet were cold, face and head flushed and hot, eyes suffused; pain in the head somewhat less acute, but still very great; the ear continues to discharge freely. Pulse 112, hard and jerking. Ordered six ounces of blood to be taken from the neck by cupping: and to take an ounce every three hours of a mixture, containing two grains of tartar emetic, in 8 ounces of water.

12th. Was much relieved by the cupping, passed a good night, no return of rigors; skin generally warm; pulse 120, and hard; tongue white and moist, bowels confined.

Repr. mistura. Habeat Haust Rhei.

13th. Much better, slept well, skin warm, face pale, and the head much cooler; pulse 116, but softer, tongue white and moist, gums slightly sore, bowels opened by draught.

14th. No visible change.

15th. Has had vomiting and hiccup apparently caused by the antimony, is now quite free from pain in the head, or intolerance of light; no discharge from the ear; pulse 120, and soft. The antimonial was suspended, and diaphoretic mixture given.

16th. Passed a sleepless night, but has dozed a little this forenoon, makes no complaint of pain, but says he is weak, and seems torpid, and a little inclined to rave; bowels not opened since yesterday, tongue cleaner, but red at the tip; has slight hiccup this morning; pulse 128, soft and compressible; ordered to continue the diaphoretic mixture, and to have six ounces of wine.

17th. The nurse states that he lay awake a great part of the night with his eyes fixed on the ceiling in a vacant stare, raving a good deal, with occasional hiccup; this morning he is delirious, and answers incoherently. There is great tremor of the tongue and lips; his hands are also tremulous, and there is an appearance of extreme prostration about him; pulse 144, small and compressible, respiration natural, tongue red, and rather dry, bowels free, urine normal.

A blister was applied to the nucha, and the wine continued, with the following bolus every third hour.

Calomel and Camphor, of each 3 grains; James' powder 2 grains; Opium half a grain.

18th. Last evening he vomited, and had much hiccup; vomited again during the night, had but little sleep, and raved a good deal, leaving the bed occasionally, and walking about the room; answers incoherently with a fatuous smile; appears much prostrated; pulse 116, and improved in character, tongue white in the centre, and moist, tremulous when protruded; has much subsultus; skin cool, feet cold. Ordered to continue the bolus; to have 8 ounces of wine, and a draught containing 20 drops of the acetum opii, and 30 of Hoffmann's anodyne liquor.

19th. Much improved. The draught was soon followed by quietness and sleep; he has since slept a good deal, and is now perfectly coherent; he has had no return of vomiting, nor for some hours has he had any hiccup; his countenance has lost the silly expression of yesterday; pulse 112, and soft; skin warm; tongue white and moist.

To have a bolus 3 times a day, and 8 ounces of wine.

20th. Passed a good night, slept almost uninterruptedly, no delirium, and but little hiccup, no headache, is perfectly rational, but very deaf; tongue steadily protruded; pulse 112, and full; skin warm, bowels open.

21st. Had a good night, and is much better, no headache, is rational; pulse 88, and soft, tongue clean, bowels regular.

23d. Has passed an indifferent night; the nurse states that he wandered a good deal in his sleep, still he appeared improved; intellect clear, countenance good, tongue clean and protruded firmly, pulse 84, full and soft, bowels regular.

To have a draught containing 12 drops of the acetum opii, at bedtime.

25th. Perfectly convalescent.

There can be little doubt, that while the state of exhaustion is sometimes produced by a depleting treatment of fever, especially by repeated purging, it sometimes owes its existence to the deficient power of the left ventricle, by which a condition of brain is produced, differing in degree rather than in kind, from the anæmic, ill nourished condition of the brain, in cases of fatty degeneration of the heart. This analogy is suggested by Dr. Stokes, in his researches on the condition of the heart in typhus.

One source of difficulty in recognizing this state, is that the weakness of the heart may be said to be the rule rather than the exception in fever; another is the similarity of symptoms in the opposite states. It is true, that in many, the exhausted brain is indicated by pallor of the face, by a clear eye with dilated pupil, cold, white, or mottled skin, a weak whispering voice, limpid urine, tendency to faintness in the erect posture, &c.; but the eye may be suffused, and the pupil contracted from want of sleep, and other signs mentioned may be wanting. I could not give you a better illustration of the difficulty and importance of this diagnosis, and its bearing upon treatment, than the following case which I have slightly condensed from Dr. Ormerod's work on fever.

The case is entitled, "Fever with papular rash; affection of the brain; symptoms relieved; death by exhaustion."

The patient was a previously healthy, temperate man, admitted into hospital on the seventh day of fever. He had taken purgative medicine previous to his admission. His face was pale and anxious; eyes glassy; head hot and perspiring; feet cold; tongue moist and furred; pulse 74, small and soft; bowels freely open from medicine; urine free and high-coloured; skin moist, with a papular rash about the chest and abdomen; pain in the head much increased by cough.

Eighth day. No sleep from cough and headache; face flushed and perspiring; eyes suffused and glassy, skin hot and moist; pulse 92, tongue thickly coated with white fur; cough, and rusty expectoration.

Ordered to be cupped to six ounces from between the scapulæ; to take two grains of calomel to-night, and half an ounce of castor oil to-morrow morning.

Ninth day. The cupping gave immediate relief; cough and headache are much less severe, but he was wakeful and restless during the night; his face is flushed and hot; tongue dry and furred; pulse 98, with more power; three copious evacuations from the bowels.

Ordered to have the head shaved, and to be cupped to four ounces from the temples; to have two grains and a half of mercury with chalk thrice a day, and a saline draught with camphor mixture every six hours.

Tenth day. Restless, with incessant pain shooting from the forehead to the temples and occiput; pulse 108, small; face hot, flushed, and anxious.

Ordered to be cupped to six ounces from the temples; to take half an ounce of castor oil immediately, and to continue the medicine.

Eleventh day. Head hot; face flushed; skin hot and dry; tongue dry and furred; pulse 110, small; three copious, liquid ochrey evacuations from the bowels; he breathes hurriedly and with difficulty, but auscultation detects nothing more than slightly prolonged expiration.

Ordered ten leeches to the forehead; to take a little more nourishment, and to continue the medicine.

Twelfth day. Delirious; restless; two loose, ochrey evacuations from the bowels; abdomen tender; pulse 120, small and soft.

To continue.

Thirteenth day. Two scanty liquid evacuations; pulse 120, small and soft.

Fourteenth day. Lies on his back, unconscious, breathing with difficulty; unable to protrude the tongue; stares wildly at any one who addresses him, but does not move a feature, and returns no answer.

Ordered to be cupped to six ounces from the temples immediately; to continue.

Fifteenth day. He lay in the same way for two hours after being cupped, then, soon after, bursting into a profuse perspiration, he recognized and spoke to his father; had some sleep during the night, and is now quiet and free from pain; pulse 98, small and soft.

Ordered to have six ounces of wine and more beef-tea; to continue the saline draughts, and cold lotion to the forehead.

Sixteenth day. Noisy and restless during the night; when visited, trembling and hardly sensible.

Ordered a blister to the nucha, and a common enema.

In the evening he was lying on his back, perspiring profusely; trembling, and muscular rigidity; dysphagia; perspired all night, and died at 6 A. M. of the seventeenth day.

Post-mortem Examination.—The brain and membranes were in all respects natural, only there was a little old thickening of the arachnoid, and no more than the ordinary quantity of fluid in the meshes of the pia mater; the lungs were slightly gorged with blood; the heart healthy, containing fluid blood, &c. &c.

Well may Dr. Ormerod say: "If we looked to morbid anatomy only as a guide to future treatment, there would be a decided condemnation of all the depletive treatment here adopted."

You will sometimes meet with nervous derangements at or immediately following the period of crisis. Thus, occasionally the "exacerbatio critica" is marked by a sudden increase of delirium, subsultus, or tremor; the delirium being impulsive, the patient restless and agitated, jumping out of bed, leaving his room, striking his attendants, &c. After a time this agitation subsides; he lies down, and falls into a calm and prolonged sleep. Sir D. Corrigan has pointed out that the nature of this nervous excitement may be recognized "by the absence of coma; by the hysterical character of the agitation, and by the copious secretion of urine loaded with lithates." Occasionally some attention is required to distinguish the profound sleep which follows this delirium from coma. This, however, can be always done, by noting the equal distribution of heat; the state of the pulse, respiration, pupils; and also noting the consciousness, which is perfect in the former, while it is lost in coma.

We also often see other nervous symptoms occur as modes of crisis, and as consequences of imperfect crisis. The first are met with chiefly in the young and in females. In women whose menstrual period corresponded with the commencement of fever, hysterical symptoms—sometimes hysterical convulsions—are apt to occur on the eve of crisis. In the young of either sex a fit of convulsions is not uncommon, even when no cerebral symptoms have marked the earlier period of the fever, beyond, perhaps, a more than usual amount of headache. I once saw a case of this kind in a boy aged 14, who complained of headache, without suffusion of the eyes, or any other symptom of cerebral inflammation. Towards the termination of the fever a degree of stupor came on; after this had continued about a day, he had a severe convulsive paroxysm, followed by a return of stupor, which continued for two days, and was again followed by an attack of convulsions; after which he recovered without any further bad symptom. I ascertained that this boy never had epilepsy previously. The urine at the time of the fit was copious and limpid, like the urine of hysteria.

A very similar case is given in Henderson and Reid's Report of the Fever of Edinburgh, in which a single paroxysm, preceded by stupor, was followed by immediate recovery. In some of these cases the cerebral circulation seems to become much deranged, for the first time, on the eve of crisis. This appeared to be the case in a girl aged nineteen, who was admitted into hospital under my care on the eighth day of fever, and had no cerebral symptoms until the fourteenth, when she was suddenly seized with convulsions; the attack continued for fifteen minutes, and recurred after five hours, the last fit being followed by much stupor, pain in the head, and distortion of the mouth. The evident cerebral congestion was treated by cupping the neck, calomel, and small doses of tartar emetic. On the next day the patient was free from fever, and her subsequent recovery was rapid: the facial paralysis remained for several days, and gradually disappeared.

Of the less serious nervous lesions following crisis, sleeplessness with delirium is frequent in those whose cerebration has been active previous to the fever, or too soon after its termination. One of the most obstinate cases of this kind which I ever witnessed, was in the person of a young man of studious habits, who after a favourable crisis, passed sixty hours in delirium without sleep. He was at length quieted by repeated doses of muriate of morphia and tartar emetic. A similar and most interesting case is narrated by the writer of a review on Dr. Graves' Clinical Medicine, in the twenty third volume of the Dublin Medical Journal, First Series. The occurrence of these symptoms from emotional excitement of various kinds during the early stage of convalescence, is not unusual. (*Appendix VII.*)

Sometimes we meet with the more serious affections of fatal convulsions, or of paralysis, as immediate consequences of crisis. The former are mentioned by Cheyne as having thus occurred in his own and in Dr. E. Percival's practice. Paralysis after fever is met with either in the form of true hemiplegia, or paraplegia, or in that form described by Dr. H. Kennedy, who gives the following example of general paralysis from the poison of fever : " A woman aged 44, was attacked by severe fever attended with petechiæ, and recovered from it but slowly. When the time came round that she might be supposed to exert herself, it was found that she had lost the power of moving both the upper and lower extremities; in fact, she was completely paralytic, and no means seemed to be of the slightest use. The arms presented exactly the appearance of limbs paralyzed by lead. In truth, no one could see this patient without having the conviction forced on his mind that the shock of the fever had literally destroyed the nervous agency. The patient lived about one month in the state described, never having given any sign of rally whatever, and it is particularly worthy of note, not even a threatening of stripping. The brain only was examined ; it was particularly healthy; there was a very small quantity of effusion into the ventricles, and a slight degree of venous congestion."—(*Dublin Quarterly Journal,* v. 20.)

LECTURE XI.

CEREBRO-SPINAL LESIONS.

WHEN epileptiform convulsions concur about or soon after crisis, they are usually due either to reactive irritation, consequent upon long continued passive congestion, or to blood contamination, as in the uremic convulsions after crisis of the relapsing fever. An interesting case of the former kind came under my observation some years since, in a lady who had passed through typhoid fever caused by malaria. The patient, aged 45, was of an excitable, hysterical temperament and jealous disposition; her habit of body was full, and she was accustomed to live well. The brain had not been apparently engaged during the fever, and she had therefore been allowed by her medical attendant to retain a mass of hair at the back of her head, of which she was very vain. She had also continued to take wine after crisis, and up to the time of my visit.

I was consulted in consequence of the supervention of severe headache with some degree of stupor, and constant distressing vomiting and retching, followed, after about a day, by an epileptic fit, which was ushered in by the usual scream. The fit recurred after a few hours, and again at intervals during the two following days. The treatment employed consisted in withholding wine, cutting off the hair, applying cold to the head, and leeches behind the ears; also in the exhibition of grain-doses of calomel at short intervals. She had no fit after the leeching, and the vomiting ceased after the first or second dose of calomel. As is usual after fever, ptyalism was very rapidly induced; when all unfavourable symptoms disappeared.

This lady had never previously had epilepsy, nor has she since suffered any return of it.

I have directed your attention to the important influence of previous states of nutrition of the cerebro-spinal centres upon the nervous lesions of fever, and I have given you some examples of this influence in the cases of prior disease or injury, of sunstroke, overexertion, &c. I have to remind you that mental and emotional states have an equal influence; as is seen in those cases in which the patient, at the access of fever, labours under some depressing emotion, whether it be fear, shame, grief, despondency, or anxiety.

In accordance with the law that the portion of the organism undergoing disintegration is that most subject to the operations of a morbid poison, the emotional centres of the nervous system are, in such cases, those portions whose functions manifest peculiar derangement. The intellect is often little affected, the patient retaining volitional control over his thoughts, but not over his motor actions. If delirious, he raves on the subject of his previous emotion, or perhaps manifests general depression of spirits, and a hysterical tendency. I believe that it is to this class we should refer the pseudo-hysterical cases which Cheyne, and after him Graves, have mentioned, under the name of dangerous, hysterical complications of fever, of which Dr. Graves says: "In every case of fever where symptoms resembling those of hysteria come on, you should be apprehensive of danger. I do not recollect having met with a single instance of this kind which did not terminate in nervous symptoms of the most formidable nature." It is probable that a form of blood-contamination to be hereafter noticed, exists in many such cases; but in others the character of the nervous symptoms is determined by the previous emotion. These cases are usually marked by the greatest disturbance of the respiration; the low temperature of the surface; the early and general subsultus and tremor, especially of the facial muscles; dysphagia; and by the early loss of the sensational consciousness, all of which symptoms, point to the great centre of emotion, the medulla oblongata, as the seat of lesion.

Some years since I witnessed and took notes of a very striking example of this form of nervous lesion, due to the influence of depressing emotion. My patient was a member of the profession, between fifty-five and sixty years of age, enjoying a large practice, wealthy, beloved by his family, but unhappily of a peculiarly nervous and sensitive temperament. Having to appear as a witness, in a cause in which he had taken a more zealous interest than was prudent, he was subjected for several hours to a cross-examination of the most painful nature, by a barrister notorious for performances of that description. He returned home broken down by shame and vexation. He could not sleep or eat; he absented himself as much as possible from the society of his family; and fever being then prevalent, in the town in which he resided, he soon contracted the disease.

The symptoms of profound nervous depression appeared at once; he began to sink from the first day; and on the eighth day the stupor, difficulty of breathing, coldness of surface, and failure of the heart's action, were so marked that he was to all appearance moribund. He was, however, kept alive till the fifteenth by the free employment of stimulants, especially of wine and brandy, the sinking of the temperature continuing throughout, with most painfully laborious breathing, and dysphagia; occasionally profound stupor and stertorous breathing; universal subsultus, tremor of the jaw, lips, and tongue, together with paroxysms of tremulous rolling of the head, &c. Near the end of the second week, the skin, which had been yellow, inclining to bronze, improved in colour, and the maculæ faded, but the nervous lesions increased daily; hiccup became constant, the voice sank to a whisper, the tongue became retracted, the breath cold, he had involuntary discharges, a bed-sore formed, and he died at length completely exhausted.

In this case, there had been scarcely any delirium throughout the illness, except upon the subject of the trial, but the aspect always indicated extreme despondency and want of desire to live; thus, when offered wine, it was usually refused with the expression: "It is of no use; wine cannot work miracles." The nervous lesions seemed specially referable to the medulla oblongata, the nerves of which were all profoundly affected.[1] The sinking of temperature, tremor of the facial muscles, disturbance of respiration, and dysphagia were the most remarkable I have ever witnessed; the last

[1] The nerves which take their origin from the medulla oblongata, mesocephale or crura-cerebri, are especially apt to be affected by emotions. The choking sensation which accompanies grief, is entirely referable to the pharyngeal branches of the glossopharyngeal and vagi nerves, which come from the olivary columns. The flow of tears which the sudden occurrence of joy or sorrow is apt to induce, may be attributed to the influence of the 5th nerve, which is also implanted in the olivary columns upon the lachrymal gland, or of the 4th nerve which anastomoses with the lachrymal branch of the 5th. The more violent expressions of grief, sobbing, crying, denote an excited state of the whole centre of emotion, involving all the nerves which have connection with it, the portio dura, the fifth, the vagus and glossopharyngeal, and even the respiratory nerves which take their origin from the spinal cord, as the phrenic, spinal accessory, &c.—(*Todd and Bowman's Physiological Anatomy of Man*, vi. p. 355.)

was at times complete. Flying blisters, sinapisms, &c., produced no effect on this latter symptom, but I could usually remove it, for the moment, by dipping my fingers in water and smartly dashing the face, when a few mouthfuls would be freely swallowed. He retained his intellect to the last; and, though unable to speak, he exhibited to me signs of recognition, a few moments before he died.

In the following instance, it is doubtful whether the fatal result was more due to depressed feelings and dread of death, or to some pre-existing disease of the brain. As in the case just narrated, the amount of muscular tremor was most remarkable:—

J. Carolan, aged 24, was admitted into hospital under my care, on the 10th day of typhus contracted by infection. The symptoms which attracted observation, were severe headache, of which he constantly complained, extreme despondency, and fear of death.

It was stated, that he had suffered for years from obstinate headache, for which he had been treated in hospital, without relief.

The pulse, on admission, was 100, tongue moist, covered with a white fur; heart's action natural.

On the 11th and 12th days he had low delirium.

On the 13th, he complained of great pain in the head; had frequent vomiting, rolling of the eyeballs, and fits of general tremor, continuing about ten minutes each time. Tongue dry, pulse 124, heart's action stronger, and the sounds louder than before.

Leeches were applied to the temples, and afterwards to the mastoid process, with temporary relief of the headache.

14th. Had passed a sleepless night. Vomiting continued; the occasional tremor was much increased; heart's action jerking and violent; pulse small and rapid; extremities inclined to be cold; tongue dry. Towards evening the tremor became so violent as to require the nurse to hold him in bed during some of the paroxysms.

No vomiting during this day, but he lies on his back incapable of speaking, or protruding his tongue.

15th. Fits of tremor return every hour, and are stronger than before. Complete loss of consciousness; heart's action variable; some times strongly excited, at other times more natural; involuntary evacuations; profuse sweating.

16th. Died at 8 A. M. No post-mortem examination allowed.

Each day's experience increases our estimate of the importance of those secondary contaminations of the blood, which occur to a greater or less extent in the advanced stage of all fevers; as well as of their influence in the production of secondary lesions, more especially of the excito-motor centres of the nervous system.

These contaminations arise from the presence in the blood of—

(*a*) The products of increased metamorphosis of tissue.

(*b*) Of resorbed sanguineous congestions or exudations.

(*c*) Of retained or suppressed excreta.

Several causes usually concur whereby the products of regressive metamorphosis are present in the blood in excess. Of these indulgence in animal food and vinous or spirituous liquors; muscular exercise during the latent period of fever; exposure to cold at the same period; and the increased disintegration of tissue which occurs in all fevers, without proportionate increase of elimination, are among the most influential. The well-known influence of a highly animalized diet, in producing, or increasing, a tendency to epilepsy, is an illustration of the special affinity of these products of metamorphosis for the excito-motor centres of the brain, and spinal axis. We have another, in the existence of the same affinity, in the case of retained urea, itself a product of disintegration. The tendency to peculiar nervous lesions in the wealthy and well-fed has long been observed. Thus, Sims gives a graphic description of them as prevailing in his time "among the middle ranks of the people, who use much flesh in their diet, and whose prevailing foible is an indulgence in spirituous liquors." And I cannot better detail the symptoms, as they have been usually observed, than in Dr. Cheyne's sketch of a form of fever, frequently met with in the upper ranks of society during the epidemic of 1817. "Persons in this class of society, more especially if they had passed their thirtieth year, who had been accustomed to live fully and luxuriously, were liable to a fatal form of the disease, which, to the inexperienced, was often very deceptive: the intellect was clear; the manner rather hurried, but otherwise natural; the patients declared themselves without pain or uneasiness, unless what arose from great weakness, which they were astonished at, so little seemed to themselves to ail them; and this at a time when their skin felt greasy, and was covered with dun petechiæ, when their eyes were glassy, their countenance somewhat suffused, and their breathing quick. *Such patients were very liable to convulsions*, in which case death (which in the upper ranks frequently took place on the eleventh day) was seldom distant," &c.

Of the second class, you have examples in the furious delirium sometimes occurring, with other nervous symptoms in typhus complicated with pneumonia; and in the advanced stage of typhoid with intestinal ulceration.

It has been frequently noticed that patients, presenting the delirium of typhus with pneumonia, present also a remarkable yellow, or sallow, discoloration of the surface, which Hasse has shown to depend on the tinging of the serum by broken up blood corpuscles and exudative material. There is therefore, good reason to ascribe the nervous derangements of this complication, to blood contamination.

The active delirium which occasionally arises in the advanced stages of typhoid, has been explained by some, as the result of the occurrence of similar contamination produced by absorption from the surface of the intestinal ulcers; and this opinion is supported by the fact, that this delirium is frequently associated with a low condition of these ulcers, and with hemorrhage and other signs of a depraved condition of the blood. Dr. Murchison, however, has proved that these symptoms are often the effect of the deficient elimination of the products of disintegration of tissue. I shall have to notice his important observations hereafter.

Of the third class, you have striking examples in the peculiar nervous symptoms produced by suppressed menstruation and suppression of urine, or deficient elimination of urea. The female whose menstruation has been arrested by the exciting cause of fever—whether cold, or mental shock—usually displays serious blood lesions, associated with nervous symptoms of an emotional or hysterical character, such as Cheyne, and after him Graves, state to have been almost uniformly fatal in their experience.

Some of you had an opportunity of seeing an example of this, in a woman named Leeson, who was admitted into our fever ward about a year since.

This woman contracted fever when menstruating. The catamenia were at once checked, and as you witnessed, the nervous symptoms in the fever which ensued, were of a highly marked hysterical character; obstinate mutism being one of these, and remarkable sinking of the temperature another. Inasmuch as very similar symptoms not unfrequently follow the suppression of the catamenia without fever, I think we may fairly ascribe them in both cases to a common cause; and this cause appears to be the poisonous action of a retained or suppressed excretion upon the emotional centres of the nervous system.

The most frequent, and by far the most serious, nervous lesions arising from blood contamination are those arising from deficient excretion of urea, the effect of one or other of the several causes, to the study of which I directed your attention in a former lecture.

The great importance of this complication will justify me in dwelling at some length on its study.

The first effects of the poisonous action seem to be felt by the spinal axis, as evidenced in the remarkable tremor and subsultus; but it also manifests its action on the cerebrum, producing peculiar psychological phenomena. The patient seems to preserve his intelligence, but becomes peevish, dispirited, and fretful; he is occasionally drowsy, but not equally so at all times, and before the stupor becomes profound he awakes from his doze suddenly, with a start, or the appearance of fright; sometimes with a scream. I recollect being led to the diagnosis of this complication in the case of a young lady in typhoid fever, from her mother mentioning that she always awoke from sleep in a state of alarm, and with a faint scream. When addressed, the patient seldom looks at the physician, and answers the questions addressed to him with averted head, and in a dreamy, careless, or sullen manner, as if looking at and thinking of something else; occasionally he is mute, or perhaps, instead of replying, he mutters as if in conversation with another person.[1] The eye is usually red, as if from want of sleep; and the pupil, previous to the occurrence of coma or convulsion, is contracted as if from opium. The face and forehead usually present a flush of a brick or copper-red colour, diffused and permanent; in some cases, however, it inclines to purple; and in patients affected with chronic renal disease, the face is occasionally pale and bloated. As the disease advances, if we watch the patient for a few moments, we observe the eyelids to close, and the eyeballs to roll convulsively

---

[1] Dr. Walsh describes a manner closely resembling that which I have noticed in many cases of uremia, as occurring in tubercular meningitis. He says: "I have now observed at least six cases of this affection in the adult, in which a peculiar form of mutism formed a striking symptom. The patients when questioned, looked steadily in the speaker's face for a few moments, and then, without making the slightest effort at speech, deliberately, but without any sign of petulance, turned their heads away."—*On Diseases of the Lungs and Heart*, p. 392.

under them; at the same time a spasm passes rapidly over the face, causing an expression of pain, and he utters a low, plaintive, tremulous moan; the facial convulsion becomes gradually more marked, and may sometimes be excited by touching the lips with a spoon.

I have frequently questioned these patients as to the cause of the peculiar moan, but could never obtain an answer from any one of them.[1] I believe it is not a voluntary act, but a simple reflex phenomenon of which perhaps the patient is not conscious. I lately witnessed something like this in a case of fatal apoplexy to which I was called very shortly before the patient's death. Although perfectly unconscious, she cried hysterically each time that any one spoke loudly in her ear. It appears to be always accompanied with facial spasm, or rotation of the eyeballs, and though different in its character from the scream which ushers in the epileptic fit, is probably, like it, a true convulsion.

Another characteristic sign of this complication is, the muscular tremor affecting every part of the body and extremities. This is painfully evident when the patient, who is usually obstinate, insists on being raised in the bed to take a drink from his own hand. Sometimes there is a droop in one eyelid, sometimes squinting, occasionally double vision. After a few hours the muttering subsides into stupor, from which the patient can for a time be aroused by speaking loudly—exactly as in narcotism from opium—but which passes gradually into coma, with nasal stertor. Very generally he awakes from this state, and becomes noisy and unmanageable; to this delirium convulsions soon succeed, followed either by coma or a second convulsion—occasionally, but not often, by a return of consciousness—and death.[2]

---

[1] Dr. Cranfield makes the same remark in the case of patients in the secondary fever of cholera : "Those that died moaned almost continually, and when questioned as to the cause of it, they could not assign any. They became comatose, and fell into collapse."—*Practical Treatise on Cholera*, p. 138.

[2] I might adduce many examples from the older writers on fever, to show that they recognized the peculiar symptoms of this complication, though not its cause. Thus Stoll describes the sullen character of the delirium as, "Silentium ad interrogata sermonicatio cum absente;" and Armstrong thus sketches the group: "Sometimes acute inflammation of the brain is to be recognized by a glary, bloodshot eye, a contracted pupil, an agitated expresssion of the countenance, and a peculiar species of moaning, which scarcely ever ceases for a moment; to these indications confusion of mind, tremors of the muscles, and coma rapidly succeed, and the patient expires at last with a bloated, pale face, and laborious breathing." These are the symptoms, not of acute inflammation, but of uremic poisoning of the brain.

The course of the symptoms will sometimes differ from the above according to age, constitution, and different complications. Thus, in the old and enfeebled subject, lethargy and coma will supervene quickly, so as to take the place of other symptoms; and in those who labour under a more acute form of kidney complication, vomiting and bloody urine will be superadded. In most of the cases which I have seen, vomiting occurred once towards the close, a sudden, copious vomiting, like that sometimes seen after narcotism;—in the following case for example:—

A gentleman, aged 45, of full habit and purplish complexion, with marked arcus senilis, was much exposed, at the period of the invasion of fever, and for two or three days subsequently, to cold, and severe bodily and mental exertion. His habits of living had been previously luxurious, and he was daily accustomed to drink a large quantity of wine.

The fever came on gradually, and I was not consulted till the eleventh day, the patient even then obstinately refusing to see any physician. I found him propped up in bed, with a wild, angry expression of countenance, but with perfect intelligence: breathing rapidly and with effort, as if after strong muscular exertion; his face was not flushed, but bloated, dusky, and haggard; the lips and ears purple; the eyes suffused; the lower jaw and lips agitated by a constant tremor; the tongue dry, and tremulous when protruded; the surface of the trunk and extremities cold—the latter particularly so—mottled and slightly livid; a few dark maculæ were scattered over the abdomen, which was tumid and tympanitic; there was moderate diarrhœa, of an ochrey character; the heart's action was weak; the pulse small, weak, rapid, and most irregular, its rhythm being entirely lost.

There was no change in the symptoms till the thirteenth day, when the facial convulsions and peculiar moaning I have described set in, and recurred frequently in the course of the forenoon; followed, at 2 P.M., by a severe and prolonged epileptiform convulsion, after which, for twelve hours, neither spasms nor moaning returned. The urine, on examination, was now found to be smoky in appearance, and coagulating on the application of heat.

In the course of the evening he had frequent fits of loud yawning; and at midnight, while I was giving him a spoonful of wine, he suddenly vomited a large quantity of fluid.

Early on the morning of the fourteenth day the moaning and

facial spasms returned, the latter being excited by putting a teaspoonful of fluid to the lips; then occasional stupor, merging in coma supervened; and about 1 P. M. a second and fatal convulsive paroxysm.

It may be useful to recapitulate shortly, the principles which I have endeavoured to bring before you in the foregoing lectures.

I. You will bear in mind, that in your study of the cerebro-spinal lesions of fever, you are engaged in the investigation of a problem involving:—

(*a*) The mode of action of the fever poison upon the great nervous centres. (*b*) The states of nutrition—present and pre-existing—of these centres which predispose them to become the seats of the special action of the fever poison (*c*) The derangements of their capillary circulation which may arise from previous injury or disease; from the influence of the exciting cause; from peculiar epidemic influence; from stasis the effect of long-continued decubitus; from passive congestion caused by obstructed pulmo-cardiac circulation, or weak left ventricle, or anæmia, consequent upon depleting treatment. (*d*) The influence of the various forms of secondary blood contaminations, which concur to set up at the later periods of fever, what is called "the typhoid state"—examples of these being met in septicæmia, and in the different retained excreta. (*e*) The influence of the depressing emotions. (*f*) The perturbations of crisis, and the cerebral excitement which occasionally comes on during the early periods of convalescence.

II. That the cerebro-spinal functions will be variously deranged according as one or other of these influences may operate in combination with the fever poison.

III. That the symptoms will vary according to the portion of the great nervous centres which may be the seat of impaired nutrition, or of deranged states of the capillary circulation. That accordingly, different groups of nervous symptoms will be set up by deranged states of the cerebral hemispheres, of the mesocephale, or of the upper portion of the spinal axis.

IV. That "ceteris paribus" the derangement of the cerebral hemispheres will seldom be marked by more than mild delirium: the motor functions remaining unaffected; while, in proportion, as the important organs at the base of the brain become engaged, a more impulsive form, "delirium ferox," usually manifests itself, attended with jactitation, rolling of the head, tremor and subsultus, rigidity of muscles, &c.

V. That the upper portion of the spinal axis being the special seat of the attractions of the products of disintegration of tissue, the most serious nervous derangements referable to the medulla oblongata are set up in those cases in which these products have produced secondary blood poisoning, as in uræmia.

VI. That besides the poison of urea—which it is well known acts upon this portion of the nervous centres—the following cases may be included in the above category:—

(*a*) The excessive accumulation of the products of metamorphosis, in the blood of persons living luxuriously, up to the invasion of fever, in whom a set of symptoms arise undistinguishable from those of uræmia. (*b*) A similar accumulation from the rapid disintegration of muscular tissue in patients who have been exposed to severe muscular exertion during the earlier periods of fever, and who accordingly manifest the same group of nervous symptoms. (*c*) The blood poisoning resulting from an arrest or suppression of the menses. (*d*) Certain depressing emotions, as grief, fear, shame, anxiety, remorse, and above all despair of recovery, with dread of death; all of which appear to make these—the emotional centres—the seats of the operation of the fever poison in a peculiar degree, and manifest symptoms referable to the medulla oblongata, such as derangement of the respiration and of the circulation; sinking of the temperature; convulsions; tremor; subsultus; and dysphagia, &c. &c.

VII. While you must endeavour, on the one hand, to avoid the mistake of ascribing the nervous derangements caused by the toxic action of the fever poison, to cerebral inflammation; you must, on the other, avoid the opposite error of ignoring the existence of true inflammation, when this is indicated by the groups of symptoms enumerated above, whether it occur in the form of acute congestion of the brain, as in some epidemics of typhus; of arachnitis, the effect of previous injury, of epidemic, cerebro-spinal arachnitis, or of reactive irritation and inflammation, set up in the advanced period of typhoid fever.[1]

---

[1] That true inflammation of the brain and membranes does occur from time to time in the course of fever, was proved in the case of two patients, admitted into the Meath Hospital since this lecture was delivered.

For the following notes of the first of these, and of the post-mortem appearances, I am indebted to Mr. J. Moore:—

"George James, æt. 40, a cab-driver, was admitted into the Meath Hospital on the evening of the 2d of August, in maculated typhus.

"He attributes his illness to cold caught by sitting upon a wall, exposed to a cold wind, while perspiring after exertion. While thus sitting he felt suddenly chilled. On the following morning, the 28th ult., he arose from bed complaining of cold and shivering. Epistaxis came on repeatedly during the next few days.

"When I saw him, on the morning of the 4th inst., he was profusely maculated, the spots being very dark and indelible on pressure. His breathing was hurried, his eyes heavy and congested; pulse, 116; respirations, 30; temperature, 103–5°; heart's impulse and sounds weak; tongue thickly coated with a dirty white fur.

"August 6 (11th day), and for two following days, he continued in much the same state. The heart's action becoming gradually more excited, and the patient's manner more sullen and obstinate. He had also frequent rolling of the head upon the pillow; and for the last two days of his life he roared loudly from time to time.

"On the 15th day the temperature rose from 101–6°; on the day previous, to 102–2°; at the same time he perspired to excess. He died on the morning of the 16th day.

"On post-mortem examination, the dura mater appeared slightly but sensibly thickened; the quantity of subarachnoid fluid much increased; the vessels of the membranes and of the substance of the brain much congested; both lobes of the cerebellum softened; the right more especially, which was reduced to a pulpy consistence; the medulla oblongata was also softened to some extent; no deposition of pus was observed, and the cerebro-spinal fluid was clear, though of a deep yellow colour."

The other patient was a young man, who was unable to give any account of himself; but from the time of his admission lay in a state of gradually increasing stupor, with ptosis of one eye, strabismus, retention of urine, and dysphagia (toward the close).

Upon examination post-mortem, Mr. Todhunter found intense venous congestion of the cerebellum and upper part of the cord; and on the lower surface of the cerebellum a circumscribed collection of yellowish-green pus included between the arachnoid and the nervous substance, and extending from one lateral lobe to the other, across the middle lobe.

There was no appearance of pus elsewhere, nor any exudation of lymph; although the arachnoid covering the cerebellum bore traces of inflammation throughout most of its extent.

Several interesting cases, bearing some analogy to the above, are recorded by Sir D. Corrigan in the proceedings of the Pathological Society of Dublin.

"The first was that of a female admitted into the Hardwicke Hospital on the 17th of March, 1841, labouring under maculated fever of a low character. She was irritable, weak, and restless; complained of severe headache, and could not sleep; the feces and urine were generally passed involuntarily.

"At the time of her admission, she was in the eighth month of pregnancy. On the 29th, without any premonitory symptoms or any labour pains, the child was expelled alive, but died a short time afterwards. The lochia also appeared, but after some time became suppressed. The delirium under which she had laboured for several days was now replaced by coma, and she died on the 1st of April. In addition to the coma, she became universally jaundiced a short time before death.

"On examination the following conditions were observed—

"In respect to the brain, it was exceedingly firm. The whole surface of the organ was very vascular, and presented a very remarkable bright red tinge at various points. A portion of this redness was really the result of increased vascularity: but a portion was also produced by effusion of blood on the pia mater. The base of the brain in front of the pons Varolii was covered with lymph. The principal features of the case were the extreme hardness of the white substance of the brain, the brilliant shading of red on the surface, and the effusion of lymph on its base.

"In the second case the woman, who was a patient in the Hardwicke Hospital, had fever and bronchitis, then the intestinal mucous membrane became affected; and along with this were the usual symptoms of the common adynamic fever of the country. On the fifteenth day she was apparently getting well, when she was observed in the evening to be heavy and somewhat confused; in an hour afterwards she was comatose, and on the following morning when he visited the hospital, he found her lying on her back, with her arms paralyzed and contracted across the chest. She died within twenty-four hours after the accession of the cerebral symptoms. On examining the contents of the cranium, there was observed sub-arachnoid, infiltration of pus, which was also very remarkable in the sulci of the convolutions. Dr. Corrigan made a section of the brain in the presence of the Society, and found that its tissue was healthy, except, perhaps, a slight increase of vascularity scarcely perceptible.

"In another case, mentioned by Dr. Corrigan at the same meeting, a man, convalescent from fever, became suddenly violent and refractory during the night. Dr. C. saw him in the morning; he was lying on his back, with the arms firmly folded across the chest; the teeth set; he would not reply to any question, but spat through the teeth when any person spoke to him; he rapidly became comatose, and died in that state. A considerable quantity of lymph was found effused at the base of the brain, extending from the origin of the optic nerves to the pons Varolii."

With these, and numerous other recorded cases of typhus before me, presenting post-mortem evidence of true inflammation, I cannot subscribe to the following dictum of Dr. Murchison. "When the rash of typhus is present, it may be always concluded that there is no cerebral inflammation; for *post-mortem* examinations show that inflammation of the brain or its membranes, rarely, if ever, occurs, even as a complication of typhus."

Nor, consequently, can I accept, without reservation, the axiom of Dr. Stokes, adopted by Murchison, "that those symptoms which indicate inflammation of the brain under ordinary circumstances, do not indicate inflammation when the case is one of typhus fever." I have in the text pointed out the extreme difficulty of the differential diagnosis between the toxæmic symptoms of pure typhus and the condition compounded of this and of acute congestion or inflammation; still the diagnosis can in many cases be made, and, whenever possible, should be made, as the treatment and prognosis are essentially different in the two conditions.

11

LECTURE XII.

DIAGNOSIS OF FEVERS.

HAVING sketched in outline the functional derangements and visceral lesions that occur in fever, I have now to invite your study of this disease from another point of view; viz., as having a more or less definite course and duration; conforming in both of these particulars, more or less perfectly, to certain laws of periodicity; as manifesting pathological changes peculiar to each of the several forms; as presenting different modes of resolution or crisis; and exhibiting certain differences in the manner of fatal termination, and in the post-mortem appearances in various forms of the disease.

The practical object or result of such an inquiry is, first, the diagnosis of fever, considered as an essential disease, set up by the action of a morbid poison; from several forms of pyrexia, or other morbid states bearing more or less resemblance to it; and, secondly, the differential diagnosis of the various species of continued fever.

The diagnosis of fever, as an essential disease, will be simplified by our determining, what is its special and invariable characteristic phenomenon.

This, I believe, you will find to be its paroxysmal character, or conformity to the law of periodicity—all fevers consisting essentially, of a succession of paroxysms, and the termination of the fever being but that of the last paroxysm;—and you will find that the diagnosis of essential fever will be more or less difficult, in proportion as local complications or other conditions may interfere with this conformity.

The diagnosis of the various species will be derived from several specific characters differing in each; one of which is, the fact, that they present the paroxysmal character, and conform to the law of periodicity in different degrees.

In reviewing the several affections more or less liable to be confounded with essential fever, I am sorry that I must confess my inability to give you any certain guides to a differential diagnosis in these cases. Patient and accurate observation and prolonged experience will alone furnish you with the power of discriminating between asthenic pneumonia and typhus; between delirium tremens and the delirium of typhus; between the fever attending acute tuberculosis, whether meningeal or pulmonary, or abdominal (tubercular peritonitis), and the remittent form of typhoid of early life; between the urinæmic condition and the adynamia of advanced typhus, of which it so often forms an important constituent; and between other forms of the typhoid state, and the advanced stages of fever. And if you should ever have the charge of a fever hospital, you will have abundant opportunities of realizing this difficulty as experienced by others, in the numerous cases of these affections which are sent to the fever wards by medical practitioners, under the mistaken idea that they are those of essential fever. I confess I have frequently treated as fever the pyrexial condition ushering in pneumonia, not becoming aware of the true nature of the case until one or more days had elapsed. In a similar manner I have mistaken tubercular peritonitis, and acute phthisis for remittent typhoid; and I still consider the differential diagnosis of these, and tubercular meningitis, from typhoid as, perhaps, the most difficult in the practice of medicine.

Your differential diagnosis will be much aided by a knowledge of the exciting cause of disease, whenever this can be obtained; by the presence of the distinctive characters of one or other affection; and in some cases, by the fact of a pre-existing state of disease, or of a constitutional tendency to it. Thus, in a certain case, if you discover that the patient has been exposed to its contagion, the diagnosis will incline to typhus. If the typhus-rash is present, this diagnosis will be confirmed. If, on the other hand, the exciting cause was cold—typhus not being prevalent at the time—and if cough and rusty expectoration existed, with the rational symptoms of inflammation of the lungs—you would diagnose pneumonia, even although the physical signs were not as yet discoverable.

You will also be assisted, by closely and carefully noting the features of resemblance which your more accurate and mature observations may convert into those of disagreement. Thus the face of a typhus patient and that of one labouring under pneumonia present at first view a strong resemblance; but if more closely studied, you will recognize a marked difference between the supine sleepiness of the one, and the painful, anxious expression of the other. Neither is the colour the same; one being dusky and dirty, while the other presents either a bright suffusion over face and brow, or a darker and more circumscribed flush mingled with the icteroid tinge of pneumonia, in the stage of hepatization. I can give you no better idea of this difference, than by repeating what I have so often pointed out to you at the bedside; that while the typhus countenance is accurately and happily likened by Dr. Jenner to that of a man awaking out of a drunken slumber, the face of the patient in the early stage of pneumonia exactly resembles that of a man who has set for some time before a strong fire. The mode of respiration presents another feature of difference. The breathing of the fever patient has been aptly compared to that of a man who pants after exertion. It is breathing with effort; semi-voluntary breathing. The respiration in pneumonia is constrained and painful; in a large proportion of cases interrupted by pleuritic stitch; and above all, presents an alteration in its ratio to the pulse, peculiar to itself.

The nervous symptoms of fever, most resembling those of delirium tremens, are those of exhaustion, which come on at a late period, and being preceded and attended with other symptoms of fever, are not likely to be mistaken. An error is more to be expected from confounding delirium tremens with commencing fever; but even this is not probable, if care be taken to ascertain the previous history and mode of life of the patient in delirium, and the causes which led to the attack.

The diagnosis of typhoid from acute phthisis, and *vice versa*, is one of the most difficult in medicine. I lately attended a young lady through typhoid fever of forty-eight days' duration, in whom, for the last three weeks, the signs of consolidation of the depending portions of both lungs were present, with cough and expectoration, hurried breathing, nocturnal exacerbations of fever, and morning sweats. Resolution seemed to be effected by gradual crisis, at this advanced period. On the other hand, I have, in several instances, witnessed the same symptoms and signs, herald the rapid and extensive deposition of tubercle; so that, in the absence of the rose-coloured spots of typhoid, I regard the differential diagnosis in such cases as one of extreme difficulty, sometimes, even, impossible. The diagnosis of tubercular peritonitis from typhoid is also, sometimes, very difficult. In both you will meet with vomiting, diarrhœa, periodical exacerbations, wasting, pain and tenderness of the abdomen; and certainly my experience does not lead me to the conclusion of Dr. Murchison, "that in most cases of tubercular peritonitis, the abdomen, unlike that of typhoid fever, is retracted." I have more frequently observed the contrary to be the case.

With regard to tubercular meningitis, Dr. Russell Reynolds[1] observes:—

"No greater difficulty of diagnosis can occur than that which is sometimes presented by a case in which the question arises, whether the symptoms are due to meningitis with fever of a low (or typhoid) type, or to typhoid fever with cerebro-meningeal complication. The question is not so much whether inflammation is or is not present (for it may exist in the latter), but whether that inflammation (cerebro-meningeal condition) is primary or secondary; in other words, whether fever is the result or secondary product of the inflammation, or whether the inflammation is one of the many secondary phenomena of the fever." Dr. Reynolds justly observes that the difficulty is greater with regard to typhoid than to typhus fever. I may add, that the younger the patient, the greater is the difficulty.

We endeavour to arrive at this diagnosis: 1st. By ascertaining the cause, when this can be satisfactorily done. 2d. By the presence of the special symptoms of one or other condition; or by the absence of some essential characteristics of one or the other. Thus, if ochrey diarrhœa, with tenderness and gurgling over the cæcum, are present, more especially if rose spots and the bronchial affection of typhoid are superadded, the diagnosis of fever may be made with certainty; but if these or the two first be absent, and vomiting not only ushered in the attack, but continued for some days, arachnitis may be diagnosed with nearly equal certainty; more especially, if there is also persistent intolerance of light and sound, and increased cutaneous sensibility. Moreover, a close observation of the patient, will detect differences in the mode in which certain features are affected in the two states; for example, the eye in the typhoid patient is usually clear, and the pupil open and dilated; while in arachnitis, the contrary occurs; this sign, however, is subject to ex-

---

[1] Reynolds on the Diagnosis of Diseases of the Brain.

ceptions, and the same may be said of the appearance of the tongue, &c. One sign, I think, you will find all but unerring when present, as it is peculiarly characteristic of cerebral mischief. I refer to that deranged condition of the circulation, in which we meet with a jerking impulse of the heart, with irregular, intermitting, pulse, of unequal volume and uncertain rhythm. You may occasionally diagnose the existence of tubercular meningitis, and impending effusion into the ventricles by this sign, when the intelligence is perfect, and scarcely any other symptoms are present.

Any difficulty in the diagnosis of fever from the exanthemata is most likely to occur in the early periods of these diseases. Thus, I have more than once known the efflorescence which in typhoid precedes the appearance of the rose spots, mistaken for scarlatina—an error more likely to occur if sore throats exists, as it sometimes does at this stage of typhoid. I have occasionally seen a rash in measles, which bore an unusual resemblance to the measly symptom of typhus, and *vice versa;* and I remember a case, which occurred in this hospital some years ago, regarding which opposite opinions were entertained. The rash which resembled both affections being, in fact, the eruption caused by balsam of copaiba.

Elderly persons attacked by bronchitis or pneumonia, erysipelas, or, most of all, by epidemic influenza, frequently slide into a condition resembling typhus, and hence denominated the typhoid state. In popular language you may sometimes hear that such a person's illness has run into typhus. Of couse I need not tell you, that although their condition may present several of the characteristic features of typhus or typhoid fever, it is essentially different from either. This state, has been well defined as—consisting in an excess of that general derangement of the assimilation, which can be traced in every fever—as usually presenting a combination of the asthenic and septic types; a poisoned state of the blood, with a depressed state of the nervous system.

Sometimes the blood poisoning may have its source in the original affection, as in erysipelas or a resorbed pneumonic congestion; or it may be the product of deranged assimilation; in other words, of the accumulation of products of disintegration which the secreting organs cannot eliminate, lastly it may be, and according to my experience often is, uremic poisoning. In other instances it has arisen from septicæmia, produced by the entrance into the circulation of putrid matter from the disease of the internal ear, or absorbed from an ill-conditioned intestinal ulcer, or from resorbed sanguineous congestion; and it should be observed, typhus and typhoid are not unfrequently themselves thus complicated, septicœmia being superadded to the original poison. In all these cases, the patient lapses into what is termed the *typhoid state*. The colour becomes dark and muddy, ecchymotic petechiæ appear on the depending portions of the trunk, and perhaps purpuric spots on the extremities; the tongue has the appearance of typhus, the mouth is dry, and the teeth and lips are covered with sordes; the odour of the breath is often urinous, the respiration sighing and laborious; the abdomen tympanitic; the cerebral functions are oppressed; the patient is drowsy and apathetic, has low delirium and subsultus, and finally lapses into coma. (*Appendix VIII.*)

I have several times observed this group of symptoms to coexist with partial or complete, retention of urine, in old men; and I would caution you to attend carefully to the state of this excretion, in all cases of disease or injury you may be called on to treat in persons of advanced life.

The diagnosis of the various forms of this typhoid condition must be derived from the previous history, primary affection, and existing symptoms; and can only be arrived at by the careful study of all these. We have no short and ready method, of distinguishing such a condition as I have described from the advanced stage of typhus, of which it is often a complication. Notwithstanding, however close may be the resemblance, the two forms of the disease are essentially different, the one not being convertible into the other.

We shall study the specific characters of the several forms of fever under our consideration, as they present themselves, at the periods, of invasion or access, of the fully formed disease, and of termination or crisis;—denominated by writers, the stages of depression, reaction, and subsidence, glancing for a moment at the post-mortem changes and sequelæ presented by each form.

The access or initiatory paroxysm of all fevers is in many respects the same. In the language of Fordyce, who has described the "fever paroxysm" more graphically and minutely than any writer with whom I am acquainted, " a fever of about eight, ten, or twelve hours' duration may present all the phenomena and may complete its existence as perfectly as if it had taken as many months." That this idea is not founded upon any inadequate conception of the febrile phenomena is evident from his comprehensive and oft-quoted definition of the disease, as well as from the accuracy and minuteness of his description of what may be termed a microcosm of fever—the fever paroxysm.

The first thing to be noticed is suddenness of access—a suddenness not peculiar to any form of fever, but more marked in typhus than in typhoid, and still more so in ephemera and relapsing synocha than in either of the others. Thus, a man may sit down to dinner with a good appetite, and not be able to eat a morsel, being at one moment in perfect health, and in the next stricken with fever. To the undefined feeling of indisposition succeed other subjective and objective symptoms; general uneasiness, restlessness, præcordial oppression seeking relief by sighing and jactitation, frontal headache, pain in the back, giddiness, and confused perception, nausea, bodily languor and mental hebetube, horripilation, and a feeling of chilliness amounting frequently to shivering.

If you are called to see the patient at the moment of access, he presents the objective symptoms of change in colour of the face and general surface; he appears shrunken, dingy, and perhaps yellow, or of a bronze hue; his eye dull and heavy; his tongue soft, blanched and tremulous, perhaps with a commencing dry streak on its centre. At this stage, too, the kidneys secrete a large quantity of limpid watery urine, the bladder retaining but a small quantity, and micturition being accordingly very frequent.

To the above stage of depression succeeds that of reaction; marked by heat and dryness of skin, excited action of the heart, and rapid bounding pulse, increased headache, and pain in the back, and not unfrequently vomiting. The bowels are usually costive, and the urine becomes scanty and high coloured.

If the fever is, to use the words of Fordyce, completed in one paroxysm, as in ephemera; or if it assumes the remittent character, as in relapsing synocha; perspiration more or less profuse will follow the above stage, sometimes even occurring in the latter affection, Dr. Cormack observes, without the intervention of a hot stage, and commencing while the patient is in his initiatory rigors. But this well-marked and complete initiatory paroxysm is peculiar to that form of fever, being less constant in typhus, and often escaping notice altogether in typhoid.

As a rule, the typhus patient can fix the date of access with tolerable certainty; but frequently, the lassitude, pain in the back, loss of mental energy, and slight chilliness, alternating with heats and perspiration, which mark its invasion, pass unheeded for a day or two, until his attention is arrested by a marked shivering, together with increase of headache and prostration; compelling him to take the bed. At the same time he breathes with effort, and has a short cough, which increases his headache; he complains of want of sleep, or his sleep is disturbed and painful, and he now presents to the observer the flushed, dusky face, with sleepy expression and congested eye, characteristic of the disease; and usually from the fourth to the sixth day of fever, the characteristic rash which I have described in a former lecture, makes its appearance, and confirms your diagnosis.

As I have already mentioned, the invasion of typhoid is not so well marked or definite with respect to time as that of typhus.

A patient—generally a young person—will present himself to you having walked to your house, perhaps, from some distance, complaining of diarrhœa, with probably some abdominal pain. You are struck with the bright, but languid eye, pale complexion, with, perhaps, a circumscribed flush on the cheek; his skin is hot, and his tongue has a white fur, and is red at the tip and edges; you note the wavy dicrotous character of the pulse, which is somewhat accelerated. He will make light of his illness, but if closely questioned, will tell you that he has more or less frontal headache and pain in the back, alternate slight chilliness and heat, want of appetite, thirst, and debility; that his sleep is disturbed and unrefreshing; and that he usually awakes thirsty, hot and restless, at an early hour (from 2 to 4) in the morning. Perhaps this patient's illness is of nine or ten days' duration; if so, on making an examination of the abdomen you will find, in addition to the meteorism, tenderness, and gargouillement, described in a former lecture, a number of round or elliptical rose-coloured spots, scattered over the abdomen and chest, and your diagnosis will be thereby confirmed. Such a mode of access is by no means uncommon; and even when the commencement is marked by a slight shivering, this, and the subsequent fever, is usually so little attended to, that unless compelled by the severity of the diarrhœa, the typhoid patient—if an adult—seldom seeks medical aid until a comparatively advanced period of the fever.

If we trace the onward march of these fevers, we shall find that the stage of reaction in relapsing synocha, rapidly succeeds to that of depression, and is usually of highly marked character. The headache is severe, the delirium active, the pain in the back and limbs continues during the course of the primary attack, the pulse is, as a rule, far more rapid than it is in typhus, the heat of the skin also higher. In Dr. Cheyne's able report of the epidemic of 1817-18, he gives the results of thermometrical observations in 250 cases; and in 214 of these, the temperature was from 100° to 105°. The tongue usually continues moist throughout, being white, with vividly red tip and edges. The aspect of the patient during the primary attack is very unlike that of typhus; being flushed, without the dusky dirty look of that disease, or else presenting that sallow subjaundiced colour, which has been remarked in all the epidemics of relapsing fever. In a few instances, there is presented a true jaundice, which resembles the intense yellowness of tropical yellow fever. I say resembles, because I cannot consider our epidemic, relapsing synocha in the least indentical with yellow fever; presenting as it does this "specific difference," as pointed out by Dr. O'Brien, " that in the latter the yellow colour of the skin is pathognomonic and general, though perhaps not a universal symptom. while in the endemic fevers of this country it is a rare and accidental occurrence, arising from an accidental cause."

This cause I shall have to refer to presently, when speaking of the complications of this form of fever.

Usually on the evening of the fifth or seventh day, when the fever is at its highest, and the pulse ranges, perhaps, at 140° to 150°, crisis suddenly occurs, ushered in by a well-marked *exacerbatio critica*. The several stages of this change were so accurately described by Dr. Cheyne, that I cannot do better than give you his graphic sketch. "The most perfect crisis," says Dr. Cheyne, "consisted of three stages. First, a state of restlessness and anxiety, with flushing of the face, rapid pulse, frequent laborious breathing, and increased heat of the surface, with great distress at the pit of the stomach, with heat, tenderness, or pain; which distress was not unfrequently relieved by vomiting. The patients were in the state of universal uneasiness, which would have been truly alarming had we not known its tendency; but this is well understood, even by the servants of a fever hospital, who soon come to know by these symptoms that the patient is near 'the cool.' This state sometimes lasted for the greater part of a day, during which one of our experienced nurses, who was fond of figurative language, would generally remark that 'the cool was hovering round' the patient. Secondly, a rigor or tremor not unlike the cold fit of an ague; the patient shivered and complained of excessive cold. I never, save in two instances, was able to measure the temperature during the rigor of crisis, in both patients the thermometer stood at 105°, even while the patient was shivering and complaining of excessive cold, and asking for an additional blanket. In one of these patients the thermometer in the evening stood at 100°, although the rigor was not followed by sensible perspiration; next morning the thermometer stood at 97°, the tongue was clean, the pulse 88°, and the patient convalescent. The rigor of crisis seldom lasts long, perhaps only a few minutes, perhaps half an-hour, or an hour. Thirdly, warm perspiration flowing from the whole surface of the body; this, which in general completed the salutary effort, the nurses in the Hardwicke Hospital call the 'cool,' being aware of its efficacy in reducing the heat of the body."—*Dublin Hospital Reports*, vol. ii.

The convalescence now commenced, is often marked by voracious appetite, and seems for a few days to be perfect. But in a great majority of cases on the fifth or seventh, or it may be the fourteenth day after crisis, there is a return of the initiatory rigor, and of the symptoms of the first attack. This occurrence is so constant, as to give the name of relapsing synocha to the disease. It has been observed in numerous epidemics, as for instance, those described by Dr. Rutty in the last century—the Irish epidemics of from 1801 to 1847-8, and the Scotch epidemics of 1817-19 and 1843—and long ago gave rise to a proverb, that "Short fevers are the most prone to relapse." Such observers have noticed a correspondence between the duration of the primary paroxysm and the interval preceding relapse. Thus Dr. O'Brien records a case in which "the patient suffered two relapses,—the interval between each attack was five days, and was synchronous with the febrile period, so that the whole duration of her fever, not including the intermissions, was fifteen days. The same thing occurred in one or two other cases." All observers seem to agree, that as the epidemic advances, the tendency to relapse becomes less marked, so that they are then the exception rather than the rule.

The progress, duration, and mode of termination of typhus differ materially from the above description. Less sudden and violent in its invasion, it displays less marked paroxysms in its course, and, totally unlike the former, its termination is often by an insensible crisis, which, however, differs from the still more gradual lysis of typhoid. I shall not enter into any detailed history of the disease, as such is no part of my object. But I may remind you that while its paroxysmal character is faintly marked in comparison to that of typhoid, there is a close conformity of its course to that of the exanthemata, to which class it properly belongs. The nervous symptoms occur as primary phenomena at an early stage; the prostration generally compelling the patient to betake himself to bed about the 3d or 4th day, and preventing his leaving the bed without assistance on the 6th or 7th day. The characteristic rash appears from the 4th to 6th day, fading from the 14th to the 17th day, unless in cases in which, owing to a depraved state of blood, the exanthem becomes converted into, or is replaced by, ecchymotic petechiæ; the exanthem and specific fever usually subsiding together, and crisis being followed by desquamation of the cuticle and falling of the hair as in scarlatina.

While the crisis of typhus is silent and gradual, as compared with that of relapsing synocha, it is abrupt and defined in comparison with that of typhoid. "I think," says Dr. A. Stewart, in his admirable essay on the two fevers, "I may appeal to the experience of every physician, and more especially of every clinical clerk in a fever hospital (for they have more constant opportunities of observation), whether they have not often been struck at seeing, during their morning visit, the glassy eye, the haggard features, the low muttering delirium, the stupor approaching to coma, the tremor, the subsultus, the carphology, the rapid, thready, tremulous, and intermitting pulse, of the previous evening, the formidable array of symptoms, in short, which seemed to indicate a speedy and fatal termination, exchanged for the clear eye, the intelligent countenance, the steady hand, the comparatively slow and firm pulse, and the returning appetite of approaching convalescence." (*Edin. Med. and Surg. Journ.*, v. 73, p. 305.)

The nature of the process by which this wonderful change is produced is a mystery. We know not how the silent transformation is effected. Of one thing, however, you may be certain it is not necessarily through the agency of evacuations; which often do not occur at all—at least sensibly—or when they do are frequently injurious. There are many cases, moreover, in which evacuations are evidently post-critical rather than critical; the consequences, not the cause of, crisis.

An obvious practical inference from this fact is, that although in special complications, and under exceptional circumstances, measures tending to elimination by some excreting organ may be indicated, any attempt to determine crisis by evacuation, any form of factitious crisis in short must be fruitless, as from experience we know it will probably be fatal, in its result.

The progress, duration, and termination of typhoid or enteric fever differ in many particulars from the other forms. When the period of invasion has been determined—not always an easy matter—it has been remarked that fever of a remittent character becomes established about the 3d day. There is usually a morning and afternoon exacerbation—more marked in young patients—and while the diarrhœa of the initiatory period continues or returns in an aggravated degree, there is a complaint of frontal headache and giddiness, sometimes of tinnitus, epistaxis, wandering pains, and restless and disturbed sleep. In some cases vomiting continues for a few days. There is meteorism and pain on pressure over the ilio-cæcal region. The urine at this period is high coloured, containing an excess of urea, and I have often observed, throughout this stage, a copious deposit of lithates every alternate day. I have already fully described the eruption which now appears in successive crops. At this period the diagnosis of typhus fever, from typhus of the same duration, is not difficult; but it becomes more so as the disease advances, and nervous symptoms and the phenomena of toxæmia, bearing a close resemblance to those of typhus make their appearance. During the third week, in many cases the noisy delirium, subsultus, and deafness closely resemble the same symptoms in typhus; and to these may succeed, in the typhoid or septicæmic stage, a group of symptoms resembling those of the worst forms of secondary blood contamination in that disease.

These nervous symptoms, as was long ago proved by Louis, do not arise from any inflammatory affection of the brain, and we are, therefore, obliged to seek their cause in some other condition. Two have been assigned; absorption of septic matters from the intestinal ulcers—septicæmia—and "retention in the blood of those products of tissue metamorphosis which ought to be eliminated by the kidneys." You will find the former theory stated by Dr. Todd in his clinical lectures, edited by Dr. Lionel Beale, page 107. (*Appendix*.) The latter is the explanation put forward with much force by Dr. Murchison, who argues from the want of proportion of the cerebral and other symptoms of the typhoid state, to the amount of intestinal ulceration; adding this most important observation, "on the supervention of the typhoid state it is found that the urinary solids, which have previously been so much in excess of the normal amount, diminish; and in several instances I have ascertained that the quantity of urea excreted in 24 hours diminished on the advent of cerebral symptoms, and increased again on their cessation. In one case, the quantity which was 292 grains, when the patient was delirious and unconscious, rose to 964 grains, when the delirium abated, and the consciousness returned. In another case, the quantity which at first was 422 grains fell to 352 grains, on the appearance of delirium and stupor, and rose to 490 when these symptoms ceased."

With regard to the termination of typhoid, it differs in a marked degree from that of relapsing fever and of typhus.

The termination of typhoid is not usually by crisis, but by lysis or insensible resolution. If you visit your patient in the morning in the commencement of the 3d week of the fever, you will mark a decided fall in the temperature, and will probably observe a copious deposit of lithates in the urine. You may, perhaps, flatter yourself that the fever is resolved; but if you pay an evening visit, you will find that the temperature has risen to the average of the week preceding, and this alteration between morning and evening will, perhaps, go on for 6 or 7 days, the average temperature subsiding to the natural standard usually about the commencement of the 4th week. This change is attended with remarkable deposit in the urine as compared with the crisis of typhus.

While the primary attack of relapsing synocha runs its course in a week, or even less, and typhus usually terminates in 14 to 17 days; typhoid, as a rule, continues from 21 to 28 days, and not unfrequently much longer. A comparison of the average duration of large numbers of the two fevers, justifies the conclusion, " that even apart from complications, and the changes of a relapse, typhoid is a much more protracted disease than typhus."—(*Murchison.*)

We have next to review the characteristic differences, presented by the pathological changes which occur in the course of each form of fever.

With regard to relapsing synocha, it may be said, that on the reception of the poison into the circulation, a change in the blood is set up, which first manifests itself in the sallow, yellow, bronzed, or purple colour of the skin; the tint varying with the amount of intensity of the blood poisoning. Secondly, in the formation of congestions, more or less in degree, in the right side of the heart, the vena-cava, the liver, stomach, and spleen. The symptoms and signs indicating a marked degree of this condition being, a feeling of weight and anxiety at the præcordia, restlessness, continued vomiting of green fluid—which, in extreme cases, becomes black —epistaxis, and hemorrhage from bowels. The skin becomes deeply jaundiced, sometimes, perhaps, in consequence of gastro-hepatic congestion; sometimes, also, there is much reason to believe, as a direct consequence of the introduction of the poison into the blood. At least, a very large proportion of jaundiced cases occur, in which there are no other symptoms of more than the ordinary amount of congestion.

The most serious pathological condition in this fever, is no doubt, the presence, in an appreciable quantity, of urea in the blood, in consequence of the marked excess of disintegration of tissue over the urea excreted; causing the symptoms of uræmic poisoning of the nervous centres to become evident.

If this is the most fatal condition, congestion and enlargement of the spleen is certainly the most frequent, in relapsing fever; and I have long entertained the opinion, that it is to this, more than any other cause, that the characteristic tendency to relapse is owing. In my report of the epidemic of 1847–8, I mentioned that in every case in which I observed the persistence of splenic congestion and enlargement after crisis, relapse followed. The cause of the relapse I believe to have been the gradual commingling with the circulating mass, of a large quantity of blood, which lying by, so to speak, in the congested organ, did not share in the depuration of the mass during crisis.

For the purpose of comparison, we may consider the pathological phenomena arising in the course of typhus and typhoid; as due to the elective affinity of the typhus or typhoid poisons;—to pre-existing conditions of the blood, or of the organs which become the seats of complication; or to accidental and extraneous circumstances, as season, locality, and treatment.

Of these, the first group only are uniform in their appearance, and characteristic of the fever; and it is with respect to them therefore that a comparison may be best instituted.

For this purpose I shall select, 1. The eruptions; 2. The order of the occurrence of the pathological phenomena, viz., of the pulmonary lesions; the urinary and the cerebro-spinal lesions.

1. I need add nothing to what I have said of the first in a former lecture, except, that the difference in their appearance, in the date of their occurrence, in the continuous persistence of the one as compared with the successive crops of the other; and lastly, the frequency of the conversion of typhus rash into ecchymotic petechiæ, while no such change is observed in that of typhoid; constitute in the aggregate what appears to me to be, a specific difference between the two eruptions

2. All observations tend to show that the pathological changes in typhus as compared with typhoid occur in a different order. In typhus the bronchial mucous membrane seems to become the seat of irritation in the early stage; and in certain seasons this irritation passes into true inflammation, which continues throughout the fever, constituting the formidable catarrhal typhus of this country. Pneumonia may occur at any time during the progress of typhus, but, as well as hypostatic congestion, is most frequent at the early periods.

The cerebro-spinal lesions, when due to the direct action of the typhus poison, also set in early; those of a later period being either inflammatory, or due to secondary blood contamination or to exhaustion.

Of the intestinal lesions, it may be said none belong to typhus; constipation is the rule; and if diarrhœa occurs spontaneously, it is at an advanced period, and, in most cases, is colliquative or critical.

The urinary lesions do not, strictly speaking, arise out of the operation of the fever poison in either case; but inasmuch as the balance between disintegration of tissue and excretion is most disturbed in typhus, uræmic poisoning is of more frequent occurrence in this disease than in typhoid.

If you study the order of the corresponding lesions in typhoid, you will find it strikingly different from the above. Unquestionably, the seat of the primary operation of this poison is the intestinal follicles, which have been found swollen and ulcerated within twenty-four hours of its invasion. The diarrhœa characteristic of this lesion is often the earliest symptom in the fever.

Far from the cerebro-spinal system being early engaged in typhoid, it is, in a large proportion of cases, not at all, or but little affected; and when delirium, subsultus, and other nervous symptoms do occur, it is usually at a late period; when they are either due to accidental complication, as cerebro-spinal arachnitis; or to reactive irritation set up in the brain, which has suffered passive congestion from long-continued decubitus; or from this combined with secondary blood contamination.

You will find a similar difference in the period of the occurrence of bronchial and pulmonary lesions. When true bronchitis occurs in typhoid, it is at a late period, seldom earlier, I should say, than the third week; while the longer the duration of the case, the greater the risk of hypostatic congestion as well as its accompanying lesion, lobular pneumonia.

Pre-existing conditions of the blood, or of the nutrition of the different organs will, of course, be alike influential in both diseases. Septicæmia or uræmia, for instance, will lead to the same typhoid condition in both, and any viscus in a state of special predisposition at the access of either, will suffer from the special operations of the poison in a disproportionate degree during its course. But we may observe this difference between the results of the two, that whereas after typhus—a blood disease—the visceral health, as a rule, continues good; typhoid, which is rather a disease of the blood-making organs—the function of sanguification being consequently deranged—is more frequently followed by tubercle or some other form of diseased nutrition.

Of the differences arising from the influence of such causes as locality, season, and treatment, little need be said. I may remind you, however, that the effects of season in generating and aggravating the two forms of fever, are exactly the opposite in each. As to the influence of treatment, it has been proved that typhoid patients will not bear to have wine pushed to the same extent as those in typhus; while on the other hand, their tolerance of opium is much greater. Purgatives may be safely administered in the one, at a stage when they would be highly dangerous in the other; and while a nutritious and stimulant treatment after the one, generally has the effect of hastening the progress to recovery, the safety of the patient demands a more guarded, cautious, and abstemious treatment in the early convalescence of the other.

With regard to the differences presented in the critical and post critical periods, you will observe—first, the most striking difference in the tendency to relapse. As you are already aware, relapse is the rule in epidemic relapsing synocha. It is very much less frequent in typhoid, and when it occurs, may be ascribed, not in all cases, to a law or necessary condition of the disease, but frequently to mismanagement, more particularly to errors in diet, by which irritation and deposit has been set up anew in the ulcerated glands of the intestine. It is to be remarked that the lenticular rose spots may reappear in such a case, as well as in those rare instances in which typhoid, replaced for a time by typhus, again sets in, on the subsidence of the latter (as in the case of the boy Morritt, before referred to).

In typhus it is all but unknown. This may, and often does, relapse into typhoid, and typhoid into typhus; but a relapse of typhus into typhus, is as rare as that of measles into measles, or as a second attack of scarlatina, during the convalescence from the first. As each of these has been known to take place, I would not deny the possibility of the same occurrence in typhus; but of its extreme rarity, I need no other proof than that among many thousand cases, in the course of twenty-five years' hospital experience, I have met with no instance.

The difference in regard of emaciation presented by the convalescents from typhus and typhoid is very remarkable, wasting being scarcely observable in many cases of the former, while it is often extreme after the latter.

The sequelæ of the several forms of fever cannot be said to present specific distinctions, or to justify us in considering any of them peculiar to any one form, with one exception to be presently noticed.

As a detailed description of these sequelæ is no part of my plan, and as they will necessarily claim our notice when considering the subject of treatment, I shall merely enumerate those which are the most frequent, and of the greatest importance.

The first in frequency is bedsores, which are undoubtedly more prone to occur in typhus than in typhoid fever. The same remark holds good of pyæmia, of diffuse cellular inflammation, erysipelas, and œdema glottidis, cancrum oris, and parotid bubo.

I think phlegmasia dolens is as often a sequela of typhoid as of typhus; and this form of fever is more frequently followed by ulceration of the larynx; acute phthisis; and disease of the intestinal glands, with marasmus; mental imbecility is also of much more frequent occurrence after typhoid than typhus, more especially when the fever has been prolonged beyond the usual duration; but if any morbid condition can be said to be peculiarly a sequela of typhoid, it is the intractable form of ulceration of Peyer's glands with enlargement, softening, and occasionally suppuration of the glands of the mesentery, attended with ochrey diarrhœa, retracted abdomen, griping pain and tenderness, and with a dry, harsh skin, and extreme emaciation, which occasionally supervenes upon a temporary convalescence, and too frequently resists our best treatment.

With regard to the modes of fatal termination:—

That of simple, uncomplicated typhus occurs by syncope or coma; failure of the power of the heart, and failure of nervous energy inducing the one, the conjoined action of the fever poison, and of the accumulated products of disintegrated tissue the other. Any of the complications may end fatally, each in its own peculiar manner. Those which most commonly influence the result are the cerebral, bronchial, pulmonary or uræmic complications. The laryngeal, though of rare occurrence, may be said to be uniformly and directly fatal.

The first and most striking feature of difference in the fatal termination of typhus and typhoid fevers is in the period at which this occurs. While a large proportion of the deaths in typhus occur between the 11th and 15th days, and in uncomplicated cases, the patient's chances may be said to improve with the duration of the disease, the contrary holds good of typhoid, in which the dangers increase with the duration of the illness; this contrast continuing, even increasing in the stage of convalescence, in which the typhus patient may be said to be safe, while the typhoid is still exposed to great risk from the existence of intestinal ulceration.

Typhoid fever may prove fatal;—by exhaustion or asthenia from the long-continued fever, or by this causing accumulation in the blood of the products of disintegrated tissue, and consequent blood poisoning and coma as in typhus; or it may be the result of one or more local lesions, as intestinal ulceration—which may lead to fatal hemorrhage, or to intractable diarrhœa and consequent inanition, or to perforation and fatal peritonitis; general bronchitis of the suffocative character; hypostatic pneumonia.

A very striking difference is presented by the rates of mortality of the two fevers at different periods of life. Dr. Murchison gives the tabulated results of 3506 cases of typhus, and 1820 cases of typhoid, admitted into the London fever hospital during ten years. The average mortality of typhus cases, from 5 years to 25 years of age, was as follows:—

From 5 to 10 years	7.65	per cent.
" 10 to 15 "	4.95	"
" 15 to 20 "	4.76	"
" 20 to 25 "	9.05	"
While from 40 to 45 years it was	29.79	per cent.
" 45 to 50 "	39	"
" 50 to 55 "	52	"
" 55 to 60 "	51	"

Taking the same periods of life, the average mortality of typhoid was:—

From 5 to 10 years	14.43	per cent.
" 10 to 15 "	12.08	"
" 15 to 20 "	16.18	"
" 20 to 25 "	20.03	"
While from 40 to 45 years it was	17.39	per cent.
" 45 to 50 "	25	"
" 50 to 55 "	25	"
" 55 to 60 "	55	"

According to Dr. Murchison, relapsing synocha observes the same law as typhus, the rate of mortality increasing as age advances. The proportion of fatal cases has, in all epidemics, been very low in the primary attack. Death has sometimes occurred at this stage from intense splanchnic congestion, the direct effect of the action of the poison. "Thus, in one case," says Dr. Law,. "from its commencement to its fatal termination, the disease lasted only four days. The case to which we allude was a man aged forty years, of a full habit, who exhibited a stupefied aspect like one in a state of intoxication, his eyes were deeply suffused. The surface of the body was of a greenish-yellow hue, the colour of tallow; the pulse was scarcely to be felt at the wrist, and the stethoscope indicated an extremely feeble action of the heart. No stimulants, either applied externally or given internally, appeared to produce the slightest result or response. The effect in this instance resembled that of a strongly concentrated poison acting with overwhelming powers on all the energies of the system, and deadening them to every stimulus. Here, if the morbific impression was made on all the system alike, all was equally overpowered by it."

Another case reported by Dr. Law has been already referred to, in which the patient, at first slightly jaundiced, rapidly became deeply so, with pulse 180 in the minute, lethargy and coma ending speedily in death. "Examination after death exhibiting the liver and spleen much enlarged in size, and congested, and so softened in structure that they seemed as if they had been soaked in blood."—(*Dublin Quarterly Journal*, vol. viii.)

At other times death has occurred from syncope or exhaustion, owing apparently to the violence of fever; while in many cases the same cause has produced such excessive disintegration of tissue, as to set up a fatal form of uræmic poisoning, due to the accumulation of the products of this metamorphosis in the blood. The idea once entertained that the fatal result was sometimes caused by jaundice seems to be without much foundation. "Henceforth," as Dr. Henderson observes, " before any one can assert that he has lost a case of this fever from jaundice, he must be prepared to show that there was no urea in the blood."

By far the greater number of deaths, however, occur not during the primary attack or the period of intermission, but during the relapse. In some cases, with symptoms closely resembling those of typhoid fever, as obstinate diarrhœa, pain and tenderness, gargouillement over the cæcum, and not unfrequently fatal peritonitis, with or without perforation. In others, with dysentery equally obstinate and fatal. From observation of the epidemic of 1847-9, I am convinced that two causes operate in different cases to produce these pathological conditions. One is the frequent coexistence of the poison of endemic typhoid, with epidemic relapsing fever, the one disease replacing the other in the same patient. The other is the influence upon the blood of the resorption into the circulation of a quantity of depraved blood laid up as it were in the congested spleen, and commingling with the circulating mass after crisis. I shall be mistaken if future observations do not prove this to be an important element, not only in the production of relapse, but also in determining the enteric lesions which occur during the second and third attacks in this form of fever.

I need not add anything to what I have already said on the anatomical lesions of relapsing synocha; it may suffice to contrast those of typhus and typhoid fevers.

The former may be comprised in a single sentence. Dark fluid blood and general congestion, with softening of the tissues, are the only appearances found in uncomplicated typhus. Upon this point the observations of pathologists of all countries are strikingly uniform. They are thus summed up by Dr. Murchison—

"The most extensive results of *post mortem* examinations of typhus yet published are those of Messrs. Gerhard and Pennock (50 cases); A. P. Stewart (22 cases); John Reid (147 cases); Thomas Peacock (31 cases); William Jenner (43 cases); F. Jacquot (41 cases); and Barrellier (166 cases). My own observations, amounting to 54, entirely confirm the results arrived at by these authors, as do also the dissections of many hundreds of dead bodies at the London fever hospital during the last fourteen years."

"Speaking generally," continues Dr. Murchison, "the appearances found after death from typhus, are of a negative character, there is no constant or characteristic lesion."

While this testimony is given with regard to typhus by observers of different countries in Europe and in America, a similar uniformity characterizes the reported dissections of cases of typhoid in all countries.

Wherever it is met with, the same affection of the agminated and solitary glands of the intestines and of the glands of the mesentery is invariably met also. These morbid appearances are, to use the words of Dr. Murchison, "constant in and peculiar to pythogenic fever.

The two diseases are thus forcibly contrasted by Dr. A. P. Stewart. "If asked to describe shortly the pathology of typhus, I might sum it up in these words: general congestion; no prominent local disease—a congestion so general and so excessive, as is rarely, if ever, met with in typhoid fever, or any other disease—a congestion singled out by most authors as one of its leading characteristics—a congestion that is evident during life, by the livid skin and petechial eruption, and is found after death to have affected more or less every organ in the animal economy—a congestion so constant as to be often passed over as almost valueless, but, which future researches may prove to be the grand peculiarity of typhus; and which, in common with many other considerations, directs attention to the blood as the essential seat of the disease.

"If required, on the other hand, to give a brief account of the pathology of typhoid fever, I should be inclined to sum it up in these words: prominent local lesion, comparatively little general congestion."

We have now arrived at a position from whence we can review

the evidence upon which we found the division of typhus, typhoid, and the relapsing synocha, into three distinct species.

Unwilling as I am to enter upon questions of a controversial character in these lectures, or to occupy your time with anything which is not essentially practical, I cannot ignore the question of the classification of fevers, or lightly regard its relations to the more practical part of our subject.

Besides, I have a strong personal interest in a question which engaged my attention and occupied much of my time many years ago; and if I then was led by seven years investigation to the conclusion that all the evidence was in support of a diversity of origin of the typhus and typhoid fevers, while, "according to the doctrine of identity of species we must believe, that the same poison shall, at the same time and place, and among the same collection of individuals, produce two diseases totally dissimilar in their modes of access, symptoms, pathology, treatment, and modes of transmission," I can now say that twenty-five years' subsequent experience has confirmed my belief in the truth of these views; and I am more than ever convinced, that whether we regard the genesis and modes of diffusion of these fevers on the one hand, or their clinical history and pathology on the other, we find the conditions under which they arise, and the laws by which they are governed, so distinct and diverse, that we cannot but regard them as constituting different species.

I shall not occupy your time by any detail of evidence, but merely recapitulate the results arrived at from the considerations referred to.

I. The first conclusion at which we arrive is, that whereas in a multitude of instances exposure to the contagion of typhus has been followed by typhus, we can find no satisfactory instance of typhoid following such exposure. In other words, typhus generates typhus and typhus only. Neither is typhoid capable of communicating typhus. The two diseases are not intercommunicable.

II. I must consider as scarcely less important, the fact proved by unquestionable evidence, that while typhus confers a remarkable immunity (fully equal to that conferred by other exanthemata) from future attacks, it confers no such immunity from subsequent attacks of typhoid; and conversely, that typhoid does not exempt from attacks of typhus, as our daily experience in this hospital abundantly proves.

I have already adduced facts to show that the same rule applies to relapsing synocha.

III. That while the occasional occurrence of a hybrid disease is unquestioned, this is so infrequent and exceptional as really to prove the rule. That the coexistence of typhus rash with rose spots and dothiénentérite, no more proves the identity of these diseases than does rubeola—the hybrid of scarlatina and measles—prove that these are one and the same poison.

IV. That the conditions in the individual, and those external to him, which predispose to each form of fever, vary with regard to each other.

. The great predisponents to typhus are ochlesis and the continued imbibition of the poison—the most seasoned by constant exposure are the most highly predisposed.

The reverse seems to hold with regard to typhoid, as was long since pointed out by Louis, and subsequent observers who have accepted his definition of typhoid as a disease—" Propre aux jeunes sujets, *principalement à ceux qui se trouvent depuis peu de temps au mileau de circonstances nouvelles pour eux.*" The other peculiarity remarked by Louis, that typhoid is prone to attack the young, by no means belongs to typhus, which is rather a disease of middle life. Thus, according to Dr. Murchison, while the average age of 1772 cases of typhoid admitted into the London fever hospital was 21—25 years, that of 3456 cases of typhus was 29—33 years of age, or about four years above the mean age of the population.

Of another powerful predisponent—famine—it may be said that it has not appeared to influence the occurrence of typhoid in any marked degree; on the contrary, persons in the higher ranks of life are far more subject to this disease than to typhus; while with regard to the latter, famine appears to be one of the circumstances which, with overcrowding, want of clothing, and want of fuel, exercises a powerful influence as a predisponent.

. The evidence in favour of its influence is, however, much stronger in the case of relapsing synocha; and although I have maintained in a former lecture that not famine, but epidemic influence, is the true exciting cause of this disease, the constant though by no means invariable pre-existence of famine at the epidemic periods, fully justifies us in regarding it as the great predisposing cause of the fever. The other predisposing causes, such as cold and moisture, intemperance, fatigue, depressing emotions, &c., seem to act pretty equally with reference to all forms of fever.

V. That the laws which regulate the prevalence and diffusion of these diseases vary. As a rule, both do not prevail together to the same extent; and in hospital we usually observe typhus to prevail in the winter and spring, and typhoid in the autumn. The three forms may, however, exist together; but in this case we observe a remarkable difference in their diffusion. Typhus and relapsing synocha spread over the country; the former being diffused apparently by deportation, a single case furnishing the germ of the epidemic of a town or village; relapsing synocha no doubt spreads in the same manner as typhus, and also by atmospheric or epidemic influence; typhoid is never epidemic, in the true sense of the term, but is strictly endemic, the disease of the locality in which the fecal miasm, which is its cause, exists.

It follows, as a corollary, that the diffusion of typhus and typhoid is to be arrested by different measures. No one doubts that the segregation of the typhus patient will accomplish this object, while I have over and over again witnessed its failure to prevent the occurrence of other cases of typhoid in houses exposed to fecal miasm.

What the separation of the infected has failed to do, a heavy shower of rain, or a change from hot and dry, to cold and wet weather, or frost, will at once accomplish; while these have no such influence on typhus, but rather operate in the opposite way, by leading to confinement within doors and over-crowding.

VI. It may be mentioned that whereas the greatest number of fatal cases occur in the outbreak of an epidemic of typhus, the fatality of typhoid increases as the season during which it prevails is prolonged.

To enumerate the arguments derived from the pathology and symptomatology of the fever would be to repeat a large portion of the preceding lectures. I shall merely recapitulate the conclusions at which I have arrived, in the fewest possible words; trusting that your memories have retained the facts and considerations upon which these are founded.

I believe, then, that those diseases must be of distinct species which differ so remarkably.

1. In their invasion and early history.
2. In their conformity to the laws of periodicity.

3. In their duration and modes of termination.

4. In their external phenomena, more especially their eruptions.

5. In the special affinities of the poisons, the order of symptoms, and their internal pathology.

6. In the mode of fatal termination.

7. In the anatomical appearances.

8. In the law regulating the mortality with regard to age.

9. In the comparative rapidity and security of the convalescence in typhus, and its complete exemption from relapse; as contrasted with the comparative certainty of relapse in epidemic synocha; its frequent occurrence in typhoid; and the other risks and uncertainties of the convalescence in this disease.

Finally, you will meet with abundant proofs in your clinical study of fever, that this distinction is not merely theoretical or speculative, but that, on the contrary, it has an immediate bearing upon important questions of prognosis and treatment.

LECTURE .XIII.

PROGNOSIS.

WE have now to enter upon the consideration of fever, regarded from another point of view, namely, as presenting data for a rational prognosis. There is no portion of the clinical study of fever of more importance than what is termed its prognosis; by which I wish you to understand, not merely an acquaintance with what are called prognostic signs, and that just estimate of their value in particular cases, to be acquired only by prolonged experience, and that incommunicable tact which such experience alone can give—but rather the study of the tendencies of a given case of fever to one or other termination, as inferred from the past and present condition of the patient, the history of the disease, its complications, and all the conditions which can influence its course or affect its result. It is in this sense the term is used by Hippocrates, who says—"The physician must be able to tell the antecedents, to know the present, and foretell the future; must meditate on these things, and have two special objects in view with regard to diseases—namely, to do good, or to do no harm." Taking this axiom as a guide, you should endeavour to attain to that foreknowledge which is based upon a comprehension of everything related to the disease on the one hand, and an accurate estimate of the constitutional power, the proclivities and individual peculiarities of the patient on the other; thus you will be enabled to anticipate all tendencies of whatever nature, and to provide against all emergencies, foreseeing what is about to take place, and the measures which will be necessary; arranging, beforehand, the means to guard against all accidents, and indicating the moment when this or that remedy should be administered or withheld.

Considering prognosis in this comprehensive sense as the knowledge of tendencies, whether of the disease or of the treatment, you will make it a part of your daily study at the bedside. Each day you will carefully estimate the tendency of the case to run a certain course, to develop certain complications, or to terminate in one or other of the modes I have already described; deriving your data from everything in the individual, the disease, and the treatment which can possibly influence the result.

Before entering on this investigation of the conditions of prognosis in a particular case of fever, it is well that you should be reminded of certain facts or laws which may be asserted of fevers in general; and which are derived from the characters of the epidemic period, the type and history of the disease, the age, sex, previous history, health and habits of the patient.

One of these facts is that the rate of mortality in fever will vary considerably in different epidemics, at different periods of the same epidemic, and throughout the whole continuance of the same epidemic at different places. Thus it has been observed, that at a time when a low or asthenic type of scarlatina or erysipelas prevails, a fatal form of fever will prevail also. At other times the prevailing diseases being mild, cases of fever of a severe form will be less fatal than usual.

Of epidemics of typhus, it has been remarked by Gerhard and others, that the earlier cases were by far the more fatal. The same has been observed in epidemics of cholera. While the epidemic typhus generally becomes less fatal, the contrary is rather the tendency of the endemic typhoid, of summer and autumn, which gradual auguments in severity, the poison becoming more and more intense, as decomposition advances, until a change in its character follows upon a change of season. The marked difference in the rate of mortality at various periods of the same epidemic, is sometimes due to the prevalence of several forms of fever at these periods.

I have already in a former lecture, drawn your attention to the fact, that three forms of fever have prevailed in various proportions at different periods of the same epidemic. This has been observed by Armstrong, Cheyne, and others, in the epidemic of 1817, and by many observers in that of 1847. As the average rate of mortality of these various forms of fever varies, the death rate at different periods of the epidemic will consequently vary also.

There is another point of view from which we should regard these different forms of fever, with reference to rational prognosis —viz., the difference in the complications which occur, and the several accidental lesions which require to be guarded against in different species of fever. While in typhus the nervous and pulmonary complications are the chief causes of anxiety in the progress of the case, and the uncertainties of crisis render imperative a guarded prognosis, up to that period; the same care and caution are called for in typhoid, by the more prolonged and indefinite duration of the disease, the great danger of imperfect crisis, and the peculiar risk of a latent and insidious affection of the intestine, often continuing for some time after the termination of the fever; and the existence of which may perhaps be first revealed by a fatal hemorrhage, or peritonitis from perforation.

In short relapsing fever the dangers of the primary illness are very much less, but death seems sometimes to occur from the excessive disintegration of tissue, caused by the highly marked crisis. Relapse, with its dangers, is infinitely more frequent than in the other forms; and the greatest caution is required to avoid mistaking for convalescence, that period of remission which so often ushers in a second, or even a third attack of fever.

The age of the patient has a marked influence on the result of fever. While typhoid fever is much more fatal than typhus in early youth, it seems to be an ascertained fact, that after twenty years, there is a steady increase in the rate of mortality in typhus, up to fifty; about which age, it amounts to one-half of those attacked. After this, the increase is still more rapid, until at eighty years, it is almost uniformly fatal.

We find the explanation of these facts—First, in regard to typhoid, in the great demand upon the formative energy during adolescence, when all portions of the organism are undergoing constant molecular change. The danger from this source is, of course, in proportion to the rapidity of growth, and is much augmented by anything causing increased waste, such as violent exercise, exposure to cold and damp, dissipation, &c.

Secondly, the increased fatality of typhus in advanced life is no doubt due to the gradually diminishing energy of the function of nutrition; progressive metamorphosis becoming less active, and the products of regressive change being less perfectly eliminated by the excreting organs.

The greater mortality of adult males than of females, as well as the disproportionate mortality of the pregnant and puerperal females attacked by fever, admit of a similar explanation.

The occupation and mode of living have each an important influence on prognosis in fever. Take two patients of the same age, say twenty-five or thirty years; one a healthy farm labourer; the other, a hardworking student, a professional man, or an anxious man of business. There can be no question that the chance of recovery of the latter is far less than that of the former; owing to the greater severity of the cerebral complications, sure to arise in the man whose brain is undergoing the most active metamorphosis at the time of seizure.

The peasant has another advantage over his neighbour in the comparatively less amount of effete organic matter in his blood, the more healthy condition of his excreting organs, and the consequent diminished risk of the dangerous secondary blood contaminations to which the luxurious are so liable.

You will occasionally meet with personal and family peculiarities of constitution, of a nature not, perhaps, ascertainable, but which have much influence on the result of fever; as I have before remarked, these frequently determine the course and complications of the disease. The members of some families seem not only to have a proneness to become the subjects of fever, but also exhibit a tendency to succumb under it. Of course, you will, in such cases, endeavour to ascertain from what special condition the danger is to be apprehended, so far as this can be learned from the history of previous fevers occurring in individuals of the family. It may be a peculiar predisposition, or a family tendency to some complication; or some peculiar physical condition not in our power to obviate; or it may be connected with residence in a particular locality, or some other preventible condition.

Residence in an infectious locality has often a most unfavourable influence on the result. If, as in the instance of a fever nurse or of a clinical clerk, it has been in the atmosphere of the fever ward, the accumulation of the typhus poison too often manifests itself in a malignant form of fever; while on the other hand, prolonged residence in a malarious atmosphere, by its influence on the blood, renders an attack of typhus, septic in its type; or, it may, together with the typhus poison, generate a hybrid disease, presenting the unfavourable combination of the cerebral complications of the one, with the abdominal lesions of the other form of fever.

The previously existing diseases of the patient, which have been observed to have the most injurious influence on the course and termination of fever, are, diseases of the brain, of the heart, of the respiratory organs, and of the kidney.

You might naturally expect, that a brain suffering from malnutrition, will not resist the disintegrating action of the typhus poison; and that the tendency to pulmonic congestion in fever will be greatly augmented by pre-existing chronic bronchitis, by mitral valve disease, or by a weak dilated condition of the right ventricle. Again, if your patient is suffering from chronic disease of the kidney, and deficient elimination of urea, nothing will probably avert the most formidable secondary contamination of the blood, and death by uræmia.

A knowledge of the history of the disease, including the circumstances of the seizure, and the phenomena of the stage of incubation, or latent period, is necessary; as well as that of the management pursued during the early stage of the disease.

"Physicians of experience," says Dr. Cheyne, "pay particular attention to the circumstances under which fever commenced, and to its first symptoms." This axiom you will do well to remember. With regard to the first, the tendency of a case of fever to death or recovery, may be influenced by the condition of the patient at the time of seizure, and the peculiar predisposing cause; or, by the special exciting cause. If the access of fever has occurred when the nervous system was exhausted or depressed by excessive exertion, fatigue, long watching, want of food, or by the depressing emotions of grief, anxiety, shame, or fear, the prognosis, will, *cæteris paribus*, be unfavourable. The high rate of mortality among the better classes, during the epidemic of 1847–8, has been ascribed to these causes by many writers. To these may be added the continued expenditure of nervo-muscular force after the seizure by fever, during its latent period and earlier stages, for the history of too many valued members of our profession has proved that, after a short struggle, the patient is struck down with more than the ordinary prostration of strength and energy, dying not unusually several days before the ordinary period. Some of you have witnessed in our fever wards, the fatal influence exercised upon the nervous system and the blood by the prolonged bodily exertion of the peasant, who has continued to labour during the earlier days of his fever.

Such cases rarely indeed recover, but always manifest severe and peculiar nervous symptoms, and the worst forms of blood dyscrasia. Ever remember that fever has a special affinity for those portions of the organism which are undergoing disintegration or waste; that under its influence, disintegration of tissue is increased; and you will appreciate the value of the practical rule, never to allow any exercise of the nervo-muscular function, which can be avoided, after the access of fever has occurred.

The influence of grief, anxiety, and despondency, are not less fatal. It has been often observed, that a presentiment of death, at the commencement of fever, is almost certain to be realized. If the patient has not a desire of life, he seldom survives. Shame and remorse often destroy their victims by the agency of fever.

The conditions in which the patient is placed during fever, exercise a powerful influence upon the result.

I have already alluded, in a former lecture, to the spontaneous generation of typhus by overcrowding and want of ventilation; and to the power that this ochlesis has of intensifying, and rendering more malignant, an existing fever. You will therefore readily imagine, that under these circumstances the rate of mortality is much increased, and accordingly our hospitals, barracks, and prisons, furnish many examples of the fact. Take the following from Dr. Dillon's account of the fever of 1847: "In March, 1847, our county gaol (Castlebar) was crowded to more than double its capability, those committed being in a state of nudity, filth, and starvation. The prison hospital, calculated to hold 16 patients, had, early in March, eight cases of low typhus. From the character of the fever, the condition of the prisoners, and the fearfully crowded state of the gaol, I clearly saw we must, before long, have a full visitation of bad typhus; and accordingly applied for immediate and further hospital accommodations. Before temporary sheds could be erected, we had fully one-fifth of the inmates in bad maculated typhus. Our Roman Catholic Chaplain, deputy governor, deputy matron, and a turnkey, fell victims; every hospital servant was attacked, and from our wretched overcrowded state, the mortality was fearful; fully 40 per cent."

While the poorer classes are usually the most exposed to this influence, the more wealthy are not without their dangers, arising from over care and injudicious nursing. I shall have to draw your attention to this subject in my next lecture.

It is generally believed, that if fever is sudden in its invasion, and this is followed by proportionate reaction, the patient being at once laid aside; the chances of recovery are greater than when a period of slight indisposition preludes the apparent commencement of the fever, and renders this insidious and uncertain, wanting the evidence of reactive resistance of the constitution. Statistics furnished by several observers support this belief. Thus, according to Chomel, of 76 attacked suddenly, 26, or less than one in three died; while, of 39 in whom fever commenced with preliminary indisposition, 20 died, or more than one-half. According to Dr. Ormerod, of 115 cases of different types, in which the mode of aggression was noted in St. Bartholomew's Hospital, the fatality of the cases in which it was sudden, was one in five; of those in which it was gradual, one in three.

I think these results may in a measure, be owing to a circumstance of which I have already spoken—the expenditure of nervomuscular energy during the earlier periods of fever, which is obviously more likely to occur in the insidious, than in the sudden mode of invasion. But apart from this it can be readily imagined, that the insidious access may be generally the more dangerous in both typhus and typhoid fevers, though, perhaps, for different reasons. In the former, it would seem that a prolonged and insidious period of incubation is sometimes associated (whether as cause or effect I cannot say) with more than usually profound nervous depression. In the latter, the most remarkable cases of insidious aggression have been found associated with the most extensive disease of the glands of Peyer. A short time since, a young woman was admitted into this hospital, labouring under symptoms of peritonitis, which proved fatal after a few hours. Her history was, that for 7 or 8 days she had felt weak and languid, but had continued her work as a housemaid, until the day before her admission. On examination of the body, the glands of Peyer were found swollen and infiltrated, several were in a state of ulceration, and one of these ulcers having perforated the bowel, had given rise to the fatal peritonitis. The mesenteric glands were also enlarged and infiltrated, and the spleen congested and soft. My friend, Dr. Carl Martius, has informed me, that an eminent surgeon of Munich, performed a capital operation in the hospital theatre the day before his death, which was caused by a perforating typhoid ulcer of the intestine.

Instances of this kind have been recorded by many writers. According to Chomel, as quoted by Dr. Murchison, ten out of twelve instances of perforation occurred in a latent form of fever;

and Dr. Bristowe accounts for its greater frequency in males (11 to 4) by their continuing to work up to the very instant of perforation. I have myself witnessed so many cases terminating fatally by perforation or by hemorrhage, that I make it a rule in private practice to apprise the patient's friends of the risk there is of one or other occurring, even in the mildest form of the disease. I may add, that this warning may sometimes save the patient, by leading to increased care and caution in the regulation of the diet and management during convalescence.

But if the insidious accession of fever is often attended with danger, it by no means follows that cases of sudden and severe invasion are exempt from their own risks. On the contrary, typhus, like scarlatina in a concentrated dose, occasionally kills, as has been said, *uno ictu*, deadening the energies of the cerebro-spinal centres, destroying the contractility of the heart and the consistence of the blood, and causing general capillary congestion; the patient passing at once into a state of stupor or coma, as if under the influence of a narcotic poison; the blood, after death, being found fluid, or semi-fluid, and stagnating in the heart, lungs, brain, liver, and spleen.

I have already, in a previous lecture, quoted Armstrong's graphic sketch of this congestive typhus. I may here refer to two examples of it observed by Dr. Cullinan, of Ennis, during the epidemic of 1847. He says: "There were presented to my observation some remarkable cases of fever, which ran a fatal course in a period of from six to twenty-four hours. They were not cases of cholera of any kind. At first, I supposed that their rapid progress was influenced by previous privations, but now I am of opinion that their course and issue were determined by the agency of some potent aerial poison. I remember particularly two such cases which occurred in the county Clare gaol, where the food was always abundant and wholesome. A young man had eaten his breakfast, and began to complain. In two hours afterwards I found him so weak that he could not stand without support; his head was giddy, his eyes glassy and without expression; his pupils contracted and vision dull; his skin was cool, pulse very weak, almost imperceptible; the action of the heart rapid and indistinct; secretion of urine scanty; tongue rather clean, but clammy. He was laid in bed, got some wine and chlorate of potash, and had a sinapism applied to his epigastrium. He sank gradually, and died in eight hours.

"The second patient died in little more than six hours from the time of his first complaining. His symptoms were very similar to those detailed. The gaol, at this time, was greatly crowded, and fever and dysentery were very prevalent and fatal among the prisoners."

Passing on from the initiatory stage of fever to the study of those indications of unfavourable prognosis arising in its progress, we shall find that, so far as these belong to the disease, they may be usually referred either to want of conformity to the type of the particular species of fever, or to the unusual severity of the complications, especially of such as arise from different forms of secondary blood contamination.

As an example of the first, I may mention, that if the delirium and slight stupor, which are normal phenomena, and, as such, excite no apprehension in the earlier stages of typhus, occur at the same period in typhoid fever, they are most unfavourable prognostic signs. Again—if in typhus the pulse rise to the same rate of frequency as is usual in relapsing synocha, it is a bad sign; and if, at a comparatively early period, say on the ninth or tenth day of typhus, there occurs a critical effort marked by copious sweating, such as ushers in recovery in relapsing synocha, the result may be said to be almost uniformly fatal.

Another example of a symptom favourable or fatal, according to the form of fever, is rigor. In relapsing fever, this frequently marks the commencement either of favourable crisis, or of a change into the intermittent type; while in typhoid it too often indicates the commencement of fatal peritonitis, or some other inflammatory complication. In estimating the prognostic value, of the deviations from healthy conditions in the course of fever, you should not merely measure the amount of derangement of function, but also compare this with the typical amount of such derangement at a corresponding period of the same form of fever, and as far as possible this should be done with reference to the cause of the deviation, and in the case of severe nervous lesions to the seat of the lesion; those referable to the medulla oblongata being infinitely the most important of the cerebro-spinal lesions in fever. Headache, for example, is a normal deviation from the feeling of health in the first week or more of typhus, and we know that it alone, although severe, is not a sufficient cause for alarm at this period. But headache, coming on at an advanced period of the disease, is an abnormal symptom, and of more or less serious importance, according to the special cause, as indicated by the associated nervous symptoms—such are cutaneous dysæsthesiæ, muscular rigidity, convulsions, &c., in inflammation of the brain or membranes.— The same remark holds good of vomiting, which is salutary at the invasion of fever, and almost certainly fatal at its close; and which, occurring at different periods during fever, may be pronounced more or less unfavourable, according as it has a cerebral, renal, enteric, or bilious origin.

Hysteria is another example. It occasionally ushers in a favourable crisis, while, on the other hand, it sometimes occurs as one of a group of most unfavourable nervous symptoms, and accordingly has been pronounced a fatal one by such high authorities as Cheyne, and subsequently by Graves. Lastly, hemorrhage from the bowels, so serious, often so fatal a sign of intestinal ulceration in typhoid, has been observed to be, occasionally, a favourable crisis in typhus by H. Kennedy and others. These examples may suffice to convince you that you cannot safely pronounce on the value of a sign or symptom, regarded singly, without reference to its cause, and to its conformity to type.

But it is in your daily clinical study of fever, that you will derive the most important prognostic indications from the course and progress of the disease. It is by close and daily observation you will be able to measure the patient's power of resisting the influence of the disease; to judge of the conformity of symptoms to the type of the special form of fever; and to ascertain the nature and extent of secondary complications, and the effects of treatment. Following the plan of investigation of each function successively, you will first collect the indications which present themselves on a review of the external condition of the patient. You note the colour of the surface, the expression of the face, the eye, the decubitus, the eruption, the temperature, each of which afford indications favourable, or the contrary, according to their conformity or otherwise to the type of fever, or according as they indicate a healthy state of the blood and vital organs, irrespective of fever; or, on the other hand, give evidence of secondary blood contamination (septicæmia), of severe visceral congestion, or of an unusually depressed condition of the nervous system. You have examples in the value of sudden jaundice as a sign in septicæmia in gastro-hepatic congestion, and in low forms of pulmonic congestion; and you read in the half-closed injected eye—dry or coated with a film of mucus—one of the most unerring signs of sinking of the vital powers and approaching death. A glance will show you the condition of the capillary circulation, whether the eruption is bright and healthy, dark, becoming prematurely ecchymotic, or, as in some of the worst cases, if it has appeared several days before the usual time, and passed rapidly into that condition. You will note if the patient lies tranquilly on his side, with his limbs moderately flexed; or, as in unfavourable cases, on his back with the limbs stiffly stretched out; or, as in equally hopeless cases, with the legs firmly flexed, and the heels pressed against the buttocks, the arm being, perhaps, suddenly flexed on your touching the wrist, or falling passively and heavily when lifted. You note, moreover, in serious cases, occasionally, a tendency to jactitation, rolling of the head on the pillow, and sliding out of bed.

The prognostic signs derived from observation of the respiratory and circulating systems, are: Respiration favourable, if easy, and not hurried or attended with frequent sighing. Unfavourable, if hurried and laborious, irregular, and interrupted by frequent sighing. Also unfavourable, if it be that form of ascending and descending breathing, described by Dr. Stokes as occurring in cases of fatty heart. Cerebral breathing, if noisy and of a nasal character; and stertor, if persistent and accompanied by coma, or nasal in character, as in renal disease. If, on physical examination of the chest, you find dulness on percussion of the depending portions from hypostatic congestion, well marked at an early period of the fever, and varying with change of position; especially, if this does not disappear during the early period, but continues till a more advanced stage, indicating failure of the pulmonary capillary circulation. Or if the signs of extensive and extending capillary bronchitis or pneumonia are present, conjoined with bloated, livid face, and injected eyeballs, and with a tendency to stupor. Aphonia from laryngitis or œdema glottidis, when occurring in typhus or typhoid, is also a prognostic sign of the gravest possible import, as well as an indication for prompt and decided treatment.

The signs derived from the circulating system of a favourable character, are, a moderately strong impulse, with healthy sounds of the heart; a perfect correspondence between the heart's impulse and the radial pulse; the latter neither very much above or below the natural standard; an active state of the capillary circulation;

and, lastly, absence of congestion of the right cavities of the heart at an advanced stage of the fever.

On the other hand, few recoveries from typhus take place with a pulse over 120 in the early period, more especially if a daily increase is observed in its rate—(the pulse may rise considerably on the eve of crisis)—you may also draw an unfavourable prognosis from a daily progressive increase in the frequency of the pulse, more especially if this coexists with a want of correspondence with the heart; the impulse of the latter, previously weak, becoming strong and jerking, and the pulse weak and thready. In such a case, you will often observe the same disproportion between the action of the carotids and the radial pulse. If there be a want of correspondence between the number of beats of the heart and pulse, with other signs of mitral disease; or if the cardiac sounds and impulse become so faint as almost to disappear; or if there is a distinct cardiac intermission, with a marked slowness of pulse, conjoined with stupor or coma, the prognosis is most gloomy, as it also is when the heart falters, and becomes irregular in its rhythm, and unequal in the force of its beats.

The favourable signs derived from the digestive system, are: A tongue not too clean, but conforming to the type of the fever, protruded readily on being moistened, and withdrawn promptly; normal thirst; willingness to take nourishment and wine, and even a relish for them. Bowels not too free in the advanced stage of typhus, or in the third or fourth week of typhoid.

The signs of danger, are a black and dry tongue, which the patient cannot protrude, or one morbidly moist, clean, and tremulous throughout the fever, or shrivelled, dry, and black; or, worst of all, a cold and retracted tongue.

Absence of thirst and disgust of food, with disinclination to swallow wine or other drink, are of bad omen; still worse, if there is dysphagia. Vomiting at a late period, is a fatal symptom, more especially if it has been preceded by meteorism, and by loud gurgling on swallowing liquid. "In nonnullis," says Hoffman, "peculiare hoc funesti ominis symptoma visum, quod liquidum assumptum et in stomachum delatum, accidente leviore corporis commotione vel erectione tantum, iterum in os cum murmure ac borborygmo ascenderit." And Burserius mentions an epidemic of typhus in which "one would have thought the patients were affected with hydrophobia. Such was the tendency of the stomach to spasm, that they fell into convulsions on touching anything liquid; and everything they drank was immediately rejected." A change from diarrhœic to dysenteric stools; hemorrhage from the bowels; obstinate hiccup, sudden and severe pain in the abdomen, with tympanitic distension; green vomiting, and other signs of peritonitis. In very protracted cases of typhoid, a retracted abdomen with diarrhœa is a sign of the worst possible augury; and in the advanced stage of typhus there is sometimes a sudden and copious liquid evacuation by stool—a true colliquative discharge—shortly before death.

The normal urine of fever is of high colour, and high specific gravity, containing a larger proportion of urea than in health; whenever this gives place in the progress of fever to a limpid colourless fluid of low specific gravity, the prognosis is *pro tanto* unfavourable. The other unfavourable signs derived from the urine, are, prolonged retention; albuminous urine of low specific gravity; suppression; bloody urine; the two last, if associated with cerebral symptoms, are, I may say, invariably fatal. I can add nothing on this subject to what I have already advanced in my lecture upon the secondary blood contaminations, especially the uræmic.[1]

The perversions of the special senses are signs of great prognostic value, more especially those of sight, hearing, and the cutaneous sensibility. Taste is sometimes morbidly acute, or rather perverted; but more frequently it is totally lost in serious cases. It is always a favourable indication when the patient retains his relish for wine and beef-tea; recovery has, in many instances, been heralded by an urgent demand for a draught of cold water. On the other hand, the obstinate refusal to take wine under circumstances demanding its administration, is one of the most unfavourable symptoms.

Loss of sight and double vision are most unfavourable signs; loss of hearing has been generally considered the contrary. It may be questioned if it is so, unless regarded as a contrast to morbidly acute hearing; which is decidedly unfavourable. There can be little doubt, that deafness coming on at a very early period of typhus, is also decidedly unfavourable.

Cutaneous anæsthesia and hyperæsthesia are both unfavourable prognostics when occurring in combination with other symptoms

---

[1] Burserius enumerates among the fatal signs "thin crude urine not depositing a sediment, and not turbid; or black urine; or having a black or reddish sediment, in consequence of its mixture with blood. Likewise, permanent suppression proceeding from defective secretion."—*Institutes*, vol. 3.

indicating lesion of the nervous system, as when hyperæsthesia is accompanied with green vomiting and muscular rigidity, in cases complicated with cerebro-spinal arachnitis; or, when anæsthesia is present with absence of thirst, dorsal decubitus, dusky skin, stertorous breathing and stupor. In the worst cases, we sometimes meet with a combination of both anæsthesia and hyperæsthesia; the patient being insensible to cold or to the action of the bowels or to that of the bladder, as well as unconscious of the presence of surrounding objects, yet screaming when an attempt is made to turn him in bed.

In order to draw any prognosis from the alterations in the consciousness, or the intellect, you should carefully distinguish between those which are due to the simple operation of the fever poison on the brain, and those—far more formidable—which arise in cases of active congestion, or in secondary blood contaminations. Much and careful observation will be required to enable you to recognize these several states; however, I shall not now repeat what I have already said on nervous complications, but merely enumerate some of the most unfavourable symptoms of this class. I may remark, then, that heightened states of the consciousness, such as excited manner, watchfulness, acute hearing, &c., are considered of unfavourable omen. The same may be said of delirium, if combined with cerebral respiration, with cerebral vomiting, spasmodic rigidity of muscles and cutaneous dysthesia; if it be violent in the early periods of typhoid, or the later periods of typhus, if associated with convulsions, or with stupor—a frequent combination in some of the worst cases—or with prolonged watchfulness, more especially when the patient has been previously subjected to some severe mental or emotional excitement, or depressing influence; when the delirium is of an hysterical character, in the case of a female in whom menstruation has been arrested or suspended at about the period of access of fever by mental shock, or exposure to cold; or, lastly, most unfavourable when associated with symptoms of lesion of those important centres which preside over the automatic movements of the heart, the diaphragm, and the stomach, and which regulate the temperature. I need not tell you that you will recognize these latter in the sinking of the temperature of the body, the irregular and sighing respiration, the weak, fluttering, and irregular pulse, with, probably, a jerking tumultuous action of the heart; and, in the tympanitic belly, hiccup, and dysphagia of the advanced stage of fatal typhus—all pointing to lesion, more or less profound, of the automatic nervous centres, rather than to the less serious derangements of the cerebral hemispheres; and justifying the observation, that of all the local lesions in typhus, the most fatal is that involving the medulla oblongata.[1]

The signs which portend a fatal termination from a septic condition of the blood, are: musky or cadaveric odour of the body; changes in the colour of the surface, to yellow, livid or leaden tint; or a mixture of these. Conversion of the fever rash into petechiæ, often after the rash has disappeared, giving rise to the idea of a fresh eruption; subcutaneous extravasations of blood; hemorrhages from mucous membranes; sudden discharges from the bowels of copious fetid liquid stools; extreme tympanitic distension of the abdomen, hiccup, &c. &c.; all point to a fatal end; and whether occurring in typhus or typhoid, the prognosis is equally lethal.

The prognostic indications derived from the effects of treatment, past and present, may be briefly mentioned:—

If the patient has been subjected to active lowering treatment, more especially to continued purging before he comes under your care, be very guarded in your prognosis. I have very often known a fatal result quickly produced by a large dose of salts, sometimes by a violent emetic, administered several days after the commencement of fever. The indications which you obtain from observing the effects of your own remedies, will be chiefly derived from those of wine, or other stimulants; opium, or other narcotics. If wine agrees with your patient, as evidenced by a fall in the pulse, increased warmth of the general surface, tendency to sleep, &c., he will, in all probability, do well. If, on the contrary, it produces poisonous effects, such as increased determination of blood to the head, the extremities becoming cold and the head hot, the heart's impulse strong and jerking, and the pulse weak and thready, it is a bad omen. The same may be said of opium. I never remember to have seen a patient recover in whom furious delirium resisting

---

[1] "Although the patient should be insensible to all external objects; although he should sleep very little, or scarcely at all; yet if the deglutition and respiration remain unimpeded, the patient is not to be despaired of; it happens even most commonly that he recovers. But if he respires with great difficulty, or hardly at all; or if the deglutition be almost totally prevented; or if attempting it throws the patient into convulsive contractions, he rarely survives."—*Fordyce, 3d Dissertation on Fever*, p. 111.

the calmative influence of opium, passed into stupor resembling that caused by its poisonous action, realizing Hoffman's description of the effect of narcotics in fever: *ad mortem aditum parant, placidum quidem, sed certum.*

Another indication which deserves mention is the inefficacy of blisters, which is sometimes observed in states of profound insensibility. Lastly, I may again mention, the fatal sign of rejection of nourishment or wine, by vomiting, at an advanced period of typhus.

It only remains for me to enumerate a few prognostic signs observable at and subsequent to the period of crisis.

In typhus, while we bear in mind that the result can never be determined before decisive crisis, in a disease so variable in its accidental complications, and so subject to exacerbations in the critical period; we augur well of the result of the crisis, if it occurs at a normal period, if the eruption has previously faded, if the heart has not become abnormally excited. Prognosis is also favourable, if, at the period of crisis, no excessive evacuations by the skin and bowels occur; if the patient falls asleep, and continues to sleep with little or no interruption for many hours, showing in his waking intervals, a sensibility to thirst, to cold, and other external impressions; and if he at the same time swallows freely. A desire for cold water at this period is a most favourable prognostic sign; another is the voluntary drawing-up of the bedclothes over the shoulder. Even before the patient has become conscious of the sensation of cold, a favourable augury has sometimes been drawn from the contraction of the dartos muscle upon exposure of the surface.

The following are also favourable signs at this period. Should the patient, on awaking, be found to have lost the expression of drunken slumber characteristic of typhus, the eye being bright and intelligent, the features placid, and the muscles no longer tremulous; if the heat is natural, the skin soft, the respiration easy, and the pulse soft, full, and less frequent, the tongue protruded steadily, expanded and moistening at the edges.

Similarly, in typhoid, we anticipate a favourable change when delirium subsides, and sleep becomes tranquil and prolonged, diarrhœa ceasing at the same time; during the third week to the fourth the daily eruption of rose spots gradually subsiding. When the bronchitic affection of this period is moderate, and the pulse loses its dicrotous character, and becomes distinct, full, and soft. When the urine presents well-marked deposit, the tongue becomes clean at the edges, and the skin becomes gently moist. Lastly, if there be no fulness or pain of the belly, or tenderness over the ileo-cæcal region, and no tendency to return of diarrhœa or vomiting.

But in typhus, should critical evacuations occur prematurely, or to an excessive amount, without resolution of fever; should visceral complication, cerebral or pulmonary, remain after a decided critical effort; or obstinate pervigilium and impulsive delirium, with tremor or subsultus, or obstinate diarrhœa, with increasing emaciation, follow upon crisis; or should petechiæ appear after crisis, the prognosis is unfavourable.

It is, in like manner, unfavourable in typhoid when the intestinal affection continues severe after the fever. Perforation is then too apt to occur, but even without this lesion the symptom is most formidable, especially if accompanied by a harsh dry state of skin, and wasting of the body. Should the catarrhal affection after having yielded, reappear, with rapid pulse and a tendency to hectic, the supervention of acute phthisis is to be apprehended, more especially if the physical signs of dulness on percussion, sibilant and crepitating râles, &c., present themselves.

Of the prognosis in the two forms of fever, at this period, it may be said, that of typhus is simple, easy, and certain, as compared with that of typhoid; which involves, besides other risks, the conditions and termination of an affection so latent in its symptoms, and so little influenced for good by treatment, as is ulceration of the intestinal follicles. Perhaps more cases of perforation occur during convalescence than during fever, and then when least expected; thus, of twelve cases reported by Chomel and Louis, two only occurred in severe fever, ten in the milder form. I remember a fine, previously healthy country girl, who had passed through mild typhoid fever in hospital. She was one day sitting in bed combing her hair, and was in the act of laughing at something said by a fellow-patient, when she was suddenly seized with a sharp pain in the abdomen, indicating the occurrence of perforation, and the commencement of peritonitis, of which she died on the day following.

Dr. Anderson relates that, on one occasion, there was a patient in the Glasgow infirmary who appeared almost convalescent. He was visited by the doctor in the evening, and made no complaint;

the first person who entered the ward in the morning found him dead. On inspection, his death was found to have arisen from peritonitis, in consequence of a perforation of the bowel, which, says Dr. Anderson, must have happened after the evening visit.

These, and similar facts, should make you most guarded in your prognosis in all cases of typhoid fever, as well as most careful in your treatment, and strict in your injunctions with regard to food and management. I should not omit to mention that you will sometimes derive valuable assistance to your prognosis from the thermometer. Thus, Dr. Aitken states, on the authority of Dr. Parkes, "that a sudden and marked reduction of temperature has indicated the occurrence of hemorrhage from the sloughs of Peyer's patches, in typhoid fever, several days before blood appeared in the stools." The same writer also warns us that the morbid process in fever cases has not terminated till the normal temperature of the body returns, and remains unchanged in the evenings and throughout all periods of the day. The persistence of a high evening temperature in typhus or typhoid fever, and its incomplete subsidence after crisis, indicates incomplete recovery, supervention of other diseases, unfavourable changes in the products of disease, &c. The onset of even a slight elevation of temperature during convalescence is a warning to exercise careful watching over the patient and a due control over his diet and actions.

Of the prognosis at and after the critical period, in relapsing synocha, it is not necessary to say much. I may remind you of the risk of fatal uræmic poisoning, from what may be termed excessive *perturbatio critica*, causing the accumulation in the blood of effete material—the product of excessive disintegration—to an amount which the excreting organs cannot eliminate, and this notwithstanding a perfectly healthy condition of the kidney. You should also remember that no matter how complete the crisis may appear to be, relapse may occur again and again. I believe that you will not, in any case, be able to say it will not occur, but that there is a condition in which you may say with all but certainty that it will do so. I refer to the continuance, after crisis, of congestive enlargement of the spleen. The prognosis in the second attack is, of course, more doubtful than in the original fever; and the data will differ, if, as not unfrequently happens in the epidemic periods, the relapse assumes the character of typhoid fever.

I think I have now enumerated most of the data upon which

you should found your *estimate of the tendencies* to one or other termination in any given case of fever; and, in conclusion, I would urge you to devote much of your attention to the study of prognosis in this sense, rather than in that of *predicting results.* You may collect, from observation or books, a large number of empirical prognostic symptoms and signs, and, by employing these judiciously and fortunately, you may acquire an ill-founded reputation for prescience and sagacity. On the other hand, the study of rational prognosis, as I have endeavoured to set it before you, will enlarge and render accurate your knowledge of the phenomena of the disease; of its laws, and of the conditions which influence the patient's resistance to it; and with the development of that knowledge will be certain to increase, your care and caution in the treatment of the emergencies arising in the course of fever, and in calculating the chances of death or recovery.

LECTURE XIV.

TREATMENT.

In entering upon the last portion of our subject, I have again to remind you, that these lectures are not didactic descriptions of fever or of its treatment. Therefore you must not expect from me a code of directions for the management of each species of the disease. In other words, you must not expect that I shall offer you suggestions for every emergency that may arise, or lay down a routine system of treatment such as you will find in class-books. All that I propose to do, is to offer you a guide to the study of treatment at the bedside, by placing before you the principles upon which the management and general medical treatment of each form of fever is founded; and the special indications which are presented by each secondary lesion or complication.

Thus far only can definite rules of practice be observed, for in no disease will you be more frequently compelled to depart from routine, and to accommodate your measures to the exigencies of the case, taking these and principles as your only guides.

As a preliminary step to this study, you must divest your minds of the idea that we possess any means of curing fever in the strict sense of the term.

It has been justly said, "You can no more cure a fever than you can quell a storm; but you can guide a case as you would guide the vessel through the danger."

It may be your duty in some cases, to attempt to cut short the disease in the period of access by a shock, while in others, the patient may be saved by active interference during the last paroxysm. But these are exceptional instances; and during the course of the fever, your duty will be watch its development, and to wait upon nature, abstaining from all injudicious interference with her operations; while you carefully note any aberrations from the normal type of the disease, and meet such complications as may from time to time occur; always remembering the axiom, that in the treatment of fever "the physician has constantly to bear in mind, not only what he *now has* to do, but also what he *may have* to do."

The general management of a case of fever seems based upon the consideration of its nature as a morbid poison: its conformity to the laws of such; and its tendency, under certain conditions to be aggravated by secondary blood contaminations during its course.

The special treatment of any case will depend upon (*a*) the special affinities of the poison; (*b*) the nature and extent of the secondary lesions; (*c*) the tendency to one or other mode of termination—which will peculiarly influence the treatment of the last paroxysm, or period of crisis. The management and treatment of the period of convalescence also demands our special consideration.

The term poison naturally suggests the idea of an antidote. As, however, the action of the fever poison is not, like inorganic poisons, definite and limited in extent and duration, but is, as it were, the result of catalysis—setting up a change in the blood, which, extending to every molecule of the body increases and perpetuates itself—there can be no antidote in the true sense; in the manner, for example, that lime-water is for oxalic acid, or white of egg for corrosive sublimate. The only antidotal remedy in fever is pure air: the oxygen of which, combining with the products of regressive metamorphosis, prevents their accumulation in the blood, and thus averts the consequent secondary contamination of that fluid, which constitutes the great source of danger in the advanced periods of the disease.

But admitting that we have no antidote, the question arises, have we any means by which the operations of the poison, once set up, may be arrested, shortened, or modified, if these measures are employed at an early period.

In others word, does the consideration of the primary actions of the fever poison, as I sketched them in my first lecture, and of its tendency to be eliminated by certain excreting surfaces, suggest any measures, which, by producing a shock, shall thus aid the efforts of nature to unload the right cavities of the heart, remove the stasis in the capillaries, restore the balance of the circulation, and at the same time, arouse the moderating nervous centres from their state of partial paresis—thereby restoring the impaired functions of respiration and molecular nutrition—while the elimination of the poison by its natural outlets, shall be at the same time facilitated.

Obviously, if it were possible to effect these objects, fever would be arrested; or at least so modified as to become slight in amount and short in duration.

I believe that hundreds of cases of fever have been thus arrested chiefly by three measures which medical fashion—rather than reason or experience—has long since consigned to comparative neglect. These are, bloodletting, emetics, and cold affusion.

With regard to the last, which may be said to have passed into complete disuse, I think no one can read the statements of Dr. Currie, and entertain a doubt that fever can be, by this means, arrested in its initiatory stage, however much we may doubt the prudence or propriety of the treatment under ordinary circumstances.

Of the efficacy of bloodletting and emetics in arresting fever, Dr. Graves wrote thus: "If I were called to visit a patient who had been attacked with shivering, headache, quickness of pulse, increased temperature of skin, and lassitude during the prevalence of an epidemic or after exposure to contagion, and happened to see him a few hours after the attack, I should certainly bleed him and give him an emetic, and I think he would have a very good chance of escaping the disease.

"I have myself witnessed many cases in private practice of medical men and students, who had been attacked with symptoms of fever after the exposure to contagion, and who escaped by taking an emetic, and being bled in proper time."[1]

I can also state, that in the course of my life, I have employed these measures in the instance of persons suffering the symptoms Dr. Graves here enumerates, and in whose family, fever at the time existed, and I have, as a rule, found them to succeed, if employed within the first 24 or even 36 hours after seizure.

I think they should be employed together: first a moderate bleeding, 6 or 8 to 10 ounces, according to age and constitution, and immediately afterwards the emetic. The bleeding should never be employed alone. But, if only one of the measures is chosen, it should be the emetic, and neither ought to be attempted after the period mentioned, as they are then both useless and unsafe.

[1] Clinical Medicine, vol. i. p. 138.

Before I enter on the question of the modus operandi of these measures, I should acknowledge, that high authorities have discountenanced their employment. Thus Sir T. Watson—"in the few instances in which emetics or cold affusion have seemed to arrest fever, or to check its progress, that effect has always occurred at the very commencement of the complaint; so that we cannot be sure (and the probability is the other way) that they were really cases of fever at all, or that they would not have ceased, even although nothing had been done for them."

The late Dr. Todd holds similar language: he says, "all the cases in which it has been said that typhus has been cut short by a very large bleeding at the outset, or by free vomiting, or by some other means, are fairly open to the strong suspicion, if not to the charge of erroneous diagnosis.

"It is plain, if you think on the subject but for a moment, that without an exact diagnosis, this question of the early curability of typhus cannot be settled. Now those who have seen most of this and other maladies, know best how difficult, nay, how impossible, it often is in the first week or ten days to predicate with certainty of this or that case, that it is typhus fever. And therefore if you deal candidly with yourself and others, you must not affirm that you can cut short and cure typhus unless you have the most unequivocal evidence that the cases in question have been examples of that disease."

Another writer of less authority, thus carries the same argument to an absurd length.

"Emetics may perhaps have the effect of sometimes limiting the duration of simple and catarrhal fevers, or may render them milder, but in typhus, they have no such effects, for no unequivocal case of the disease, such as one having the typhoid eruption has ever been brought forward where such has been the result."

In other words, no case has been produced of typhus with eruption, say on the sixth day, which had been arrested on the eve of the second.

Our answer to the objections of Drs. Watson and Todd, is, that these would be conclusive, if medicine were an exact science, and if the physician was bound in all cases to abstain from curative measures, until the nature of the disease was ascertained. But such, unfortunately, is not the case; and inasmuch as our primary duty is to save life, we are bound to disregard reasoning so purely theoretical and speculative, and to attach due weight to considerations which we may think warrant the practice in question.

One of these is the great amount of probability that a person exposed to the contagion of typhus will contract typhus rather than simple catarrhal, or ephemeral fever. If under the circumstances I have mentioned, an individual presents the symptoms enumerated by Dr. Graves, the chances are many to one, that the subsequent illness will be that to which he has been exposed. If to scarlatina, scarlatina; if to typhus, typhus.

The other consideration is, that there is nothing in the nature of typhus which must needs prevent it being arrested, shortened, or modified, by such measures as are capable of producing the effects I have before enumerated, upon the heart, the capillary circulation, the nervous system, and the nutrition.

Of the modus operandi of bloodletting in the congestion of the right cavities of the heart, which forms one of the primary conditions in fever, I need say nothing, but that the experiments of Dr. John Reid and other physiologists, have conclusively proved that the abstraction of even a very small quantity from the overloaded auricle and ventricle arouses their contractions, when these are rendered languid, or altogether suspended by the action of narcotic poisons; and that there is every reason to believe the same remedy has a similar efficacy in the overloaded and languid condition of the right side of the heart in the early stage of typhus. (*Appendix V.*)

The action of an emetic produces, in a different way, the same effect. During vomiting, the right cavities are unloaded, the pulmonary circulation aroused, the respiration at the time rendered more perfect.[1] The influence of this change is at once felt by the systemic capillaries, the capillary circulation being aroused, and diaphoresis following. The portal circulation is also stimulated, and increased secretion of bile takes place. The nervous centres feel the influence of the active circulation of a more highly oxygenized and purer blood. I say purer blood, for there can be little doubt, that the action of an emetic, is to some extent eliminative of the poison, because of the power possessed by medicines of this class of attracting the poison to the gastric mucous membrane.

---

[1] It is thus the remarkable efficacy of emetics in the bronchitis of children, attended with atelectasis of lobules (wrongly termed lobular pneumonia) is to be explained.

The marked success of the emetic treatment in arresting the progress of typhoid fever, led Dr. Chambers to argue that the poison is carried directly into the stomach with the saliva. "At an early stage," says Dr. C., "even after the virus has begun to act upon the system, the fever may be stayed by emptying the stomach, and thus preventing the whole dose being taken up. Those who have watched my practice will have witnessed several instances of the success of this treatment; they will have seen the fever cut short, and convalescence entered upon immediately, with its characteristics of painless weakness, and emaciation gradually passing away."[1]

I need not tell you, that while I accept Dr. Chambers' testimony to the success of the emetic treatment, I entirely dissent from his views of the modus operandi of the remedy.

Of other eliminant medicines, two have been employed, namely, sudorifics and purgative.

I have long been accustomed to follow up the action of the emetic in the early stage of fever, by giving at intervals a combination of James' powder and nitrate of potass, usually adding to the first dose a grain or two of calomel.

The exhibition of three grains of the antimonial, and ten or fifteen grains of nitre, every 4 hours after the emetic, is pretty certain to cause copious diaphoresis, continuing as long as the administration of the medicine is persevered in; which should be for at least 24 hours.

I seldom have trusted to this measure alone, but I have heard of many cases, in the practice of others, in which large doses of James' powder, producing copious and long-continued sweating, have apparently cut short fever.

Once, however, the disease is established, no practice can be more injurious, than either the emetic or diaphoretic treatment, unless it be the purgative.

I think it may be safely said, that no treatment is calculated to do so little good, or so much harm, as the free exhibition of active purgatives at the commencement of fever. By these, I do not mean medicines that simply unload the bowels of their contents, but such as produce copious fluid evacuations, the greater proportion of which is derived from the blood. In and out of the profession, this abuse of purgatives prevails.

---

[1] Lectures, chiefly clinical, 4th edition, p. 75.

Among the worst cases admitted into hospital, are those of patients who have been dosed with salts by themselves or their friends. Of the practice of a certain class of practitioners, you may judge by a quotation from one of the writers, whose objection to the preventive treatment by vomiting I have already referred to. "In the commencement of the disease," says this writer, "one or two brisk purgatives, such as the compound powder of jalap, with two or three grains of calomel; or castor oil combined with the same mercurial salt, will, if repeated at the interval of 2 or 3 days, be found useful in unloading the primæ viæ, and in promoting several of the secretions; but large doses should not be persevered with, if the patient be of a weak habit of body, be cachectic, or be advanced in years. In ordinary cases, after one or two brisk cathartics, a moderate dose of castor oil, infusion or fluid extract of senna, or other mild purgative every second day, will be sufficient to unload the bowels, without causing *much watery evacuation*."

I give you this quotation as an average specimen of the practice of a certain class of medical men. The objections to it are: first, that there is perhaps no instance known of fever being arrested by purgatives. Next, that if they modify the disease, they will do so injuriously, by exciting irritation of the intestinal mucous membrane and thus producing obstinate diarrhœa and tympanitis; or by causing the draining off of serum from the blood, and thus producing in some cases, cerebral exhaustion, manifested occasionally by delirium and convulsions, in others, congestion of the kidney and albuminuria, and in others again, leading to the formation of fibrinous coagula in the heart. I have no hesitation in saying that you should never attempt to arrest or shorten fever by the exhibition of purgatives.[1]

---

[1] On this subject, Sir H. Marsh remarks, "severe purgatives at the commencement of fever, have done much mischief. How often do patients, even those of the medical profession, imagine that if they can but succeed in being well purged, all will be well. Thus is the surface of the intestinal canal often dangerously irritated, great debility induced, and the disease, instead of being mitigated, rendered far more formidable."—*Essay on the Origin of Fever.* Dublin Hospital Reports, vol. 4.

Dr. E. Percival in his admirable essay on the epidemic fevers of Dublin observes: "I am far from recommending the practice of purging severely at the commencement of fevers. Some of the worst cases which I have witnessed, have been of young medical students, who, in their eagerness to evacuate their bowels, have brought on hypercatharsis on the second or third day of their fever,

You will, however, have very few opportunities of attempting the arrest of the disease, as it seldom comes under the notice of the physician, until the period for such interference has passed, and when the attempt would be certainly injurious.

I have said that we have no antidote for the fever poison. This is doubtless, true, and yet our general treatment and management of a fever patient, is quasi antidotal; or based on the *idea* of an antidote. We carry out this idea through the following indications, which may be said to comprise most of the details of such management and treatment.

I. *To prevent the augmentation of the poison in the blood,* by removing the patient from an infectious locality, or from a situation in which he cannot obtain a due supply of pure air.

Among the poorer classes, this is done by removal from their wretched, overcrowded, and too often filthy dwellings, into the large and well-ventilated fever wards of our hospitals.

Among your private patients of the richer class, you will occasionally have to resort to a similar measure. For example, by removing them from a close, ill-ventilated, musty-smelling chamber, stuffed with furniture, from which the light and air are excluded by closed shutters and window curtains; and from feather beds covered with thick blankets; the bedsteads four-posted, canopied, and surrounded by curtains; to large and airy apartments, well lighted, free of all unnecessary furniture, there to be laid upon a hair mattress, firm and elastic; not surrounded by curtains, but supplied with firm pillows, and with but scanty bed-clothes; the latter at least in the earlier periods of the fever. I have often seen immediate improvement and rapid recovery follow the change thus made. The patient, previously both restless and wakeful, racked with headache and weariness, falling into a quiet sleep, and progressing favourably from that moment. Having thus arranged the bed-room, you have next to lay down rules for the daily management, which should in all cases be carried out, not by the patient's friends, but by trained nurses, whom you should instruct to admit air and light freely into the room; to observe perfect quiet, avoiding all conversation with, or in the hearing of, the patient, and taking care that while the daily ablutions are regularly and freely

---

with very alarming consequences, &c."—and Dr. Graves says, "the abuse of purgatives, particularly in the first stage of fever, continues, I am sorry to state, even to the present day, a blot on the character of practical medicine."

performed, and the bed linen frequently changed, yet that everything should be done without the least exertion on the part of the patient. The experience of physicians of all ages has proved the importance of this; the only difference being as to the rationale of the practice. Thus while Cullen lays down as a rule that "all motion of the body is to be avoided, especially that which requires the exercise of its own muscles," since "it is to be observed that every motion of the body is the more stimulant in proportion as the body is weaker:" physicians of the present day enforce it on the ground that during fever all exercise of the brain or muscles produces an amount of disintegration of tissue, vastly out of proportion to the amount of such exercise—probably the more excessive the more advanced the fever—by which the secondary blood contaminations of the advanced stage are rendered more formidable and dangerous.

II. You should not only take every precaution to prevent the augmentation of the fever poison, but should also use such measures as may fulfil the second indication, namely, *to moderate the increased disintegration of tissue*, which it is in the nature of this poison to set up. Thus you should supply the daily waste by suitable nutriment, and at the same time aid, so far as may be in your power, the daily elimination of the products of waste from the blood:—bearing in mind the aphorism "Impura corpora quo plus nutriveris eo magis lædes." Experience abundantly proves that wine and nutriment best produce their salutary effects, when conjoined with pure air, cleanliness, and cold water; and that they act more or less as poisons when these latter measures are neglected. Given with this precaution, liquid nourishment, wine, or brandy supply the materials of repair, and retard disintegration of tissue. I say liquid nourishment, because the stomach of a fever patient is incapable of properly digesting solid food; with the exception therefore, of a small quantity of bread given with tea, or of a cupful of panada, we do not administer solid food in fever.

The quantity of food and wine administered to a patient in fever will, of course, vary with the varying circumstances of the case. In the earlier periods, it is a good rule, as laid down by Dr. Graves, to preserve the habits of the patient; and to give food as nearly as possible at the hours at which he has been accustomed to take it.

As the disease advances, and the tendency to sinking of the vital powers increases, this rule must be disregarded, and beef-tea and wine must be given at shorter intervals. Even then, however, I think it a mistake to give the former oftener than every two or three hours; nor need the quantity in the twenty-four hours exceed two pints in the average. To this may be added a cup of tea for breakfast, and milk or whey at intervals.

With regard to wine, no precise rule can be laid down. Some cases do not require any. Some are even positively injured by it. In many others, it is our great remedy, and must be given in any quantity that may be required.

We shall hereafter consider the special indications for its administration; but I may now remark, that like food it should be given with some regard to the habits of the patient; and that in general, wine or alcohol, will produce their best effect, when given in the after portion of the day and during the night.

Like food also, it should be given at moderately long intervals, until the time arrives when its stimulant action on the nervous system is needed from hour to hour till the crisis be past.

Scarcely second in importance to nourishment in the fulfilling of this indication is sleep, of which Dr. Carpenter says: "It is a peculiar feature in the physiology of the cerebral and sensorial ganglia, that their activity undergoes a periodical suspension, more or less complete; the necessity for this suspension arising out of the fact, that the exercise of their functions is, in itself, destructive to their substance; so that if this be not replaced by nutritive regeneration, they speedily become incapacitated for further use."

Taking physiology as your guide, you will from the first endeavour to secure the periodic return of sleep; not always by the exhibition of narcotics, but by the judicious management of the patient with reference to this function.

The first and most essential measure, as forcibly insisted on by Sir D. Corrigan, is to preserve the *habit* of sleep, by marking the alternation of day and night; freely admitting light as well as air in the daytime; and only shading the patient's eyes with a curtain or blind, if the light is uncomfortable to him; avoiding over-heating the bed-chamber—a too common practice with friends and nurses, and one certain to give rise to restlessness and prevent sleep. Lastly, observing perfect silence during the hours of sleep, let there be no fussiness, no creeping about the room adjusting articles on the table, and moving furniture, no conversation or whispering, and let the patient rest for at least four hours at a time unless in exceptional cases, without disturbing him to give food or medicine.

As you will find that the anxiety of the patient's relatives will certainly interfere with the carrying out of the above injunctions, you will do well to supersede them, at all events, during the night, and to place your patient under the care of an experienced nurse.

In addition to these points of management, there are some measures likely to induce sleep, which you may employ during the earlier stages of fever. One of these is warm-sponging of the head; another, stuping the legs and feet with flannel wrung out of hot water; and a third, is the application of a few leeches to the temples or behind the ears. The latter practice was recommended by Dr. Graves, susequently by Sir D. Corrigan, and I have seen it followed by the best effects, in calming excitement, relieving headache, and procuring sleep.

The second part of the above indication is liable to misconstruction, and therefore needs explanation. While I maintain that you should facilitate the elimination of the products of disintegration, I wish you to understand that this is not to be done by the use of evacuants during the course of fever. These cannot do it; on the contrary, the class of purgatives most usually employed for the purpose, have the effect rather of increasing disintegration of the blood itself, and producing the consequences I have before enumerated. The measures you may employ for the purpose, are: first, the constant admission of pure air for respiration, and so far as may be consistent with the patient's comfort, its admission also to the surface of the body. This is a practice which has been strongly recommended by Sydenham and other old writers on fever,[1] and

---

[1] Thus Sims observes: "One of the things which had great and immediate good effects at the height and worst time of the disorder, was the application of cold air. During the first stages indeed, the cool regimen was strictly enjoined, the curtains of the bed were undrawn; the quantity of bed-clothes lessened, the drink never suffered to be in the least warmed, the linen changed as often as moistened with sweat, and as the patient often lay in small rooms not sufficiently airy, the door was not permitted to be shut, neither was the chamber allowed to be heated by a fire, nor many persons remaining in it at the same time. But in the last and virulent period, these methods, seeming inadequate to the greatness of the complaint, a farther and unusual step was taken." The patient was directed to be lifted out of bed, and with only the covering of a great coat or wrapper, not kept tight to his body; exposed to cold air for a considerable length of time, often until an universal shivering or chattering of the teeth, gave sufficient tokens of its having taken effect, &c.—*On Epidemic Disorders*, p. 211.

you may at any time witness the striking effect upon the colour of the eruption, and upon the capillary circulation, by removing a portion of the bed-clothes, and leaving the patient covered by a single sheet.

On the other hand, you will see the effects of a deficient supply of oxygen, in the dark colour of the skin and maculæ of patients confined in small, overheated, and deficiently ventilated rooms, or in crowded ships, prisons, or dormitories, and sometimes in the overcrowded wards of hospitals.[1]

Secondly, we aid the efforts of nature in this direction by the internal exhibition and external employment of pure water.

The natural instinctive desire of the patient is usually for cold water, and I need not remind you, that the experiments of physiologists have shown this to be a great depurator. Its good effect is aided, and thirst relieved, by the addition of hydrochloric or sulphuric acid, or of the chlorate of potass, of which I usually order two drachms to be given in the course of twenty-four hours. The same object is subserved by the frequent sponging of various portions of the surface with tepid water, or vinegar and water.

I shall have to refer to several medicines of the eliminant class when speaking of the complications of the advanced stage of fever.

We have next to consider the special indications which may arise in the course of fever, and which may be due to the difference of species, and to the different pathological phenomena belonging to each.

Taking these in the order of the several functions, you will find that the difference in the condition of the digestive organs in typhus and typhoid fevers necessarily involves a marked difference in the dietetic and medicinal treatment. I have already mentioned the

---

[1] I well recollect the terrible effect produced by this cause in an epidemic of typhus in the County of Meath, to which I referred in my lecture on the etiology of fever.

The unavoidable addition of three or four beds in each of my wards produced an immediate and striking change in the colour of the eruption, and a corresponding influence upon the nervous symptoms, depending on increased blood contamination.

"In every fever," says Heberden, "it is of the utmost importance to keep the air of the patient's chamber as pure as possible. No cordial is so reviving as fresh air, and many persons have been stifled in their own putrid atmosphere by the injudicious, though well meant care of their attendants."—*Commentaries*, p. 179.

fact that wine is seldom borne to the same extent by the typhus and typhoid patient. You see few of the former pass through fever without it, whilst many of the latter require none. You must also have observed that whereas the typhus patient usually takes beef tea freely throughout the disease, our nurses constantly report that owing to the occurrence of diarrhœa in the typhoid case they have discontinued the beef-tea and given milk only; broths of all kinds being found to increase such a tendency in many cases of this disease.

The special treatment of the intestinal affection of typhoid, varies according to the stage of the fever. It being an essential and normal condition in this form of fever, the duty of the physician seems to be not to check, but merely to moderate it in the earlier periods. With this view we usually apply a warm linseed poultice or a wet bandage, and give moderate doses of dilute sulphuric acid, with which I usually combine one to two minims of Battley's sedative, according to the age of the patient. At a more advanced stage of the fever, the indications may be to correct an ill-conditioned sloughy state of the intestinal ulcer, to arrest an exhaustive discharge, and to avert, if possible, the septicæmia with which these are too often combined.

Without entering into minute details, I may mention that acetate of lead and sulphate of copper, gallic and tannic acids—all have been employed by various practitioners for the two first objects. But in our experience, oil of turpentine, in doses of from 10 to 30 minims, seems superior. Yeast has sometimes been administered with excellent effect in septicæmic diarrhœa, characterized by excessive fetor of the evacuations, which latter often consist of a fluid resembling the washings of flesh, and sometimes of mucus streaked with blood, or containing shreddy particles apparently of sloughing mucous membrane. In this condition I have more than once given the bisulphite of soda or magnesia, with the effect, at least, of correcting the fetor.

At this stage, you will generally find that while the smallest quantity of animal food, in the shape of broth, will increase the discharge and the fetor, yet the patient will be benefited by the free administration of old port, which may be given with spiced water, or may be added to arrowroot or sago, in the proportion of a glassful of the former to a breakfast cupful of the latter.

If hemorrhage from the bowels occur, the indications are threefold, viz., to arrest the bleeding; to support the strength; and to correct the probable septicæmia. Ice, given internally, and locally applied—the above astringents—in addition to which may be mentioned alum, Ruspini's styptic, and ergot—are the measures to be most relied on for the first indication. At the same time, enemata of acetate of lead, or tannic acid, or oil of turpentine should be administered. For the second, wine freely administered —brandy, opium, and beef essence. For the third, if such a condition coexists with hemorrhage, little can be done; but you may try large doses of the tincture of the perchloride of iron, of dilute sulphuric acid and sulphate of iron, or bisulphite of soda or of magnesia.

You will find treatment of any kind of little avail in the event of perforation of the bowel causing peritonitis.

The indications, you should bear in mind, however, are—if possible, to arrest the escape of the contents of the bowel into the peritoneal cavity; to circumscribe the inflammation of the serous membrane; to support the patient's strength, and relieve pain when present. All of these indications are best fulfilled by opium, which should be administered in full doses, frequently repeated. Opium controls and retards peristaltic action of the intestine. More than any other measure, it limits the extension of the inflammation of the serous membrane. It relieves pain, procures sleep, and acts as a cordial, supporting the patient's strength while it lessens his sufferings.

When its good effects are most marked, it seems to produce the least narcotism; and, on the other hand, when it rapidly narcotises, it does no good.

If you are so fortunate as to succeed in limiting the peritonitis in a case of perforation, you should be careful to avoid the mistake of administering an aperient for many days afterwards. Some lives have been lost by means of this practice, which is directly opposed to the indications above enumerated.

The indications for the treatment of the thoracic complications, vary, of course, with the nature and seat of the lesion.

I have already said enough on the treatment of congestion of the right cavities of the heart. That of the weakened left ventricle consists, as you are aware, in the free exhibition of wine and nourishment.

The simple congestion of the lung, which is the rule in typhus, cannot be said to admit of any special treatment. Following, as it does, upon the altered attractions between blood and tissue, as well as upon the weakened condition of the right ventricle, its subsidence will depend upon the removal of these its causes.

Occasionally, however, this congestion merges in hypostatic pneumonia, requiring the use of suitable measures: such as dry cupping, blisters, senega, carbonate of ammonia, and not unfrequently mercury, which, in these cases, may be combined with quinine and carbonate of ammonia.

In less severe cases, turpentine epithems may be freely applied to the chest as a substitute for blisters; and when the vital powers are much depressed, bark, in infusion or decoction, with sesquicarbonate of ammonia, may be substituted for the above-mentioned remedies; and wine may require to be supplemented by brandy or whiskey punch.

These, however, are more generally indicated in the more serious complication of suffocative bronchitis, which, at some seasons, constitutes the prevailing type of disease—catarrhal typhus. In this affection the patient's life will often depend upon the energy and watchfulness of the physician. Decoction of senega with carbonate of ammonia; boluses of camphor and carbonate of ammonia; oil of turpentine, in doses of half a drachm or a drachm; and, in desperate cases, tincture of cantharides, in 30 or 40 drop doses in combination with the turpentine, have been the remedies most successful in my experience. With these should be conjoined the most liberal allowance of brandy or whiskey, in the form of punch; and the use of every possible mode of external stimulation, as sinapisms, turpentine epithems, and blisters—more especially flying blisters, as recommended by Dr. Graves. I recollect a patient whose life was saved by my pouring boiling water upon his legs, after all the above-mentioned means had failed. He was fast sinking into the fatal coma in which these cases terminate, and could not be aroused by any measures which had been previously employed. The pain produced by the scalding water, however, did so most effectually, and under the use of the turpentine and cantharides mixture he eventually recovered; the only bad consequence being a large superficial sore on the calf of each leg, which, however, healed rapidly.

LECTURE XV.

TREATMENT—CONTINUED.

The indications derived from the urinary functions are referable to the bladder, or to the kidneys.

Your daily attention is required to prevent the ill consequences arising from retention of urine, and with this view you will occasionally find it necessary to introduce the catheter twice or three times in the course of each twenty-four hours.

It is equally important that you should, from time to time, observe the qualities of the urine excreted, since daily increasing observations show, that the most serious symptoms of the advanced stage of fever are frequently owing to deficient elimination, by the kidney, of the products of disintegration of tissue.

There can be no doubt that the uræmic poisoning which occurs manifestly, and as a rule in the stage of reaction, or secondary fever in cholera; also occurs exceptionally, and for the most part, in a more latent form and milder degree, in many cases of relapsing fever, in typhus and even in typhoid fever; the difference being, that while total suppression of secretion occurs in the one, the danger in the others is more frequently connected with a change in the quality of the fluid secreted. The observations of the older writers, of the great danger of pale limpid urine in the advanced periods of fever, and its connection with the most formidable nervous symptoms, are thus fully confirmed and explained, and the proper treatment of such cases is at the same time indicated. This, however, is by no means the only form of deranged urinary secretion you may be called to treat, in fulfilling the indication to restore the healthy balance between increased disintegration of tissue and elimination of its products.

I may here remind you of the conditions under which the symptoms due to the retention of urea in the blood, may occur. They are (1) acute inflammation or congestion of the kidney, arising either out of the exciting cause of the fever, such as cold applied to the loins; from the direct action of the fever poison on the blood and on the capillaries of the kidney, as in congestive typhus; or from septicæmia. In these cases the urine will be either suppressed, or bloody, smoky from the presence of hæmatin, or albuminous.

(2) Pre-existing disease of the kidney; in which case the urine will probably be of low specific gravity or albuminous.

(3) The disturbed balance may be due, less to any affection of the kidney than to sudden and marked increase of disintegration, with diminished elimination. The urine being not in all such cases albuminous, but sometimes simply deficient in urea.

This condition is met with more especially after the crisis of relapsing fever; or in the course of typhus or typhoid, when the patient has either lived luxuriously or has been exposed to excessive muscular exertion at or about the period of the invasion of fever,[1] or as I have seen in several cases, has had excessive bilious evacuations by stool in the early period of fever.

(4) Any two of the foregoing conditions may coexist in the individual; indeed, in no case can any one of them be certainly expected to occur as a simple or solitary phenomenon, since the complex and varied pathological conditions of fever will certainly influence them in various ways.

In all cases, the altered relations between the blood and nervous tissue will co-operate to render the cerebro-spinal centres more susceptible to the poisonous action of the retained urea; and probably to undergo that change in their composition suggested by Dr. Oppler. (*Appendix VIII.*)

Another mode in which fever exercises an influence on the patient previously suffering from chronic nephritis, is, in the arrest of healthy molecular nutrition, by which a stop is put to the depuration of the blood by some form of vicarious secretion, by which process the fatal result is occasionally warded off for a length of time; such cases invariably terminate fatally when fever supervenes and arrests this process.

In some instances a complex morbid condition is set up, resulting, partly from the toxic action of the fever poison; partly from con-

---

[1] "The quantity of the urea secreted at any given period of life," says Dr. Carpenter, "seems to depend mainly on two conditions: namely, the degree of muscular exertion previously set forth, and the amount of azotized matter ingested as food."—*Physiology*, p. 619.

gestion or inflammation of the brain or medulla oblongata; and in part from uræmia. A good example of this is the case narrated at page 135 of this volume.

A review of the conditions above mentioned will show you that in the 3d alone, when it exists uncomplicated with any affection of the kidney, is there a reasonable prospect of benefit from any medical treatment.

In the first condition, the measures, which have sometimes been found to succeed in the suppression of urine following collapse in cholera, may be tried. These are, cupping over the region of the kidneys; turpentine epithems and linseed poultices; the vapour bath; James' powder and nitre, or other diaphoretics, followed in some cases by the administration of mercury. But, in truth, I cannot recommend such, or any other measures with confidence, as I have never seen recovery occur in fever under these circumstances. Neither have I ever witnessed "successful treatment" in the case of pre-existing chronic disease of the kidney. In the uræmia following the perturbatio critica of relapsing synocha, however; and in the deficient excretion of urea, which sometimes occurs in the latter periods of typhus and typhoid, the indications above mentioned have sometimes been successfully carried out, the secreting action of the kidney being increased, and the proportion of urea being suddenly and largely augmented, a change which has been rapidly followed by amelioration of the patient's condition.

I believe it is to Dr. Henderson, of Edinburgh, we owe the observation, of the connection between the convulsions following crisis in relapsing fever and the presence of urea in the blood; and also, the suggestion that this complication should be treated by the administration of saline diuretics, nitrate of potass in particular.[1]

I shall give you Dr. Henderson's account of this observation in his own words:—

"It was not till August that I began to entertain a suspicion that the function of the kidneys was liable to be affected, to an extent that was capable of producing the most serious consequences. The suspicion was suggested by the following circumstances: A

---

[1] Dr. Henderson seems to have been anticipated by Baglivi, who says, "In ardentibus febribus sal prunell specificum est."—p. 146. The febris ardens of the ancients was our relapsing synocha.

patient of mine in the hospital, three or four days after the crisis, of a paroxysm of fever, was seized with convulsions; and my friend Dr. Moir informed me, that in a family which he attended, two members were seized also after a paroxysm of fever had ceased, with convulsions, and that one of them died in the convulsive state. In none of these cases was there jaundice to account for the attacks of convulsion. In October I was requested to see a young gentleman who had had the crisis of his first attack of fever the day before. When I saw him at noon his respirations were reduced to twelve in the minute, and his pulse to fifty-eight; he was very languid and oppressed, and had transient attacks of confusion of mind and a peculiarly uneasy sensation in the head. As he had no jaundice I inquired into the state of his urine, and learned that he had passed none since seven the previous evening, and it was since that time that the symptoms that I have mentioned had come on. He was directed to have ten grains of nitre every hour till three doses should be taken. Whether in consequence of the medicine or not, he passed some water half a hour after the first dose, and repeatedly considerable quantities in the course of the day and succeeding night, so that at seven next morning he was found to have passed altogether at least four pounds of urine. He was much relieved by the evening of the day on which I had seen him first; his pulse had risen to seventy-two, and his uneasiness in the head and tendency to delirium had ceased. Next day he was perfectly free from every unpleasant symptom. This last case determined me to pay more attention to the condition of the urine in future, and it was not long before on opportunity occurred of making more particular researches on the subject than I had previously done, and of having the suspicion I have mentioned amply confirmed."—(*Ed. Med. and Surg. Jour.*, vol. 61.)

From considerable experience of this eliminative treatment, I can recommend you to adopt it, in those cases in which the symptoms due to retained urea may present themselves in the course of typhus or typhoid fever. The combination I usually employ is that of nitrate of potass, nitric acid, and nitrous spirits of ether. The effect I have observed, has been increased flow of the urine of higher specific gravity, with corresponding relief of nervous symptoms.

I recently attended the son of a professional friend in this city, in typhoid fever, contracted after excessive and prolonged muscular exertion. The nervous symptoms were unusually marked from the commencement; and about the end of the second week they assumed many of the peculiar characters of uræmic poisoning, such as noisy, irregular breathing, constant moaning, facial twitching, &c. On examination of the urine, which had become pale in colour, the specific gravity was found to be 1.015, and on the day following—the nervous symptoms having meanwhile become more marked—it was 1.009 only. In consultation it was agreed to give a mixture containing nitrate of potass, nitric acid, and nitrous spirit of ether.[1] On the day following, the nervous symptoms were less marked; the facial spasms in particular had ceased; the specific gravity of the urine had risen to 1.017, and it continued to increase day by day, till it amounted to 1.029. The patient's condition continued for many days to be one of great danger, but from different causes; nor was there at any time, subsequently, any return of the peculiar nervous symptoms connected with deficient excretion of urea.

The indications derived from our study of the cerebro-spinal lesions in fever are of the highest importance, and should be regarded in connection with these lesions. You may be called upon to treat nervous symptoms singly, and, as it were, quite independent of the fever in which they occur; but you should endeavour to keep in view the pathological condition from which they arise, and those with which they may be associated. They usually occur in groups, but such groups do not admit of any exact or arbitrary arrangement, as they will vary in their component symptoms in different cases.

Thus the headache of the earlier period, associated with pain in the back, weariness and restlessness, is a condition widely differing in its nature and treatment from the headache of a later period, which ushers in fierce delirium, stupor, or coma. In like manner the convulsions, which sometimes attend the crisis of fever in the hysterical female, are altogether different from the convulsions of uræmia.

---

[1] The reader will observe that this mixture is nearly the same as the combination recommended by Dr. G. Symes in tonsillitis; and I have no doubt that the remarkable success attending its employment by Dr. Symes and others, was, in a great measure, due to its eliminative action on the kidneys; the affections of the tonsils depending in most cases upon some form of blood-contamination. Scarlatina, diphtheria, syphilis, are examples of this.

You should not, therefore, regard any individual symptom as demanding a special, positive, and invariable line of treatment; in other words, as admitting of a routine treatment; but should in every case stop to consider, first, what are the pathological conditions upon which it depends, and, in the next place, what are the indications suggested by it.

These will naturally fall under the following heads:—

I. To neutralize, or moderate, the toxic action of the fever poison upon the cerebro-spinal centres, more especially of the cerebral hemispheres.

II. To arouse the failing functions of circulation and innervation, and to avert the dangers arising from venous congestion and exhaustion.

III. To moderate arterial excitement, and subdue existing active congestion or inflammation.

IV. To avert the nervous symptoms caused by various forms of secondary blood contamination by eliminative treatment.

In fulfilling, so far as is possible, the first indication, wine is our great agent; and, if the case is one of simple uncomplicated typhus, due attention having been paid to those principles of general management, which I have already enumerated—by which the balance between disintegration and elimination is preserved—you will often be struck by its wholesome influence upon the nervous system and the circulation; the pulse becoming slower and fuller after its administration, delirium abating, and sleep following—effects thus tersely summed up by an old writer: "Magna quantitate sumptum, omnem vehementiam sedat, nam qualitate narcotici colligit sensus, somnum consiliat, deliria et dolores acutissimos capitis imminuit."

The same indication is fulfilled to some extent by opium, and under the same conditions.

I have already remarked, that while wine may with advantage be more freely administered in typhus, opium is borne with safety to a much greater extent in typhoid. Many cases will, however, occur of both forms of fever, in which one or other, or both, these remedies will be contraindicated. They are so, for example, in active congestion of the brain, and when there exists a great amount of blood contamination. I think I may venture to assert that in every case in which such symptoms occur as I enumerated in a former Lecture, indicating the existence of active cerebral congestion, more especially if associated with an excited, jerking action of the heart; you will find opium—when given in the first instance —to act injuriously. I have too often seen a single dose of opium followed in such cases by stupor, coma, rigidity of muscles, stertorous respiration, and death. Nor will it be less injurious if the blood is highly charged with carbon, whether from deficient supply of pure air, or from imperfect respiration, as in catarrhal typhus; or if it is contaminated to any extent by the retained products of disintegration. In these cases opium seems to have a twofold injurious influence. It retards elimination, and so increases the mischief already existing; and it also directly aggravates the poisonous influence upon the nervous centres, of substances, to which its own action is analogous.

I have no doubt that these objections may be lessened, even sometimes altogether obviated, by the addition of tartar emetic, the eliminative action of which was conspicuously marked in many of the cases published by Dr. Graves. But even so it is a two-edged remedy, and should never be pushed to any extent, unless you are satisfied that the objections I have mentioned do not exist. If they do, you will be more safe in trusting to the less certain action of large doses of hyoscyamus in combination with antimony and camphor, given at moderately long intervals until sleep is procured.

If, however, the heart's impulse is weak, the first sound indistinct or inaudible, the head cool; if there is no muscular rigidity, no tendency to stupor; if the urine is copious, and of good colour— perhaps depositing lithates—you may safely push the exhibition of the opium and antimony, no matter how noisy or impulsive the delirium, or how rapid the pulse, and even although the pupil may be contracted, and the eye red from want of sleep.[1]

A short abstract of one of many cases in which I administered this combination in the typhus epidemic of 1839-40, will illustrate this. James Caffray, æt. 27, was admitted on the tenth day of typhus, much maculated, delirious, and swallowing with difficulty. Wine was ordered, also a blister, to the nucha.

On the eleventh day, he became violently delirious and unman-

[1] Dr. Graves mentions the pin-hole pupil as a contra-indication to the use of opium. Such, however, is by no means always the case, for it, as well as the injected eye, is frequently caused by want of sleep. Like every other symptom in fever, its true meaning can only be ascertained by viewing it in association with other cognate symptoms.

ageable. He was given half ounce doses of a mixture, containing four grains of tartar emetic and a drachm of acetum opii in eight ounces of camphor mixture.

After the second dose he slept for a short time, but on awaking the delirium returned, accompanied with much subsultus.

He was ordered a blister to the shaven scalp, and an increased quantity of wine.

On the twelfth day he had not slept; the delirium was so violent as to require restraint; he bit his nurse, swore and sang constantly; had involuntary evacuations, general tremor and dysphagia.

An enema was administered, containing 15 minims of tincture of opium, 20 grains of musk, and 5 grains of sulphate of quinine; ale was also given from time to time in addition to the wine.

On the thirteenth day he had had no sleep for forty-eight hours, and required the constant attendance of the nurse to keep him in bed; he had general tremor and subsultus; his heart was weak, and the first sound inaudible; the pupils were contracted to a point; the extremities were cold, notwithstanding the applications of sinapisms and hot water. In short, nothing could be more unpromising than his condition. Under these circumstances I determined to push the exhibition of the tartar emetic and opium to the fullest extent, and I accordingly directed an ounce of the above mixture to be given every hour. Seven doses were thus administered, when he fell asleep and continued to sleep for many hours, having been awakened at intervals to take wine or nourishment. He rapidly and completely recovered.

I may take this opportunity of reminding you that, while the patient who falls into a natural sleep at the time of crisis may be safely allowed to sleep thus for a considerable time; the patient who after prolonged watchfulness, sleeps under the influence of opium, should be awakened at intervals of two or three hours, to receive wine or beef-tea, as otherwise there is danger of sudden sinking. I may also mention that in some cases in which opium does not act quickly or in a satisfactory manner, you will find chloroform a most useful adjuvant. Some of you may remember the case of a former student of this hospital, who passed through typhus in our private ward two years since; and in whom, after the mixture of tartar emetic and opium had been given for a couple of days with little effect, the inhalation of a small quantity of chloroform sufficed to procure sound and long-continued sleep.

I should recommend you never to trust to this measure alone, but to employ it as subsidiary to opium in cases such as those I have alluded to.

It is in fulfilling the second indication, that opium in small doses, and repeated at intervals, will be found a useful adjuvant to wine and brandy. The patient has, perhaps, contracted fever after prolonged mental or bodily exertion; has had long continued headache and sleeplessness; has delirium of a low hysterical character; has been freely purged; the heart being weak, the temperature low, the urine abundant. Often the case presents many of the features characteristic of delirium tremens. Under such circumstances, wine, brandy, and diffusible stimulants being specially indicated, the influence of opium as a cordial and a hypnotic will also be found most beneficial. In these cases the frequent repetition of from one to three drops of laudanum in an ounce of cardiac mixture is the form of administration which I most prefer.

The third indication will arise under special circumstances, and at certain epidemic periods. I have no doubt that at times there exists an epidemic tendency to active congestion of the brain, just as at others there is a tendency to cerebro-spinal arachnitis, of which, as a complication of typhoid, you have witnessed examples during the last two years in this hospital; the same having been observed in other hospitals in the city.

In the epidemic of typhus of 1839-40 I observed a tendency to active congestion of the brain, and I think we find evidence of a similar tendency at different periods in the writings of Armstrong, Southwood Smith, Laycock, and, above all, of Dr. Graves, who says, in reference to the epidemic of typhus of 1836-40: "Most of the fatal cases of typhus at present die of cerebral disease, but in the majority of instances you find that these were cases in which the head was neglected, and in which the appropriate remedies were used too late."[1]

---

[1] The following observations of this great physician deserve attentive consideration:—

"For the production of cerebral symptoms in typhus there must be something more than mere congestion or inflammation of the brain; but you are not to infer from this, that there is no necessity for taking any steps to obviate or remove congestion of the head in fever. On the contrary, I am of opinion, that in typhus, one of the principal sources of danger is connected with the head, and that the cerebral symptoms should be always watched with the most unremitting and anxious attention. I am also of opinion that when there is any evidence of deter-

Apart from epidemic influence, the special circumstances under which the indication arises are, after exposure to great solar heat at the access of fever, more especially if great muscular exertion is conjoined with the exposure; fever ensuing upon mental excitement, or upon a debauch; and occasionally after exposure to cold, while the body is heated by exercise.

I do not hesitate to advise you to pursue a decided and energetic treatment in all these cases. Guided by the action of the heart—which as a rule you will find jerking and forcible, as contrasted with the feeble contractions of the left ventricle usually met in fever—you should abstract blood in a mode and to an extent suited to the peculiar nature and amount of the lesion. Thus, in cases of cerebro-spinal arachnitis, relays of leeches to the mastoid process will be the mode preferred; while in the active congestion of typhus, arteriotomy seems, on the whole, to be most useful. Take as an illustration the following case, published in Dr. Cheyne's account of the epidemic of 1817.[1] "A young man, an apothecary, whom I had been called upon to visit in typhus fever, had passed several days in a state of maniacal delirium, attended with complete loss of rest. His afflictions were considerably heightened by the warm atmosphere of a small apartment, to which he was confined, loaded with bed-clothes, and scrupulously excluded from the benefit of cool air; his mental irritation was not a little augmented by the restraint imposed on him by a pair of stout guards placed over him. I directed his room to be freely ventilated, and immediately opened the temporal artery; when a few ounces of blood had been drawn he expressed great satisfaction, hoped as he was so much relieved at that side of his head that I would open the other artery; which was done according to his desire. He became immediately composed, passed several hours in calm sleep, and speedily recovered."

mination to the head, the best way of preventing dangerous cerebral symptoms is to deplete the head by the application of a sufficient number of leeches, and then to proceed to the use of blisters. One of the best modes of doing this, is to apply six or eight leeches behind the ears, and to repeat them every six hours until relief is obtained. You should then order the head to be shaved and kept constantly covered with cloths wet with warm vinegar and water, and, at the same time, have recourse to the internal use of tartar emetic and nitre, or blue pill and James' powder. Should this plan fail of giving relief, you have a powerful aid in the application of blisters to the scalp, and this must be done extensively and at once."—*Clinical Medicine*, vol. i. p. 176.

[1] Barker and Cheyne's Report, vol. i. p. 493.

I may add, as illustrating the beneficial effects of this practice, a remarkable case which occurred to me during the epidemic period I have already alluded to. A young man, on the ninth day of typhus, had, under the use of wine, become restless and delirious; for this state his medical attendant prescribed an opiate, shortly after taking which he fell asleep. On attempting to awake him a few hours afterwards, he was found to be in a state of unconsciousness, and unable to swallow. When I was called to see him (at 2 A.M.) he had not spoken, and he then lay perfectly motionless and rigid, the eyes directed to the ceiling, the pupils contracted and immovable on the approach or withdrawal of light; his lower jaw was rigidly closed, and he manifested no consciousness when spoken to; the surface was warm, the respiration accelerated, the pulse sharp and rapid, and the heart's action strong, the sounds being loud and distinct. In my hospital practice I had recently seen several cases in which the same symptoms following the administration of opium, the patient had passed into coma and death; the evidences of arterial congestion of the brain being discovered after death. I accordingly advised the withdrawal of opium and stimulants, and the adoption of treatment calculated to lower excited arterial action—viz., the abstraction of blood, the application of cold to the head, and the administration of calomel and James' powder;—3 grains of each to be given every third hour.

The temporal artery was immediately opened; while the blood was flowing the pupil began to dilate and contract alternately, and before ten ounces had been abstracted his eye followed the movements of the attendants. A few minutes after the bleeding ceased he spoke. There was a marked increase of delirium on the following evening, for which we prescribed, in addition to the powders, a mixture containing two grains of tartar emetic in eight ounces of camphor julep. There were no further bad symptoms, and he had a favourable crisis about the fourteenth day. When he recovered his intelligence, the patient had no recollection whatever of the occurrences at the time of his being bled, but said that he was during the whole period in a state of perfect unconsciousness.

I need add nothing to what I have already said in a former Lecture, on the use of local bloodletting and mercury in the reactive inflammation of the brain or membranes, which sometimes occurs in the later stages, and during the post-critical period of fever—more especially that of typhoid.

I would earnestly impress upon you that bloodletting is a remedy which you should neither abuse, or neglect to use when occasion requires. Its indiscriminate employment in fever and its total disuse are equally to be deprecated. We have had reactions of medical opinion in both extremes at different times; and the practice has been as extravagantly lauded at one time, as it has been unreasonably censured, or flippantly sneered at, at another. "It is time," to use the words of Dr. Chambers, "that the wave of opinion, which has swelled backwards and forwards to a dangerous height, should settle down into a steady stream. We ought to know clearly *why* we bleed, and then we shall know *when* to bleed."

I entirely agree with this writer in thinking that the ill effects ascribed to bleeding have often been strictly due to the abstinence enforced at the same time; and I would caution you to repair the waste you thus produce, by administering nourishment; and also to obviate the nervous excitement and irritability, which so frequently follow the abstraction of blood, by means of opium, and even by wine or brandy. I may refer you to the case of Suttle, page 141, and to that of the woman Donnelly (Appendix VIII.), who was recently under your observation, as illustrating their good effects. I may also remind you of the remarkable case of myelitis in a man named Byrne, recently under observation in our smaller fever ward.

You may remember that this man was brought in suffering from complete paraplegia, with incomplete paralysis of the upper extremities, and rigid contraction of the fingers; that he had scarcely any thoracic respiration, but breathed almost entirely by the diaphragm; that he had also double vision, some difficulty of utterance, with turgid, congested face and eyes. That, under these desperate circumstances, he recovered; the treatment pursued having been repeated bloodletting by leeches and cupping-glasses applied over the cervical spine; blistering the spine; and mercury pushed to ptyalism; with which we conjoined full nourishment, and a liberal allowance of wine, together with a nightly dose of Battley's sedative. These latter remedies are by no means found to be antagonistic to bleeding in their influence on the nervous system; in practice I have certainly observed them to act in a more salutary manner in these cases, after general or topical bloodletting than under any other circumstances. Their effects seem to be twofold in such states, viz., to lower nervous action, and thus procure sleep, and to diminish waste of nerve tissue.

Of the last indication I have little to add to what I have already said when discussing the treatment of the urinary lesions. It has always appeared to me, that the marked success of Dr. Graves' treatment by tartar emetic, with or without opium, was in a great measure due to its eliminative action, by which, in most of his cases, evacuation was brought on in several modes; as, by vomiting, purging, and copious diaphoresis or diuresis. These generally occurred about the usual period of crisis, and were followed by the patient's amendment.

I have occasionally seen similar marked effects follow the administration of large doses of oil of turpentine at the same period. It seems to be indicated in cases in which there is great tympanitic distension of the abdomen, with nervous symptoms, characteristic of blood contamination rather than of active congestion. Such was the following:—

Michael Dunne, æt. 35, a strong, muscular man, was admitted into hospital, under my care, at an advanced stage of typhus. He was at the time in a state of profound prostration: his pulse 132, small and compressible; skin cold; tongue dry and shrivelled; belly tympanitic; he had involuntary discharge of urine and feces.

Wine and ammonia were freely administered, and for the two following days he appeared to improve; but on the third day after admission his condition seemed desperate. He lay in a state of stupor with a muddy, livid colour, general tremor and convulsive startings, incessantly rolling the tongue, the belly tumid and tympanitic, &c. &c. He had had no evacuation from the bowels for two days, and this, with the tympanitic distension of the abdomen, and the character of the nervous symptoms, led me to try the effect of oil of turpentine, as advised by Copland under similar circumstances. I accordingly gave him three drachms in a castor oil draught. This produced a copious discharge of dark liquid matter. There was no other apparent critical evacuation, but on the following day he was perfectly conscious, all unfavourable symptoms had disappeared, and he was, in fact, free from fever. For a similar instance of the striking effect of this medicine, I may refer you to Dr. Graves' *Clinical Medicine*, vol. i. p. 183. In this case, after tartar emetic and opium had been given without any beneficial effect, turpentine was administered in two drachm doses. "After the second or third dose," says Dr. Graves, "the patient had two or three full motions from the bowels, and then soon fell into a sound and tranquil sleep, from which he awoke rational and refreshed."

In the progress of a case of fever in our wards, you hear the question frequently asked: What day of the fever is this? The question has two objects, the one with reference to prognosis, which is much influenced by the conformity or otherwise of the particular case to what may be termed the normal' type of the period; the other, with reference to treatment and to the approach of the critical period, as one in which increased vigilance, and, perhaps, active interference, may be demanded.

Upon no subject connected with fever has there been more difference of opinion than with regard to critical days. The ancients evidently drew their prognostics frequently from occurrences, which, as a rule, took place upon certain days, hence named "judging days," and this is the sense in which we should employ the term. I have already pointed out to you how important it is, that the occurrences in each form of fever should conform to its type, in regard to time as well as in regard to their nature and seat. Now this rule applies to the phenomena marking the termination of fever no less than to those occurring during its course. I believe we shall find the observation of these "judging days" to be not less important with respect to treatment; and that the fate of our patient will often depend upon our recognizing the last paroxysm and treating it suitably.

There have been, from time to time, writers who have treated the doctrine of "critical days" with scepticism, sometimes with ridicule. Their objections seem to me to be founded either on a want of recognition of the difference of species of fever, or on a misapprehension of the real meaning of the term "critical days." As an instance of the former, one able writer, whose views of fever and its treatment were derived exclusively from cases of typhoid, failed, of course, to recognize in this disease any conformity to the laws as to duration, which ancient and modern writers had found to govern relapsing synocha and typhus; and accordingly he expressed his doubt in the existence of any definite period to the course of fever.

Another seems to suppose that by a critical day is meant, not —as the word implies—a judging day, but one upon which the fever *must* end; and accordingly he says, truly enough in this sense of the word—"There cannot be accorded to any one day more than another any peculiar efficacy in accomplishing a crisis." Admitting this, it cannot be doubted that there is a tendency in each form of fever to terminate upon certain days; that this tendency is well marked in synocha and typhus, less so in typhoid; and that, to use the words of Cullen, "it is by some violent and irregular cause that this tendency is obviated.[1] That typhus rarely terminates before the eleventh day, and when on this day, for the most part, fatally: and that its usual terminations are on the fourteenth, seventeenth, and twentieth days."

I commend the chapter on Critical Days, in Cullen's "First Lines," to your attentive perusal, as containing the clearest and soundest views on the subject with which I am acquainted.

Viewing crisis as consisting of two parts—namely, an exacerbation of the fever, followed by a complete remission; you will be at no loss to understand in what the duties of the physician at this period consist.

Recurring to the comparison employed by a great physician, of the treatment of a case of fever to the guiding a vessel through a storm, I would say the treatment of the last paroxysm is, in many cases, not unlike the steering of the same vessel through a narrow and shallow passage, with rocks on either side, into a secure harbour. Two dangers are to be avoided—the one by careful and skilful steering, the other by, if necessary, lightening the vessel.

This is a fair analogue of the duties of the physician in the period of crisis. He has in all cases to steer for the channel, in other words, to keep in view the main indication, to support the vital energies of his patient until the paroxysm is over. In the great majority this is his sole duty, and often no appreciable difference of treatment from that of the course of the fever will be requisite for its fulfilment; but in others, where there is an evident tendency to sinking of the vital powers, as marked by signs which I have before enumerated, he must do all in his power to support and stimulate the failing functions of circulation and inner-

[1] This is well and tersely enunciated by Baglivi: "Circa septimum, vel quarto decimum, plures acuti morbi solventur, nisi medicus intempestivo ac stulto purgantium diaphoræticorum aliorum que hujusmodi remediorum usu naturæ motus turbaverit."

vation. The mention of these functions will suggest to your minds the class of remedies to be employed. The internal administration of wine and brandy, and of the diffusible stimulants, the external use of warmth, and of revulsives and counter-irritants, are those usually prescribed. Many a patient has owed his recovery to the substitution of a more generous wine, such as Madeira, or of warm brandy and water, for the weaker wine, which had lost its power of sustaining the heart's action. In others, again, a fillip has been given to the failing circulation by external warmth, and both functions have been aroused by the free use of sinapisms, turpentine, epithems, and flying blisters. You should bear in mind that there are certain diurnal periods at which your patient is more especially apt to require additional support and stimulation, to avert the tendency to sinking which then manifests itself. Such, for example, are the hours from three to five A. M., during which a tendency to sinking most frequently occurs.

Occasionally, however, though rarely, you may meet with cases in which the dangers of crisis have been produced by the excessive or too-prolonged use of stimulants. I have sometimes known wine to be continued day after day, when it obviously disagreed with the patient, producing heat of head, restlessness, watchfulness, and a state of vascular and nervous excitement, incompatible with favourable crisis. In such a case the patient generally loathes the wine, and craves cold water, and not unfrequently the sleep of crisis has followed upon a hearty draught of the cooling fluid. It may occasionally happen that more active measures may be required to subdue the excitement thus produced.

On this point Dr. Cheyne, in his report of the fever of 1817, says: "We venture not to say that there may not be some cases to which bloodletting is applicable, as for example, the following: I found a young man, says our correspondent, who had been eight or ten days in a very low state, with all the appearance of low typhus. I ordered him to have some wine immediately, together with whatever appeared necessary at the time; but not being able to see him for a few days, and the mother of the young man not paying sufficient attention to my instructions, the wine was continued without intermission, the consequence of which was that his head became much affected, a serious determination having taken place thither. I saw immediately that no time was to be lost, and in defiance of a host of friends who beset me, told me

I was *murdering* him, I opened the temporal artery, and took away blood pretty freely. The vessel bled twice in the course of the evening, but not to any serious amount, the consequence was that he was convalescent next day."[1]

I do not know a case of greater emergency and difficulty than the one we are now considering. It is fortunately rare and exceptional, but it is one which demands the greatest caution and judgment on the part of the physician. On the one hand, if he withdraws the wine and support which have been hitherto administered, he does so in contravention of the rule, that the support and stimulation prescribed during the course of fever must be increased rather than diminished on the eve of the final struggle; while on the other hand, it is the fact that in these exceptional cases every ounce of wine does harm, and but increases the dangers of the crisis. These dangers usually arise from some secondary complication, the most frequent being some form of cerebro-spinal lesion. They are not the less dangerous, because they are usually more or less latent. As I have already pointed out in a former lecture, they may occur either in the form of active congestion, existing during the course of the fever, but masked by it, or as reactive inflammation set up at the close of fever; or they may be the effect of secondary blood contamina-

[1] Dr. Bateman gives the following example of the ill effects of continued administration of wine under these circumstances: "I found a middle aged man, who had been brought in the evening before in the lowest conceivable state of collapse, in fact, to all appearance in *articulo mortis;* the extremities were cold, the trunk bedewed with cold sweat, the pulse imperceptible at the wrist; in short a heavy respiration, and some feeble remains of the power of deglutition, were the principal signs of life. Hopeless as the case appeared, I directed him to be kept warm, and to be supplied at intervals with a teaspoonful of wine or spirits, ether, &c., as long as he could swallow. To my astonishment I found him on the following day quite sensible, and loudly demanding food. The skin was warm, the pulse firm, and he had recovered a surprising degree of vigour. Fearful of withdrawing the support of some stimulus from a person whom I had seen the day before at the point of death, I continued his supply of wine; but on the following morning he was delirious; the eyes soon became red and ferrety, his skin hot, the pulse sharp and frequent, and in a few days he died with all the symptoms of phrenitis terminating in effusion. It is manifest, therefore, that stimulants should be regulated with great caution, even when most essentially required, *and according to their present effects.*" This was, no doubt, a case of latent cerebral congestion or inflammation, not caused but rendered manifest by the wine and spirits.—*On Typhus,* p. 124.

tion. In each of these conditions, the mischief, previously latent, is often revealed by the effect of wine and stimulants, as above described.

In this, which I consider the most difficult question concerning the treatment of fever—involving as it does, not only the administration of wine, but also that of opium in certain cases—I can give you no fixed rules for your guidance. I would, however, remark, that to treat such cases successfully you must not be bound by any routine method; and you must seek your indications in the condition of the patient, not in any system laid down in books. You will look for them then in the state of the heart and pulse; of the temperature; of the respiration; of the consciousness; and of the nervo-muscular system.

Should you find the heart's action jerking and impulsive, while the pulse at the wrist is small and hard; or the heart labouring, and its right cavities and the large veins distended; should the face be suffused, the sclerotic tinged, and the head hot, while, perhaps, the extremities are cold; or should the face be bloated and purplish, and the veins of the conjunctiva congested and dark coloured; if the consciousness is heightened; the hearing morbidly acute; the eye sensible to light; the manner excited and restless; and the patient watchful; or on the other hand, if there be a tendency to stupor and coma, if there be hyperæsthesia and spastic rigidity of muscles; all these conditions, *or any of them*, existing at the period of crisis, are contra-indications to the use of wine, and may be, according to circumstances, indications for general or local bloodletting.

These measures, by relieving a local congestion which occasionally interferes with the resolution of the fever, give assistance to the vital energies, and allow the law of periodicity, as it has been termed, to come into operation.

Pneumonia, and simple pulmonary congestion, as a rule, are resolved by the crisis; but this resolution may be occasionally aided by moderate local depletion, by blisters, dry-cupping, and by the free application of turpentine.

Pleuritis is, as you are aware, an exceedingly rare complication in typhus. I have seen but three instances in the course of my life, and in one of these the patient's safety was jeopardized at the period of crisis.

This gentleman had complained of a stitch in the lower part of the right side for some weeks before fever (maculated typhus) had set in. No physical signs, however, were discoverable, with the exception of a slight frottement, nor were there any unusual thoracic symptoms during the course of the fever. On the eve of the fourteenth day, however, the breathing suddenly became hurried and difficult, and this difficulty increased so rapidly and to such an extent as to threaten suffocation. On this account I was summoned soon after midnight to see him. He was apparently dying; his breathing embarrassed to a most painful degree; the trunk and limbs perfectly rigid, so that he could not be raised to the sitting posture, and when turned over in bed, for the purpose of examining the side, it was as if we turned a log of wood. I found the right side dull to the spine of the scapula, with perfect nullity of respiration sounds.

It was evident that rapid effusion had taken place into the right pleura.

Having procured a cupping instrument I abstracted about six ounces of blood from the side. This was between 2 and 3 A. M. He soon after fell asleep, and at 6 A. M. he was breathing easily, and in a profuse perspiration. When I left the house at 10 A. M. he breathed quietly, with a soft, slow pulse; the fever had terminated by crisis, and not only the fever, but the pleuritis also, for when on the following day I examined his chest, no trace of effusion could be discovered.

But even without the existence of any definite local complication you will sometimes meet with cases in which, to revert to our comparison, it may be necessary to lighten the vessel, by relieving the blood by eliminative measures, of the products of regressive metamorphosis which have accumulated in it in the course of fever.

This is often done by unaided natural effort. In one case by a salutary hemorrhage; in another, by spontaneous diarrhœa, or by a copious flow of urine of high specific gravity; in another, the patient, after a hearty draught of cold water, falls asleep, breaks out into a copious warm perspiration, and awakes free from fever.

Without, for a moment, supposing that you can procure crisis by imitating nature's efforts at elimination, you will be able to give occasional assistance. Thus, in one case, the oppressed brain will be relieved by abstracting a little serum by means of a blister; in another, the same relief will be given by increasing the amount of urea excreted by certain diuretic medicines; in another, a timely purgative may effect the object; and in another, the hot dry skin and dry tongue with thirst, will suggest the giving a copious draught of cold water, the diaphoretic action of which may, perhaps, be aided by an antimonial.

It is not possible to lay down rules for the employment of these measures. In one case the one will be indicated; in another, the other. You must be guided in your selection by principles—keeping the indication constantly in view—and your daily maturing judgment and experience.

It is of great importance that you should not continue the exhibition of wine or brandy—so necessary on the eve of crisis—after this change has taken place. Their gradual and rapid withdrawal is suggested by theory, and justified by experience.

We know that at this period there is enormous waste of tissue, with corresponding increase of elimination of its products. By continuing our alcoholic stimulants we retard the process, thus acting contrary to the indications of nature, and to the instincts of the patient, who turns with loathing from them while he craves the more suitable nourishment. In practice we find this abuse of alcohol followed by heat and dryness of skin, rapid, irritable action of the heart, thirst and restlessness, want of sleep, and return of delirium, subsultus, &c. &c. In this state you will find James' powder, with the addition of a little camphor and nitre, to act most beneficially. Three grains of the antimony may be given every three or four hours; warm sponging of the head and face, and warm stupes to the legs, being at the same time employed. These measures will generally be followed by sound sleep and copious diaphoresis, with complete relief of the above symptoms.

The older Irish physicians were well aware of the value of antimony at this juncture, while, according to Dr. E. Percival, they prescribed its use before the eleventh day. This writer thus states his experience of the remedy: "In those varieties of fever which are protracted by imperfect crisis, where the tongue remains coated, and the skin, though temperate, is dry and torpid, I have employed antimonial powder, in small doses, with great advantage. I have usually combined with it calomel or opium (or both together), and found them highly serviceable in improving the condition of the secretory organs, and amending the discharges from the liver, the mucous membrane, and the skin. In the dysenteric fever these remedies are an invaluable resource."

I have mentioned, in a former lecture, that crisis is not unfrequently followed, after a short interval, by an access of watchfulness and delirium, especially in the young patient, or in one whose brain has been overworked previous to the fever. In this state you must be careful not to administer wine or stimulants. Neither is depletion of any kind indicated. Your treatment should be simply sedative and soothing. Opium or morphia, given by the mouth, and by enema, or perhaps in combination with antimony. In some cases, hyoscyamus and camphor will answer the purpose of calming the nervous excitement and procuring sleep. These measures will be aided by the withdrawal of all stimuli, such as light, sound, and the presence of friends; and by warm sponging to the head, together with warm fomentations to the extremities.

Sleep, thus obtained, should be kept up by the same measures for many hours, or until complete tranquillity of the nervous system is restored.

While on this subject I may mention that the delirium after crisis is sometimes produced by the pain and loss of rest caused by excessive blistering—more especially in the young patient. There can be no doubt that fatal consequences have sometimes followed from the suffering thus caused. Dr. Graves wisely cautions us against producing this *delirium traumaticum*.

I would also caution you against blistering the shaven scalp, in cases of active congestion of the brain, without having previously moderated arterial action by topical bloodletting and tartar emetic. I have repeatedly known marked increase of delirium, &c. to follow this practice.

If care is required in the management of the typhus patient after crisis, it is still more necessary during the convalescence from typhoid. You should always remember that a large proportion of cases of perforation of the intestine have occurred during this period, and you should be careful to take every precaution in your power to avert such a result.

Absolute rest must be insisted on, stimulants prohibited, and the blandest nourishment only allowed to be given. If the diarrhœa has been protracted, and if there be pain or tenderness on pressure over the cæcum, it is advisable to employ counter-irritation by means of the repeated application of small blisters in succession, or by tincture of iodine. Irritating purgatives are to be strictly avoided; the bowels being opened every third day by an enema of warm water, or perhaps, occasionally, by a small dose of calomel, or gray powder combined with opium.

Even although perforation does not occur, the patient may be exhausted by protracted and intractable diarrhœa, arising from the persistence of intestinal ulceration. You have, perhaps, treated this patient's diarrhœa during the later stages of the fever with port wine and astringents, with advantage. You may have continued these remedies after the termination of the fever, in the expectation that they will be equally beneficial; but day after day you learn that the discharge, although checked for some hours during the day, returned at night or early in the morning, and was attended with griping pain and with a feeling of sickness; the stools being thin, and brown or pea-soup colour—probably having a cadaveric odour; or perhaps, mucous, or bloody, or resembling the washings of flesh.

You note that the patient is emaciating rapidly; with a quick, irritable pulse, a red eye, and glazed tongue; the skin hot at intervals, and permanently harsh and dry; the abdomen perhaps tumid, but more usually retracted and concave in its outline. Sometimes low peritonitis sets in, with increase of pain and tenderness, green vomiting, &c. &c.; at other times death seems to be caused by inanition and exhaustion.

In this condition I have seen the use of the warm bath, aided by mild diaphoretics, of signal service. With these should be conjoined the exhibition of soothing and sedative remedies, as bismuth, hyoscyamus, and morphia. Poultices to the abdomen, and, in protracted cases, the administration of medicines calculated to improve the condition of the intestinal ulcers and of the mucous membrane. Of these, nitrate of silver and the turpentines are, perhaps, the most worthy of your confidence.

I shall not enter upon the consideration of the numerous sequelæ of fever. You will find their treatment fully described in systematic works on medicine and surgery; and these lectures have already extended to a much greater length than I originally intended.

In bringing to a close the remarks I have offered you upon the general management and special treatment of fever, there are a few principles, or general rules, which I advise you ever to bear in mind.

I. That inasmuch as fever is a disease having a definite course, duration, and termination, undue interference with it is to be deprecated, and the duty of the physician will often consist in watching, and waiting upon, the operations of nature, without interfering.

II. That such expectancy does not preclude the necessity for attention to management and regimen, together with the due regulation of the internal functions of the patient, and his external hygienic conditions.

III. That the first of these in importance—as being both curative and prophylactic—is supplying abundance of air containing its full proportion of oxygen, care being taken at the same time that the patient is not subjected to the necessity of breathing the exhalations from his own or surrounding human bodies.

IV. That equally important is the prevention—so far as may be possible—of the accumulation in the blood, of the products of disintegration of tissue. This object must be accomplished partly by pure air, the use of diluents and fluid nourishment, and also by wine or alcohol in some cases; and partly, perhaps, by the administration of the mineral acids, chlorate of potass, and other medicines of that class.

V. That unless, in case of necessity, no medicine of the evacuant, or so called eliminative, class should be administered after the period of invasion. But that such a necessity may arise during the later stages of fever, and in the event of local lesions: either with a view to avert directly fatal consequences from the lesion itself, or—in other cases—because that such lesions interfere with the natural tendencies of the fever, and frequently hinder its resolution by crisis.

VI. That whenever it may be considered necessary to employ such treatment in a case of fever, its depressing effect should be carefully guarded against by the simultaneous employment of measures calculated to repair the waste thus caused, and to support the patient's strength.

VII. That while avoiding routine, we should, as a rule, adopt a different system of management, in some particulars, in different types of fever.

VIII. That, generally speaking, with the exception of the intestinal affection of typhoid, the treatment of the primary local lesions, or those arising from the affinities of the poison, seldom demands the active interference of the physician. That the secondary lesions may probably demand active measures which will vary in their nature with the seat and nature of the lesion, and the mode in which each tends to produce a fatal termination.

IX. That the fate of the patient seems occasionally to depend upon the treatment adopted in the last paroxysm; at which period judicious interference, by assisting nature, as it were, either by removing local congestion, or by combating increasing asthenia— seems to act more beneficially than at any former stage, and that we have reason to believe salutary crisis may sometimes be thus secured.

X. That unless under exceptional circumstances, and with the above object, there should be as little interference as possible, but that as a general rule, the support and stimulants previously administered, should be given with greater frequency and in increased quantity at this period.

XI. That whereas in typhus our treatment up to the period of crisis was designed to retard regressive metamorphosis, we should be acting in opposition to the operations of nature, were we to continue it longer at a time when a rapid disintegration, and consequent elimination of its products is taking place; and that, accordingly, we withdraw a large proportion of the wine or alcohol immediately after crisis, and administer both wine and nourishment sparingly during the early days of convalescence.

XII. That this rule is doubly important in the convalescence from typhoid, on account of the danger there exists of renewal of disease of the intestinal follicles, and its consequences.

APPENDIX.

I.

I WOULD refer the student to Dr. Parkes's "Lectures on Pyrexia," published in the *Medical Times and Gazette* for 1855, for a lucid exposition of Professor Virchow's views on the proximate cause of fever. Perhaps he may obtain some idea of these from the following extract from a lecture delivered by the Professor, of which I have been favoured with a translation by my friend, Dr. Gilbert Nicholson:—

"The complicated mechanism of fever can only be understood altogether, when one regards the peculiar mechanism of the body. One must not consider the body as a dead mass, into which enters the breath—the pneuma, as the Greeks said (or, as the ancient Jews expressed it, the living order to bring into activity) one must not regard the body even as a peculiar machine, which the soul governs according to its own intentions. On the contrary, one must conceive the body to be a many-membered organism endued throughout, with life; whose single parts certainly work mechanically, but from which still each separate one has, at the same time, in itself the cause of its activity—the life. Many lives are here joined in united life; many different existences, with an independent power of life and action, are placed in a common dependence on one another, and in this dependence they are influenced the one by the other, each after its own manner and the manners of the others. Many are more highly organized, and therefore nobler and more important in the great common existence; others are weaker, small, poor, and single, seemingly of little importance, and still, in cases of difficulty, hard to be done without.

"Where now do the great regulatory contrivances lie, in the common joining together of the human body? They lie principally in the blood and the nervous system. The blood is the means of the interchange of materials; it streams to all parts of the vessels, in the capillaries, and returns after a long course to the heart, much changed, in order to be driven from that again through the lungs, that great emporium of the exchange of gases.

"It brings the oxygen with it, which consumes the materials, and thither it brings back the carbonic acid gas, which has resulted from that consumption. Each part takes from the blood its portion of materials; each gives back to the blood whatever has become useless for it. Can one still wonder that the blood can become a source, too, of general disturbance—the centre point of constitutional disease? Injurious materials force themselves into the blood in the most different ways, and whilst they arrive from that at the separate parts, they become a mighty ferment for internal decomposition. Thus the infectious fevers arise, in which the blood is rendered first impure by means of all kinds of changed substances; for the most part by chemical means, which have arisen from the decomposition of organic vegetable or animal bodies. The earth, the habitations of mankind, their nourishment, and trades, can offer the opportunity for such decomposition; but the body itself, too, can give the material, and so give the occasion to self-infection, the worst because the most secret infection. Here belong many of the so-called traumatic and inflammatory fevers, as they are seen to arise in over-filled hospitals, and therefore so often in the train of great battles.

"But every infection of the blood does not bring about fever. Cholera is one of the worst diseases of infection, and still not essentially feverish. In its heaviest forms it occasions such a reduction of heat that it has received with right the name of cholera algida. The impurity of the blood brings fever then only, when at the same time the nervous system, in its most important parts, is attacked with it, when, therefore, from the blood injurious matters make their way into a certain nervous part. But there are many channels to the nervous system, of which the circulation is only one; and there is, therefore, many a fever in which, at least at first, the blood is quite uninjured, and a pollution of it remains quite excluded. But the so-called nervous fever—typhus—does not belong to this class, since it is exactly such a decided infectious disease that, as the latest experience teaches, quite near to it lies the suspicion of poisoning. On the other hand, original fevers of the nervous system are well enough known in a popular way. One can certainly reckon among these that consuming fever which is brought on by excessive and continued exertion—it may be mental or bodily—by which the constitution is exhausted, the nervous system weakened, since in all cases of constitutional weakness, original feeble formation from defective nourishment, or from exhaustion by work, the nervous system is also inclined to febrile excitement.

"We are accustomed to say, excitement—but by this must not be at all understood that in fever, as a rule, development of strength takes place from the side of the nervous system. On the contrary, every great development of strength takes place only by fits and starts for a limited time, and where it happens it is much more in relation to an increased irritability. But such an event is much more a sign of weakness than of strength, and truly all phenomena point out that in every fever, however it has begun, the fundamental character of the nervous activity, and in particular of the regulating activity, is that of an ever increasing weakness and want of resistance. From the beginning a very expressed feeling of tiredness and want of power, frequently shows itself. The muscles obey only slowly the demands made upon them : one has no desire for activity; for any enjoyment; one shivers from the slightest breath of air; in short, one is sensible of a disturbance in all the members, which affects not so much the members in their actual being and existence, as in their relation to one another. The common equilibrium of the parts is removed, and with this is experienced a feeling of internal disharmony. This disharmony develops soon more strongly ; the beats of the heart increase, the pulse is quicker whilst all other muscles are dull:

external cold shows itself whilst the inward heat becomes even more glowing. We can easily understand why the surface of the body becomes cold while the blood is warmer than its wont; for the bloodvessels of the skin contract, grow narrow, until only so little blood can stream through them that this cannot keep up to the normal elevation the temperature of the skin, sinking from radiation. But the contraction of the vessels is still a phenomenon which points out, like the increased activity of the heart, an unusual work of the contracting parts. How shall we recognize in this a symptom of weakness? and still it is such, for the nervous system acts over all as a moderator in the natural processes of life. It is that same contrivance which mediates not only between parts in the organic collective being, but also regulates the flow of the blood, in that it changes as well the motions of the heart as the width of the vessels. If it loses the power to exert mediatory or regulating activity, it becomes palsied in its proper central elements. So many single portions of the body, even single sections of it, unfold an increased activity; yet the fact is not by this changed, that the body has experienced a dangerous debility in its most important parts; so to say in its kernel. The more clearly this conviction has been established by the modern physicians, the more has the idea found its way back, which (idea) had received the greatest attention in Germany a few decades ago; namely, that fever in itself is a wholesome reaction of the body against any disturbance which has forced itself into the body, or has arisen in it, and that this reaction conquers for itself a natural termination in the crisis, like a victory. This idea has not a little contributed to accustom physicians to the so-called expectant treatment. And this has had its good, too, since it has at last set determined bounds to the excessive quantity of compound and dangerous medicines."

II.

Dr. Stokes once informed me that at least six-sevenths of the numerous cases of gangrene of the lung, which have fallen under his observation, were caused by prolonged exposure to cold and moisture during a state of intoxication.

I think I may say that every case I have seen was thus produced. The following is the testimony of a great physician, as to the influence of similar causes in predisposing to fever:—

"Inde enim fit, quod febres hæ hieme potissimum grassentur, et eos quam maxime corripiant, qui pravo victu ac regimine usi, ac præterea, corpus intensius refrigerarunt. Vidimus id ipsa hac hieme, qua ex civibus nostris multi, intemperantiæ in baccho ac venere indulgentes, et traha diutius sub cælo rigido vecti, inciderunt in has febres, et voluptatis dederunt pœnas. Neque etiam ea propter effluvia ex iis, qui his exanthematicis febribus graviter decumbunt aut mortui sunt, omnes inficiunt promisuè, sed eos tantum, qui dispositionem ad hujus modi morbum, summam nempe succorum impuritatem gerunt, ceteris, qui ab hac immunes, a contagio etiam liberis."—*Hoffman De Febribus Epidemicis.*

My friend, Dr. Maguire, of Chapelizod, has favoured me with the particulars of an interesting case of fever, complicated with septicæmia under prolonged exposure to moisture, lately under his care. The fever, which affected four individuals in different degrees of intensity, broke out in a detached cottage in the month of May last, and appeared to him to have had its origin in a large heap of vegetable matter left to decay beside the house, "the smell of which," says Dr. Maguire, " was intolerable, so that I had it removed at once." In addition to this, his patient's bed was so placed that she was exposed during four or five days' constant rain to the dropping from the roof, by which her bed was thoroughly wetted. The consequences were similar to those in the cases detailed in the text, page 51. "About the eighth day," says Dr. Maguire, " her gums began to bleed, and on looking I found her body, particularly the lower extremities, covered with purpuric spots—distinct from the maculæ. On the day after she had also hemorrhage from the bowels. Under the use of tincture of perchloride of iron, after two days the hemorrhage ceased, the spots began to disappear, and on the fifteenth day she had a distinct crisis, and recovered perfectly without any untoward circumstance."

The cases of cancrum oris, referred to at page 54, are of sufficient interest to warrant their introduction here :—

"Ann Proudman, ætat. 7. On Sunday, December 17th, 1826 (being then in the 20th day of fever), her mother observed a foul black ulcer on the lining membrane of the right cheek, with a black spot on the tip of the tongue; this last sloughed, and presented a clean and healthy looking ulcer on the 20th. When seen for the first time at the dispensary, the right cheek was much swollen and inflamed; the mucous membrane presented a gangrenous ulceration, extending nearly to the angle of the mouth, but without any communication with the external cheek, or with the membrane immediately lining the gums; attended with an extremely fetid odour. On the 21st a speck, resembling in every respect a spot of petechial purpura, appeared on the outside of the cheek, corresponding to the ulceration on the mucous surface, which, on the 22d, had increased to an ulcer of the size of a shilling, exhibiting a dark, black spot in the centre, and a yellowish circular margin; this rapidly advanced in a circular direction to the size of a dollar, and on the 28th had made a communication with the internal surface; after this it spread more irregularly, and exposed the alveolar process of the upper and lower jaws, and extended to the orbit. The constitution sunk rapidly under a diarrhœa, and she died on the 3d of January.

"On the 13th of August, 1829, Mary Proudman, sister to the above, also aged seven, was brought to the same institution, when her mother stated that she had measles about six weeks since, previous to which she complained of pain in the left jaw, which was attributed to cold, with swelling of the left cheek; in a week after the swelling she perceived an ulcer of an extremely fetid nature on the lining membrane of the lower alveolar process, which had been gradually extending, now nearly five weeks, to the upper, and to the side of the tongue, with great flow of saliva. The progress of the disease in this case was arrested before ulceration had extended to the outer surface, by the application of the nitro-muriatic acid mixed with a little honey; with bark administered internally; and by an allowance of wine; and under a generous diet she ultimately recovered.

"The coincidence is not a little remarkable of the two cases occurring in the same family, during a febrile attack, and at the same age."—*Report of Cork Street Hospital for* 1837 *and* 1838.

III.

The fact alluded to at page 60, is mentioned by Fordyce and other older writers on fever:—

"The infectious matter," says Fordyce, "produced by a number of men living in a small space, as well as that produced by fever, may adhere to a person in perfect health, so as to be brought into another place and communicated to a whole assembly, as has been too often proved by a felon brought from gaol into a court of justice, and infecting almost the whole of the persons assembled, and that even when the felon himself was perfectly free from fever, and never had been affected by the disease.

"It often happens when numbers of persons are confined together in a small space that putrescent substances are not thoroughly cleared away, hence a person brought out of a gaol where putrescent substances have been accumulated, carries with him substances of a peculiar smell. Hence some have supposed that the infectious matter produced in this last way had sensible qualities. This is undoubtedly not the case, since the infection has arisen from a person brought out of rooms in which numbers had been confined for several months, but kept clean from all putrescent matter, so that there was no particular smell or other sensible quality. In one case that came under the observation of the author, a person under such circumstances, from whom no peculiar smell arose, or any other sensible effluvia, communicated the infection to four others with whom he was carried in a coach for about half a mile, so as to produce fevers in all of them, which fevers were violent and fatal."[1]

Dr. Ferrier includes moral causes—*e g.*, anxiety and depression of spirits —in his enumeration of the influences which generate animal poisons, "because it is not proved that the mere confinement of the effluvia of clean and healthy persons, free from mental uneasiness can become poisonous." The following remarkable case of the origin of fever from a single person under such circumstances is given by Dr. Harty.

A gentleman was suspected of having confined and ill-treated his wife. At length two gentlemen, one of them a clergyman, having obtained the necessary authority, visited the house, and examined every apartment for the wretched object of their humane search—at first in vain; but at length a small closet door attracted their notice, and having insisted on its being opened, both gentlemen eagerly entered, and as precipitately retreated. One was immediately seized with vomiting; the other (the clergyman) felt sick and faint. After a little, they recruited and called the wretched woman from her prison hole, in which she had been for weeks immured. It was a small dark closet without *light* or *air*, and in it she had been immured without a

[1] On Fevers, vol. i. p. 114.

change of clothes. At the end of a week both gentlemen had fever; both took to their beds almost on the same day. The clergyman died, and the other recovered with great difficulty, after a severe struggle. Both cases were alike throughout, except in the termination. The woman had not then or afterwards any febrile disease, and had been free from any at any period of her confinement.—*On Fevers*, p. 168.

In the following passage, Dr. Murchison adduces the arguments against Dr. Budd's theory of the propagation of enteric fever, referred to at page 64:—

"According to Dr. Budd, the poison of enteric fever, although contained in sewage, is then always derived from the alvine evacuations of an individual already suffering from the disease. The poison resides chiefly in the stools of the sick, and a drain is merely the vehicle of its propagation; or, in fact, a direct continuation of the diseased intestines. Admitting fully that the view advocated by Dr. Budd and Professor Von Gietl, offers the best explanation of the circumstances in those cases where the fever is propagated by the sick; many, if not most, of the facts adduced in favour of it are explicable on the theory of spontaneous generation, while in others the mode of communication is not so clearly established as might be desired. Dr. Budd's arguments are twofold: first, he adduces many facts to show that the disease is contagious; and secondly, he mentions many cases to demonstrate the intimate connection between its origin and bad drainage.

"Both of these positions I readily concede and have always contended for.

"But it does not appear to me equally clear that, in the cases recorded by Dr. Budd, the disease was propagated by the stools of persons previously infected. For example—in the North Tawton Fever, on which so much stress has been laid, while the facts leave little doubt that the fever was communicated in some instances by the sick to persons in health, it is not shown that the stools of the infected were the medium of communication. On the other hand, Dr. Budd records three instances, from which he argues that sewers merely transmit the poison, in consequence of receiving the excreta of a diseased intestine. In all of these instances the fever evidently arose from air or water tainted with sewage; but it is not shown that the sewage in any of the cases had first become contaminated with the excreta of a person suffering from enteric fever.

"The necessary link in the evidence, viz., the introduction of the poison, is wanting. In another instance, cited by Dr. Budd, where four cases of fever occurred in a retail establishment at Bristol, it was argued that the disease in the last three cases was due to the evacuations of the first case being thrown into the common water closet. But it was not shown that the poison was imported by the first case, which is spoken of as 'casual,' and on the supposition that the first case was due to some local cause, that cause was sufficient to account for all. In other instances, such as the Orphan Asylum at Ashley Hill, and the school in the South of England, where the fever appeared in connection with offensive latrines, it is also argued that the cases were due to the children frequenting the latrines, into which the dejections of the first patient had been thrown; but it does not appear that the first patient contracted the disease elsewhere than in the asylum or school. Even in the account of the North Tawton outbreak, although the date and locality of the first case are mentioned, it is not stated that the patient caught the disease away from the place. The circumstance much dwelt on by Dr. Budd, that extensive outbreaks of enteric fever have occasionally been preceded by two or three isolated cases, proves nothing in favour of contagion, in my opinion, except it can be shown that in these first cases the fever was contracted away from the site of the subsequent outbreak.

"Moreover, in not a few instances many persons have been attacked simultaneously in the same house, without any isolated cases preceding.

"But, on the supposition that enteric fever is occasionally propagated in the manner described by Dr. Budd, it does not necessarily follow that in every case arising from bad drainage the poison has been merely transmitted. On the contrary, there are facts which, in my opinion, demonstrate that the fever often arises independently of any such transmission. It is difficult to obtain crucial evidence on the point from what is observed in large towns furnished with a complete system of drainage, because to every instance of fever arising from the inhalation of sewer-gases, or from drinking water polluted with sewage, it would be replied that sewer emanations are 'the very quintessence of a pre-existing fever.' But even in towns, evidence is not wanting that enteric fever has arisen from bad drainage, where it is impossible to conceive that the poison was introduced into the drain. For example—in the case of the fever at the Peckham Police Station (see page 438), the cesspool had no communication with the public drains. In the outbreaks at Westminster (see page 440), Clapham (see page 442), and other places, the source of the fever was traced to the decomposition of sewage in drains which were choked up, and so, in a measure, shut off from the general drainage; while in the case of the Colchester Union (p. 443), every possibility of importation appears to me to have been excluded. On turning to what occurs in country districts, the evidence is still more conclusive. Several instances have come under my notice in which enteric fever has broken out in an isolated country-house, or in a small group of houses, miles away from where any fever was prevailing, and in which every mode of importation or of communication by drains, or otherwise, seemed impossible; and by numerous inquiries from country practitioners I learn that such events are far from rare.

"There is another view of the subject worth considering. Dr. Budd, like many former writers, regards the intestinal disease as a specific eruption, and he contends that an infinitesimally small 'dose of the poison derived from this eruption is sufficient to produce the disease.' But there is no positive evidence that the stools of enteric fever are of such a virulent character, while there are some circumstances which rather tend to an opposite conclusion.

"As already stated, since my connection with the Fever Hospital 1048 cases of enteric fever have been under treatment, but only one case of the fever has originated in the hospital; and yet the night-stools containing the excretion often remain for many hours without being emptied. Moreover, on several occasions I have known other patients sit over the evacuations of enteric cases without suffering. Surely, if these stools are so venomous, as Dr. Budd would have us believe, a contrary result might have been expected. Some years ago, I fed a pig for six weeks on the stools of patients suffering from enteric fever. They were mixed with barley-meal, and given two or three times a-day. The animal appeared to suffer no inconvenience, but on the contrary, it got very fat, and when killed its intestines were perfectly healthy.

"Again, if the stools of enteric fever are so virulent, that those of one patient can give the disease to a whole community, as Dr. Budd believes, an epidemic might be expected to be constantly raging on the banks of the Thames, which must often contain the excreta of many thousands of enteric patients; yet such is not the case.

"A careful examination of the stools in enteric fever shows that they are remarkably prone to decomposition or fermentation. In place of being acid, as healthy feces always are, they are invariably alkaline; they also contain abundance of the ordinary products of the decomposition of animal matter, in the form of ammonia and ammoniaco-magnesian phosphate—while their odour fully bears out the idea of their putrid character. Surely, such a condition would be fatal to the existence of an animal poison, such as that contained in the pus of smallpox. Dr. Budd says that the morbid material deposited in the intestines contains the specific poison, by which the disease is propagated, just as the contents of the variolous pustule contains the poison of smallpox. But if this be so, the poisonous matter is never passed from the body until it is separate as a slough from the intestine—until, in fact, it is dead and putrid. It has yet to be shown that the contents of the variolous or vaccine pustule will produce smallpox or cowpox, after they become putrid. Lastly, the fact that the prevalence of the disease is influenced by temperature (see pages 418, 419), is opposed to the idea that it depends on a specific poison derived from the sick, but is readily accounted for, on the supposition that the poison is generated by fermentation or decomposition.

"It is very probable that the stools of patients, suffering from enteric fever, are more prone than ordinary sewage to the peculiar fermentation by which the poison is produced, and even that, in certain cases, the fermentation may have commenced before their discharge from the bowels. In this way enteric fever may occasionally be propagated by the stools. But whether this be so or not, it seems to me far more probable that the poison is always the result of decomposition than that it is derived from a specific eruption, like that of smallpox.

"While, then, I entirely agree with Dr. Budd in recommending the immediate removal of the discharges from the sick, I hold that it is even more necessary to rectify without delay the escape of sewer gases into houses, and the pollution with sewage of drinking water."

The following highly illustrative case and observations, which I extract from Sir D. J. Corrigan's "Lectures on Fever," is referred to at page 65:—

"Mary Cope, ætat. 22, previously in the enjoyment of the best health, was admitted into the Hardwicke Fever Hospital, on the 23d January, ill of maculated fever. There was nothing unusual in her case. She was soon marked convalescent; when, on the 5th February, fourteen days before her death, she complained of debility and of diarrhœa; her tongue became brown and dry in the centre, but not furred; there was no tenderness of abdomen, nor tympanitis, but there was gargouillement over the cæcum; and the stools were like gruel, but neither mucous nor bloody, nor was there any tenesmus. The pulse became quicker and weaker. On the 18th February, she required wine in considerable quantity; the diarrhœa became uncontrollable. On the 19th February, fourteenth day of the attack, she died.

"*Post-mortem Examination.*—The peritoneal covering of the abdominal viscera was sound, but the ilium and a portion of the colon presented follicular enteritis in all it stages. The greatest intensity was at the ileo-cæcal valve, the entire circle of which was occupied by a depressed, jagged, grayish, irregular ulcer. In the ilium, both the isolated follicles and the 'glandulæ agminatæ,' or glands of Peyer, were attacked. The sites of the affected 'glandulæ agminatæ' were marked by oval ulcers, while around and above them isolated follicles were seen in similar ulceration, but not to such a degree. In addition to these ulcerations, the isolated follicles in both the lower portion of the ilium, and the upper portion of colon presented every stage of the disease. Some follicles were just protruding under the mucous membrane, filled with a cheesy-looking purulent matter, the mucous membrane around being swollen, red, and prominent, and from the orifices of the follicles looking like depressions, those follicles bore a strong resemblance to variolous pustules. In others the matter was in such quantity that the affected follicles presented the appearance of spherical projections, attached by pedicles, and covered by mucous membrane, hard and firm to the touch, while in others, still more advanced, the follicles had gone on to ulceration, destroying the mucous membrane, and leaving only a gray slough of cellular tissue in the place of the follicle itself. The appendix vermiformis was swollen and congested, and, on being slit up, was found distended with a tenacious purulent fluid, and some of the mesenteric glands were infiltrated with purulent matter. This was a case of great rapidity and of equal severity. In most cases the disease is bounded by the ilio-cæcal valve. In another case, occurring about the same time, which ran a longer course, the lower part of the ilium was found extensively ulcerated, the ulcers presenting different appearances; some of them presenting the striking resemblance already described to the pustules of variola, others having the appearance of white cicatrices, as if traced out by a snail traversing a zigzag course along the mucous membrane; while others again, with loss of substance in the centre, but with soft pulpy areolæ, projected above the surrounding mucous surface. In this case, as is most usual, the disease was confined to the small intestine.

"This disease may be of every degree of intensity, proving fatal within ten or fourteen days or verging into convalescence within two or three weeks; or, when there has been more extended disease, occupying even months in its progress and cure. It is in its most acute form that it bears the strongest resemblance to typhus, from the sudden prostration of strength, dry and furred tongue, maculated appearance of skin, and accompanying delirium. We have already, in our first Lecture, sufficiently shown, I hope, that there is no necessary connection between this disease and typhus. We need not revert to this. But we must not forget that this disease, like any other, may take its origin in the course of a case of typhus, or may occur quickly supervening, as in Cope's case; in other words, that while typhus exists without it, typhus gives us no surety against the complication or supervention of this disease. Follicular enteritis, of moderate degree, and typhus, cannot continue '*pari passu,*' because typhus must either terminate life, or terminate itself in crisis within a few days, while follicular enteritis, which has arisen during typhus, will continue, after the disappearance of typhus, to run its own course of weeks, or perhaps months. If follicular enteritis occur early in typhus, as one of its complications and of great severity, as in Cope's case, death may take place, from the combined effect of the two diseases, within a very few days: but still the connection between the two, typhus and follicular enteritis, is merely incidental, and is not essential."—*Lecture X.*

The following observations of Dr. Murchison on this occasional coexistence of the two eruptions, are worthy the attentive consideration of the student of fever:—

The doctrine of the compatibility of two of the exanthemata, has an important bearing upon that of the non-identity of typhus and pythogenic fever. They who maintain that the poisons of the two fevers are identical have appealed triumphantly to certain cases in which they have observed the eruptions of the two fevers to coexist. Now, allowing for a moment that the facts in all these cases have been correct, the conclusions which have been drawn, are based upon a doctrine which is utterly fallacious. The coexistence of two eruptions, no more implies an identity of the two diseases, than it does in the case of variola and scarlet fever, or of scarlet and pythogenic fever. But there can be little doubt, that in the majority of cases the facts themselves, from the manner in which they have been described, must be viewed with no small distrust. On few subjects does so much confusion prevail in the profession, as with regard to the eruptions of continued fevers. A very common mistake is to imagine, that petechiæ constitute the characteristic eruption of typhus, a mistake which has been strengthened by petechial fever—being one of the appellations applied to the disease—and it has been argued from a patient presenting both "rose spots" and "petechiæ," that the eruptions of pythogenic fever and typhus have coexisted. But petechiæ do not constitute the characteristic eruption of typhus; and they are met with in the course of pythogenic fever, in the same way as they show themselves in the course of variola, scarlatina, and many other affections. All they who have had much practical experience in studying both typhus and pythogenic fever, will admit that it is excessively rare to find the measly eruption characteristic of the one, coexisting with the rose spots characteristic of the other.

In my essay upon the etiology of continued fevers, published in the *Medico-Chirurgical Transactions* (vol. xli. p. 275), I expressed an opinion, that such a coexistence was possible; but I maintained then, as I do now, that no argument could be based upon such a coexistence, as to the identity of the typhus and pythogenic poison, any more than we should employ a similar argument to show that variola and scarlet fever, or scarlet fever and pythogenic fever were one and the same. I shall now proceed to detail the facts which testify to the possibility of typhus and pythogenic fever coexisting.

Such facts might be naturally looked for under circumstances in which a patient labouring under the one disease has been exposed to the contagion of the other, as for example, in London Fever Hospital. When the doubtfully contagious character of pythogenic fever (already alluded to) is remembered, it will not be wondered at that patients admitted with typhus have seldom contracted the former disease. During a period of ten years I have only been able to find the notes of two such cases. One was that of a female, aged twenty-one, who, with seven others of the same family, was admitted with well-marked typhus, and who, in the third week of convalescence, had an attack of pythogenic fever, the symptoms of which, however, were mild and not very characteristic. The second case will be shortly alluded to.

On the other hand it has been by no means rare for patients admitted with pythogenic fever to contract typhus during their stay in hospital. But in most cases this has been in the fifth to eighth week of convalescence from the first fever, and two or three weeks after the patients have been removed to the convalescent ward, a circumstance which is explained by the patients in the convalescent wards being thrown into more intimate relations with one another, and by typhus being avowedly more contagious during convalescence than during the height of the disease. In the following instances, however, the eruptions and other symptoms of the two diseases were almost contemporaneous.

Illustration XLIX.—A female, aged twenty-two, had an attack of pythogenic fever, which was attributed to the putrid emanations from a bad drain. She was admitted into the London Fever Hospital. The primary attack lasted three weeks. After a fortnight she had a relapse, with a return of the "rose spot;" and the day after this there was a subcutaneous "typhus mottling," along with drowsiness, heaviness, and other symptoms of typhus.

Illustration L.—A male, aged twenty-five, was admitted, with well-marked pythogenic fever.

On the 27th day there was a great aggravation of the symptoms, with much headache and stupor, and in addition to several rose-coloured spots there was a distinct subcutaneous mottling. The diarrhœa still persisted. Four days later the subcutaneous mottling had become developed into a well-marked typhus rash. The patient recovered.

Illustration LI.—A female, aged twenty-seven, was admitted on the third day of an attack of typhus. The rash began to fade about the fifteenth day, but there was no abatement of the general febrile symptoms. On the eighteenth day there was watery diarrhœa, tympanitic abdomen, and several rose spots. The latter symptoms continued for about a fortnight, after which the patient gradually recovered. It not unfrequently happens that patients are exposed to the poisons of both typhus and pythogenic fever, before their admission into an hospital. In my researches, elsewhere published, I have endeavoured to show that the poison of pythogenic fever is generated in the emanations from decaying animal matter, and that of typhus, by the respiration of an atmosphere charged with the exhalations of living bodies, although in the majority of cases the latter disease is propagated by contagion.

Now, if a certain poison can generate one group of symptoms, and another poison can generate another, surely it is but reasonable to expect that a combination of the two poisons may give rise to a morbid condition of an intermediate character, without its being necessary to conclude from the existence of such a hybrid affection that the first two morbid conditions have been merely different manifestations of the same poison.[1]

The following observations of Dr. Wilks, referred to at page 73, are highly important as bearing on the question of the value of diarrhœa as a symptom in typhus:—

"The condition of the bowels forming one of the characteristic distinctions, and, therefore, diarrhœa being an important fact to notice, as its existence or absence might be made the means of distinguishing or confounding two cases. It is important also to know the value of this symptom, or

---

[1] "British and Foreign Med.-Chir. Review," vol. xxiv.

what evidence it carries with it of an organic disease of the intestine. The occurrence of a fluid motion may, on the one hand, be denied, as constituting a morbid condition of the bowels; while, on the other hand, it may be denominated a diarrhœa, and considered evidence of organic change.

"It would be advisable, therefore, to have some experience on this point, before commencing the investigation of the present subject; and, in looking over my fever cases, reported many years ago, and before the commencement of the present series, I find I was prepared to meet a difficulty of this kind, having learned that the mere evacuation of the bowel would not decide the question of a morbid condition of the intestine, as the two following cases will show. A woman, aged 45, who was covered with a mulberry rash, and whose bowels had been generally confined, died on the 14th day of fever. She had not had any evacuation for two days before death. The necropsy showed the intestines perfectly healthy, but both small and large were filled with liquid fecal matter. My note made at the time, says, that if this patient had lived a few hours longer, she would, in all probability, have had a large liquid evacuation, and then if no subsequent post-mortem examination had taken place, there might have been strong suspicion that a diseased ileum existed. It was learned, then, that as a rule, the contents of the bowels in fever were fluid; but not always so, as another case will show. It was that of a young girl, who had been ill three weeks, and whose bowels were said to have been irregular, but generally confined, and who came to the hospital to die. The post-mortem inspection showed the intestines full of firm scybala, and on removing these, under each was found an ulcer. These cases have been before me in the present investigation, and although knowing as a rule, that regular bowels would indicate a healthy state of intestine, and a diarrhœa a morbid one, yet I was prepared to know that in fever, as in phlebitis, and other septic diseases, a fluid motion was common, and did not necessarily indicate a diseased intestine, and on the other hand, that this organ may be ulcerated without producing diarrhœa. It had also been learned that a large fluid evacuation at the end of a typhus case was not at all unusual."—*Guy's Hospital Reports*, New Series, vol. i. p. 318.

It would appear from the observations of Dr. Grimshaw and Dr. Perry, that the typical range of temperature in typhus is lower in Dublin and in Glasgow than in London. This may be best stated in Dr. Grimshaw's own words:—

"As to the typical range of temperature of the body in typhus, I am very sorry to have to differ from Dr. Aitken. I consider the range of temperature as given by that gentleman in his excellent work on the 'Practice of Physic,' as quite too high for a typical range. I am glad to say that I am confirmed in this view by Dr. Perry, in a very valuable and interesting paper read before the Glasgow Medical Society, on December 19th, 1865. By selecting twenty-five (of what appeared to me to be typical cases of typhus) from a large number of cases, and taking an average for each day, I have arrived at what I conceive to be a typical morning range of temperature for typhus fever.

"I do not at all wish to impeach the accuracy of Dr. Aitken's observations, which appear to have been made with admirable care. I think we must look to the difference in condition or modes of treatment of the patient upon which the observations were made, in order to find out the cause of this discrepancy of results. That there is some great difference in the gravity which typhus assumes in different localities is, I think, shown by the rate of mortality in different hospitals; thus 20.89 per cent.(Murchison), 14.41 per cent. (Perry), and 8.75 per cent. are the rates of mortality in typhus cases for the London Fever Hospital, Glasgow Royal Infirmary Fever-house, and Cork-Street Fever Hospital, respectively. The great difference between 20.89 per cent. and 8.75 per cent., can scarcely be attributed to difference in treatment. Perhaps the difference in the range of temperature may be accounted for by the difference in diet between the English on one hand, and the Scotch and Irish on the other, the former being much higher fed than the latter, and having, therefore, more tissue to be burned up, as it were, during a fever, and consequently a higher temperature produced. I merely throw out this as a suggestion, not having any proof of the correctness of the theory. I am afraid the temperature of the body has been relied on too much by some as a symptom of more value, than various others, which, up to the present, we have been accustomed to rely upon as our great guides. Believing thermometric observations to be of considerable value in aiding the formation of an opinion in doubtful cases, and of deep interest from a scientific point of view, I regret that these observations should fall into disrepute, owing to their being cried up by those who 'ride their hobbies to death.' It has been the fate of many valuable instruments to fall into bad repute owing to their deceiving those who relied upon them, to the exclusion of older and more thoroughly tested means of diagnosis or treatment. It is to be hoped that such will not be the case with the thermometer."—*Thermometrical Observations in Fever*, 1866.

I think it highly probable that Dr. Grimshaw's explanation of the difference is the true one.

In alluding to the thermometrical observations in relapsing fever, I omitted to mention those of Dr. Cheyne, which were made in 250 cases during the epidemic period of 1817-18. I have carefully compared these with Dr. George Kennedy's observations in 325 cases of typhus in 1837-8, and I find that while the proportion in which the temperature ranged from 100° to 105°, was slightly larger in the typhus cases, that of those exceeding 102° was decidedly greater in the relapsing fever, in 15 cases of which the temperature reached 106° or 107°.

The following case, by Dr. H. Kennedy, presents a close resemblance to that of the boy Morrit, alluded to in the text, page 77 :—

"Develin, a young man of 17, admitted into hospital during the present month, April, 1866. He had fever, but not of a severe kind, marked by the usual symptoms, and the tongue red and furred. When he was now six days ill, the spots of enteric fever appeared on chest and abdomen, and in an unusually well-marked form. On the second day of their appearance, this patient was seen by the Drs. Martin, from Berlin, who happened to be visiting the hospital. On the third day, however, the number of spots had greatly increased, and become more those of typhus, and finally the case,

beginning with she spots of enteric fever, became one of regular typhus. It was observed that, as the typhus rash declined, the typhoid spots became again quite visible, and at this period a slight attack of diarrhœa occurred."
—*Essay on Mixed Types of Fever*, page 6.

Thomas Morrit, ætat. 17, from Grand Canal Street school, admitted into Meath Hospital, with typhoid fever, February 12th, 1866. This fever was of a mild type, with few rose-coloured spots, and he was discharged apparently convalescent on the 9th of March. He was again admitted on the 17th of March, with well-marked typhus, with profuse measly eruption. On the thirteenth day of this fever the maculæ disappeared and were replaced by a profuse crop of sudamina, followed by a fresh attack of typhoid fever, with more numerous spots than in the primary attack. This terminated without any complication, and he was discharged quite convalescent on April 23d, 1866.

No. 2.—Kate M'Loughlin, ætat. 14, from Luke Street school, admitted into hospital Feb. 1, 1866, with febricula; discharged convalescent Feb. 13. Readmitted with typhoid Feb. 23d, and while in the convalescent ward, after this fever, contracted typhus, apparently from the convalescent typhus patients in that ward. She was discharged quite well on May 5th, having been seventy-one days in hospital.

No. 3.—John Hollywood, ætat. 9, from St. Peter's school, Camden-row, admitted into hospital January 28th, 1866, with typhoid fever, and discharged convalescent on Feb. 13th. He was readmitted with typhus on Feb. 27th, and discharged quite well March 27th, 1866.

IV.

According to Rokitanksy, the local hyperæmiæ of typhus—including the hypostasis in the lung—are due to the paralyzing influence of the blood upon determinate ranges of the nervous system either at the periphery or at the centres. This pathologist ascribes the occasional occurrence of tubercle and of pyæmia at the later period of fever in the one case to a conversion of the typhus crasis into the tuberculous; in the other into what he terms deuteropathic pyæmia. The following cases will give the student some idea of the changes referred to.

The first is detailed by Dr. Blakiston:[1]—

A policeman, æt. 25, had been losing his strength and appetite for three or four weeks, and had experienced "a dull, sickening kind of headache." Four days since he had been rather suddenly seized with rigors, and on going to bed and covering himself with clothes he became burning hot, and at length perspired, and was relieved. The next day he felt extremely exhausted, had much nausea and headache, and loathed food; his skin became hot and burning, and his breathing oppressed and laborious. When examined the day after his admission into the hospital, he complained of a "stupid headache" and "confusion," a nauseous taste in his mouth, and difficulty of breathing amounting at times to a sense of suffocation. His eyes were watery and congested, his face was rather swollen and livid, more particularly his lips.

---

[1] On Diseases of the Chest, p. 273.

The tip and edges of his tongue were of a bright red colour, and the middle was covered with a thick yellow moist fur. His skin was dry and hot; the bowels were confined; the urine was scanty and high-coloured, without containing albumen. The pulse was small, but rather hard, 120. Respiration short, forty in the minute. He had no cough or expectoration.

The chest sounded well on percussion; small cooing sounds were heard, more especially over the back, and now and then there was a slight trace of muco-crepitant rattle.

In three days' time he had sunk into a typhoid state, with a dark, dry tongue and great lividity of lips; a slight cough without expectoration appeared: the dyspnœa became urgent, more muco-crepitant rattle was heard, and some dulness was perceived on percussion at the lower and posterior part of both sides. He died on the sixth day of his admission, in spite of every effort to support him.

INSPECTION.—The mucous membrane of the stomach and of the lower part of the ileum was slightly congested in small patches. Both lungs were heavier and redder than usual. They gave out some little colourless serum on incision. They were studded all over with very minute granulations, none of which reached the size of a pin's head, and most of which were not a fourth of this size. They were grayish white, and opaque. A number of incisions showed that they were distinct from each other. They felt hard to the touch, but could be rubbed down between the fingers, although with some difficulty.

On this case Dr. Blakiston remarks that, while the appearance of the lung resembled ordinary pneumonic hepatization, it still differed materially from it. The lungs were pervious to air, but more friable than in such cases, whilst the granulations were much larger and less thickly clustered than in any case of pneumonia, whether sthenic or typhoid. *The symptoms indicated the march of simple typhoid fever, accompanied by a degree of obstruction at the lungs, not sufficiently accounted for by any physical sign.*

The second case to which I refer is one which occurred under my own observation during the epidemic of 1848-9:—

The patient, a fair-haired lad of 18, was admitted into hospital on the fourth day of fever. He had a sweating crisis on the eighth, and convalesced rapidly. Eight days afterwards he relapsed, and for the first time since his admission he had cough, but no expectoration. No physical signs whatever could be discovered. On the third day after relapse he had copious epistaxis, and it was discovered that the spleen was considerably enlarged; the lower posterior portion of the left side was dull on percussion.

On the fourth and fifth days the epistaxis recurred, and the dulness had extended as high as the scapula. Inspiratory murmur was feeble and interrupted on the left side, and both inspiration and expiration were accompanied with a sibilant râle. It was also remarked that the heart's sounds were heard higher than their natural situation.

He died on the morning of the seventh day of relapse.

On examination, the heart was found displaced fully two inches by an enlarged spleen. This latter organ weighed two pounds fourteen ounces. It was of a dark purple colour, coated with patches of a whitish substance, and containing internally numerous masses of the same, of various sizes. The entire left lung was of a dark violet colour externally, the upper portion crepitant, the whole lower lobe hepatized, red, and friable; part of the upper lobe, posteriorly, congested. On washing away the contained blood, the lobe became of a grayish colour, and its section showed numerous aggregated minute granules, each scarcely larger than a pin's point.

Peyer's glands were enlarged, tumid, and, in some instances, ulcerated. The mesenteric glands were congested, and contained a whitish deposit. My friend Dr. Aldridge submitted a portion of each of the viscera to microscopical examination, and informed me that in none was there to be found any product of inflammation, but that the deposit in all the organs was of a tuberculous nature.

I would direct the student's attention to some valuable observations on the natural relations between fever, pneumonia, and tubercle, by Dr. F. Zehetmayer, whose excellent memoir, translated by Dr. F. Battersby, is published "in extenso" in the 27th volume of the *Dublin Medical Journal*—first series.

The subjoined cases and observations by Dr. Stokes will amply illustrate the occasional epidemic tendency to conversion of the typhus crasis (to use the phraseology of Rokitansky) into pyæmia, and I may also refer to two interesting examples, published by Dr. Graves in the 6th volume of the *Dublin Medical Journal*, first series: of the supervention of a pyogenic condition upon fever.

Dr. Stokes exhibited a specimen of purulent infiltration of the lung, which was of interest, as being illustrative of the prevailing epidemic. It was taken from the body of a man who died in the Meath Hospital of fever during the past week.

When the patient first came under notice, the exact day of his fever could not be determined, but it appeared he had been more or less ill for a fortnight. The case was marked throughout by extreme prostration, although there was no loss of power in the action of the heart. It gradually appeared that the posterior portions of the lung were becoming dull, but the dulness was never complete, and the stethoscope did not indicate impermeability of the pulmonary tissue. The patient now began to have frequent sweats, diarrhœa, and delirium; and towards the close of the case the dulness extended, but the crepitating râle never disappeared: the pulse rose to 140, but the heart's action continued vigorous. On the twentieth day a profuse eruption of petechial spots came over the back. On dissection it was found there was nothing remarkable in the intestinal canal, beyond a slight enlargement of volume in Peyer's glands; but this circumstance is of interest, as showing that we have been, perhaps, too much in the habit of regarding what is termed follicular enteritis as an invariable concomitant of the diarrhœa of typhus fever. The spleen was much enlarged; the heart was free from disease; but the lungs, particularly the left, were in a state of sero-purulent infiltration.

Dr. Stokes observed, that these cases of purulent infiltration of the lung were examples of a morbid state clearly different from that of Laennec's third stage of pneumonia. Its antecedents; the mode of succession of its physical signs, and even its anatomical characters; pointed it out as one of that class of local disease developed under the influence of a general pyogenic state. With respect to the order of succession of the physical signs, he remarked on the want of the signs of consolidation which in ordinary pneumonia occur midway between those of the inflammatory congestion and those of purulent infiltration. In connection with the fever of this country, there had been a strong epidemic tendency to this affection lately developed. In some cases, as in that which he presented, the alteration had occurred during the existence of the fever, and while maculæ existed on the surface. In other cases, pyogenic tendency was not exhibited until after a short convalescence from the primary fever. Of this, a remarkable case had lately occurred in the Meath Hospital. The patient had gone through a severe typhus fever, and was held to be convalescent. In a very few days, an œdematous swelling attended with a slight blush on the surface appeared along the line of the lower jaw. This was followed by another swelling in the right occipital region.

The pulse became rapid; the breathing laboured, and the posterior portion of the left side showed comparative dulness on percussion and a large muco-crepitating râle. Death soon took place. Depots of purulent matter were found in the neck, and the posterior mediastinum and the left lung were found in a state of semi-solidity, and infiltrated with pus, there was no red hepatization, nor did the incised surfaces exhibit the peculiar granular aspect of the third stage of pneumonia.

In a third case at present in hospital, the patient after having recovered from a very low typhus, was attacked with pain in the right elbow joint, and within twenty-four hours, the parts presented all the appearances so well described by the late Dr. E. M'Dowell in his papers on periostitis and synovitis, in which he dwells on the characters of this purulent disease as affecting the joints, and subsequently the internal organs.—*Proceedings of the Pathological Society*, January 27, 1855.

In the following case (referred to at page 96) presented to the Pathological Society by Dr. Gordon, the coagulum was evidently formed several days before death. The case by Sir D. Corrigan, which I also subjoin, was probably connected with the action of a saline purgative.[1]

Dr. Gordon detailed the history of a case of typhus fever, in which a firm coagulum was formed in the pulmonary artery, causing emphysema of the lungs. Hugh Reilly, a dairyman, aged 20, was admitted into the Hardwicke Fever Hospital on 1st November. He was at that time fourteen days ill; from the history which he gave of his illness, and the symptoms which then existed, he was evidently labouring under an attack of intermittent fever, although I could not trace the origin of the disease to any peculiar malaria. The fever assumed the rather unusual form of the quartan ague; there were two clear and distinct days on which there was no paroxysm; but when they did occur, they were unusually severe, and were always accompanied by very sensible splenic enlargement. He was treated at first, after mild but efficient purgation with sulphate of quina in moderate doses. This plan not succeeding, a large additional dose was administered a short time before the febrile accession was expected, and thus the disease was arrested—and that suddenly: it did not yield by the paroxysms becoming gradually less severe, or the intervals becoming more distant.

---

[1] For a complete history of the formation of fibrinous coagula in the heart, I would refer the student to Dr. Richardson's Lectures in the *British Medical Journal* for 1860.

The paroxysms came on at noon; he had a severe attack at this hour on 11th November; he had another equally severe on 14th November, and this was the last. Although the attack of ague terminated, the patient did not appear to convalesce satisfactorily, and he gradually assumed the aspect of one labouring under typhus fever. The entire system became affected. The cerebro-spinal system was engaged, for the patient suffered from pain in his head and back, weariness of his limbs, soreness of the entire body, restlessness and inability to remain longer in any position; his intellect became dull, and he raved occasionally at night. The respiratory system was engaged, for he complained sometimes of cough and weight on his chest, and the existence of sonorous and sibilous râles, indicated that there was at least some congestion of the bronchial mucous membrane. The gastro-intestinal mucous membrane was engaged, for his tongue was hard, and his lips and mouth were gradually becoming parched and dry, he had lost all appetite for food, and even for drink, and the bowels ceased to act spontaneously. The principal evidences of the circulating system being engaged, consisted in marked acceleration of the pulse and its great debility; while the morbid pallor and coldness of the surface showed that the capillary circulation was not in a healthy condition. I could not assign any preponderance to any collection of symptoms, or say that any particular system was more deeply engaged than the remainder. The patient was not maculated, nor had he an eruption of any kind, at any period of his fever; neither were there present any of the other evidences of increased fluidity of the blood. The sounds of the heart were natural and perfectly distinct; the urine was scanty.

The symptoms of typhus fever continued to increase and accumulate; and on the 28th, there was superadded a loud bruit de soufflet towards the base of the heart, but not exactly in the usual situations where aortic murmurs are heard; it was confined to this situation, and did not pass into the large vessels. There was in this case no previous history of cardiac disease, nor any symptom of it during his stay in hospital. I had little hesitation therefore, in designating this murmur in the nomenclature of Dr. Stokes, as "the typhoid anæmic murmur," and I carried out the indication it afforded, by ordering for the patient an additional quantity of wine and diffusible stimulants, and extensive and repeated counter-irritation over the præcordial region.

The abnormal murmur persisted on the following day, with symptoms of greatly-increased prostration and debility. The physical signs of bronchial congestion became less evident, while the patient complained of increased dyspnœa and pain, which he referred to the epigastrium. This apparent anomaly was accounted for by the sudden supervention of extensive emphysema of both lungs, evidenced by the usual physical signs of increased resonance on percussion, and diminished intensity of respiratory murmur. He died rather suddenly, on the morning of the 30th.

The post-mortem examination showed excessive vascularity of the meninges and substance of the brain, with slight superficial softening, chiefly of its inferior surface. The lungs exhibited a perfect example of intense vesicular emphysema; the muscular structure of the heart was healthy, and the valvular apparatus perfect. The left side of the heart was almost empty: the right ventricle contained a well-formed fibrinous coagulum; it was exceedingly firm and strong; it passed into the pulmonary artery and through several of its ramifications. It was formed of concentric layers, was partially adherent to the sides of the vessels through which it passed, and terminated not very abruptly. The liver was slightly congested; the spleen was about the natural size—when incised it seemed of normal consistence; the capsule, which was slightly opaque, had a shrivelled appearance, particularly along the circumference, where was a fringe-like border, as if the parenchyma had receded somewhat from the edge of the capsule. The mucous membrane of the intestines was healthy, excepting some spots of very decided vascularity, corresponding to the situation of Peyer's glands in the vicinity of the ileo-cæcal valve.—*Proceedings of the Pathological Society.*

"Anne Rock, ætat. 68, was admitted into Hardwicke Hospital, upon the 20th November, 1851, four days ill. She had previously taken salts, which purged her violently. On admission, her countenance looked dark and sunken; the skin was dry and hot, and spotted with dark-coloured maculæ; tongue dry and brown; pulse, 108; very weak and intermittent. She lay on her back, making no complaint; and when questioned, answered intelligently. She had little thirst, and swallowed without difficulty; respiration was rapid and laborious, and bronchial râles were audible over the whole chest. On the sixth day of her illness, the maculæ were still more extensive and dark coloured, and the pulse was 116, weak and intermittent; the respiration 40 in the minute. She lay as before, and was quite collected. On the seventh day the pulse was 120; the respiration 48 in the minute; hands, cold; tongue, dry and brown, covered with sordes. She began to mutter to herself, but was sensible when roused.

"On the eighth day the pulse was 120; very weak; the hands and arms were cold, and dark purple spots of congestion appeared on her elbows, knees, and ankles, the maculæ enlarged in size, and were dark and ill-defined. She continued to sink, and died at eight o'clock next evening— ninth day.

"*Post-mortem.*—The surface of the body was of a dusky yellow colour; but where a blister had been applied it was of a dark purple. The lungs were greatly congested posteriorly, and there was effusion into the bronchial tubes. Upon cutting into the right ventricle of the heart, it appeared to be distended with a quantity of black clotted blood. Upon turning this aside, a yellow fibrinous deposit was found firmly attached on all sides in the interstices of the inner walls of the ventricle, and extending from the ventricle continuously into the pulmonary artery, and its branches to the third and fourth subdivisions. This fibrinous deposit was tough, elastic, and consistent, not more than half filling the caliber of the tubes through which it lay. Where this fibrinous formation, which was rather globular in the ventricle, passed through the neck of the pulmonary artery, it suddenly became constricted, and again swelled out in size in the trunk of the artery beyond the valves. The left cavities of the heart was full of black blood, but without any such fibrinous deposit as that on the right side. The other viscera were sound.

"In this case, it is obvious there were two stages of coagulation—the one, that forming the dark-coloured clotted mass of blood which occurred immediately at or after death, and which presented nothing remarkable; but the other coagulation, I think it is obvious, existed for some time; how long, it is difficult to say, but, probably, for one or two days before death. The extreme toughened consistence, and freedom from colouring matter of the fibrinous deposit, would render this probable; but this probability is greatly strengthened by the neck-like form of the fibrin where it crossed the valves into the neck of the artery, and which can only be accounted for by admitting, that while the polypus was in process of formation, the valves were in action, and for a sufficient length of time to produce the constriction described. Supposing this to have been the state, we can understand what a formidable aggravation such a formation must have been, and how the case must have sunk from the lesions of circulation with little comparative disturbance of any other function."—*Lectures on Fever*, page 67.

"I have observed," says Dr. J. Reid (referred to at page 99), "in several experiments on the lower animals, that disgorging the right side of the heart, when its contractions are enfeebled or suspended, by opening the external jugular, has, in some cases, a decided effect in renewing its action, and this, I am convinced, may be of considerable practical advantage in promoting the return of the circulation under certain circumstances.

"Exp. 1.—While assisting my friend Dr. Cormack in some experiments upon the physiological effects of creasote, we found on opening the thorax of a dog, immediately after it had ceased to breathe, into whose femoral vein twenty-five drops of creasote had been injected, that the heart was perfectly quiescent, and remained so even when pricked and cut superficially with a scalpel. As the right side of the heart seemed much engorged, a small opening was made into the auricle, and part of the blood allowed to escape. As soon as the blood began to flow the heart immediately resumed its contractions, and continued to act vigorously and spontaneously between two and three minutes, and only ceased after five minutes.

"Struck with the effect which disgorging the right side of the heart had in renewing its contractions under circumstances where no external stimulus was of any avail, I was anxious to ascertain if the same results would follow the unloading of the heart when arrested from other causes.

"Three dogs were killed by hanging, and as soon as they had ceased to breathe the thorax was laid open. In all of them the heart was acting pretty vigorously, particularly in one only a few months old. When the contractions had become feeble the external jugular vein was opened. This was followed by a decided but temporary increase in the contractions of the heart in two of them, which were large and full grown. The opening of the jugular vein was found rapidly to empty the right side of the heart.— '*On the effects of venesection in renewing and increasing the heart's action under certain circumstances.*'—*Physiological and Pathological Researches*, page 51.

"I have not unfrequently observed in the human species, when the respiration had been impeded for some time before death, a condition of the lungs similar to that observed in the lower animals after division of the *vagi*. In several of the fever patients I have had lately occasion to inspect at the Royal Infirmary, the lungs in the posterior and middle parts were dark coloured and gorged with blood and serum. When cut into, the substance of the lungs, in some cases, appeared at different parts denser than what could be accounted for by mere congestion of the bloodvessels, and a comparatively small quantity of blood could be squeezed from the cut surfaces, though the blood in other parts of the body was fluid; but sections of the lung generally, though not always, floated in water, and presented none of the granular appearance. In some cases the bronchial tubes contained a considerable quantity of frothy serum, in others very little. In all probability these morbid appearances are occasionally dependent on the disturbed respiration consequent upon derangement of the central organs of the nervous system. I lately saw a gentleman labouring under fever whose respirations for a short time were only eight in the minute, though the lungs at the time were unaffected, but they fortunately soon rose to sixteen in the minute, and he ultimately recovered. Dr. Allison has suggested to me that these morbid changes in the lungs are sometimes owing to another cause. He believes that in cases of fever when the heart's action is feeble, the bronchiæ often somewhat obstructed, and the blood altered, the right side of the heart is unable to propel the blood through the lungs, it consequently goes on accumulating in their depending parts, and the same results follow as when the respiratory movements are diminished in frequency. In confirmation of this view he states that he has seen these morbid changes occur in the lungs without any preceding diminution of the respiration.

"When we remember that the pulmonic circulation is dependent upon two distinct causes varying in efficacy—the right side of the heart and the chemical changes going on at the lungs—both of which are necessary for the proper propulsion of the blood through the lungs to the left side of the heart, we can easily understand how a diminution in the activity of the respiratory muscular movements, and the impaired contractility of the right side of the heart, should produce the same effect—viz., congestion of blood in the lungs and effusion of frothy serum."—Dr. J. Reid, *Physiological and Pathological Researches*, p. 205.

Many explanations have been offered of the phenomenon alluded to at page 90, none of which can be said to be perfectly satisfactory, inasmuch as it seems to occur in the most opposite conditions. First observed by Dr. Graves, in cases in which the presence of air in the pleural cavity was indicated, by other signs he regarded it as diagnostic of such a condition. My own observations and dissections proved that in some cases, at least, it existed, independently of any such effusion of air, as I found the pleuræ closely adherent in subjects in which the tympanitic clearness was observed immediately before death.[1]

Several years elapsed before my statement of the fact was believed; but the frequency of its recurrence led gradually to its general recognition.

Of those who have observed it, and offered explanations of it, Dr. Hayden deserves especially to be referred to. In his memoir in the *Dublin Quarterly Journal*, vol. 41, Dr. Hayden reviews the theories advanced by writers since the time of Dr. Graves, and adds some important observations of his own.

Having detailed a case in which a tympanitic form of resonance was present, over the seat of pneumonia affecting a portion of the lung, Dr. Hayden observes :—

"The inflammation, as is usual in typhoid pneumonia, invaded first the apex of the lung, and then slowly travelled downwards, involving, however,

---

[1] *Dublin Medical Journal*, vols. 7 and 11—first series.

only the superior lobe, and at no period extending to the posterior and inferior portion, or inferior lobe of the lung. Are we to seek, in this latter circumstance, an explanation of the tympanitic resonance, or rather that remarkable modification of dulness which might be expressed by the designation of 're-sonant dulness,' which became developed on the 3d of June, *i.e.*, thirteen days after the commencement of the attack? According to Skoda, the answer to this question should be in the affirmative: as he holds that the presence of a substratum of healthy lung, freely permeated by air, is capable of communicating a modified resonance, of a somewhat muffled character, to a solidified portion interposed between it and the chest-wall, subjected to percussion. On theoretic grounds this explanation might be deemed satisfactory. There can be no doubt that the presence of a stratum of healthy lung tissue, freely permeated by air, subjacent to a solidified and non-resonant portion, does modify, in a marked degree, the dulness of the latter, the percussion note yielded by which is of a compound character, representing in the proportions in which they are relatively present, the two elements of which it is composed. It will appear, however, on reference to Case No. II., that the phenomenon of modified tympanitic resonance in a solidified lung, cannot be attributed to this cause exclusively; in that case, although the entire lung was hepatized, the phenomenon was developed even in a higher degree than in the case now under consideration; whilst in Case No. III., all the conditions required by Skoda's doctrine for the production of resonant dulness, namely—a solidified lobe, overlying one in a healthy condition, were supplied, yet the phenomenon did not exist.

" From those cases, therefore, the conclusion to be arrived at would seem to be, that although in a partially hepatized lung, in which the solidified portion has a substratum of healthy pulmonary tissue, the percussion dulness of the former may be modified by the resonance of the latter, in such a manner as to give rise to a hybrid sound of a very remarkable character, this phenomenon may be present where the entire lung has become solidified: and again, that it may be absent where a healthy and a solidified portion of lung-tissue occupy the relative positions mentioned."

Dr. Hayden's second case presented marked tympanitic resonance over the front of the right side, with *prominence* of the side as compared with the left. The side being opened—"post-mortem"—under water no air escaped, and the lung and costal pleura were found firmly adherent. On puncturing the lung, however, bubbles of air freely escaped. "The right lung," says Dr. Hayden, " was fully distended, so as to keep that side of the chest in a state of maximum expansion. Its colour was light gray, and on section it was found solid throughout: placed in water, the lung sunk at once." On this case Dr. Hayden observes :—

" I have thought it right to give from my notes, taken at the bedside, a detailed account of the progress of this case from day to day, and likewise of the post-mortem appearances; because of the interest that attaches to it in connection with the phenomenon of tympanitic resonance in a solidified lung. In case No. I., this phenomenon likewise existed, but in a less degree, in a lung only partially solidified. In that case the hypothesis was at least warrantable, that the resonance yielded by the solidified portion was communicated from that which was in a healthy condition. In the case just narrated, however, no ground existed for such an assumption, as the lung was hepatized throughout its entire extent. Neither was it admissible to suppose that the phenomenon was due to communicated gastric resonance; for, independently of the fact that it occurred on the right side, it will be remembered that it was manifested only in the upper portion of the lung, where it could not have existed, if transmitted from the stomach, without also manifesting itself in the inferior lobe. Lastly, percussion of the lung after removal from the body afforded convincing evidence that the resonance was intrinsic. A portion of it placed upon the hand, a solid body, and percussed in the usual manner yielded the characteristic metallic note resonance of a character similar to that heard during life over the corresponding portion of the chest. The occurrence of aëriform accumulations, independently of rupture or perforation of the lung in connection with acute disease of the organs of respiration, is of great interest as introducing an element of derangement into the rules of diagnosis of intra-thoracic disease. These accumulations may be subdivided into two forms, both of which have been fully established, namely, those in the cavity of the pleura, constituting the 'simple pneumothorax,' of Graves, and the 'secondary' form of pneumothorax without perforation of Jaccoud, and those in the substance of the lung itself. In the two first cases given by Dr. Graves, there can be no reasonable doubt that there had been aëriform effusion into the cavity of the pleura. Of this, the displacement of the heart without liquid accumulation in the former case, and the compression of the lung without liquid distension of the pleura in the latter afford convincing evidence.

"The presence of a cavity in the apex of the lung in the latter case in the absence of positive proof that no communication existed between this and the pleura, may be taken as invalidating, in some degree, the evidence it affords in support of the doctrine of simple pneumothorax by secretion.

"Of this form of gaseous secretion into the pleura, another indubitable example has been admirably reported by Dr. Little of Sligo. A case is mentioned by Dr. Stokes, of typhoid pneumonia in a female, in which on the eighth day of illness, the antero-superior portion of the left side, previously dull, yielded a sonorous tympanitic sound on percussion. On the following day, resonance had extended to the postero-superior portion of the chest; but on the next succeeding it had disappeared, and been replaced by dulness.

"In this case, which is given as one of pneumothorax by Dr. Stokes, the character of the resonance described by him as being 'similar to what is produced by the stomach in the highest degree of flatulent distension,' was different from that observed in my cases, in which it was somewhat muffled.

"That there is, however, another cause than effusion of air into the cavity of the pleura, whether by secretion or otherwise, to which resonance over a solidified lung may be due, the cases reported by Dr. Hudson, in his valuable memoir on typhoid pneumonia, afford strong evidence.

"The cases now reported go to corroborate the evidence furnished by Dr. Hudson; and the second case carries that evidence a step further, by demonstrating negatively, that no air existed in the cavity of the pleura, and positively that air did exist in the tissue of the lung, in that portion of the organ which yielded resonance both before and after death.

"In Dr. Hudson's second case, as in mine, there was at first sharp pain, dulness on percussion, and absence of respiration, and of crepitus over the affected portion of the lung; hence, in both cases the erroneous diagnosis of pleuritis with effusion. Subsequently, the occurrence of bronchial respiration and of crepitus, showed the error committed, and pointed out the real nature of the disease. Morbid clearness was developed in the previously dull portion of the lung, in Dr. Hudson's second case on the fourth day of the illness; in my first case, only on the fourteenth day, and in the second, on the sixth day. Of these three cases, it continued up to death in the two which were fatal; but in that which terminated in recovery, resonance was replaced by dulness on the sixth day from its commencement. In all three cases bronchial respiration and crepitus coincided with tympanitic dulness.

"It cannot be pretended that resonance was due to a central pneumonia in these cases; firstly, because of the pre-existence by several days' dulness, in the resonant portion of the lung; and secondly, because post-mortem examination in two of them, in each of which resonance continued up to death, showed that hepatization was universal. In Dr. Hudson's seventh case, bronchial respiration and crepitus likewise coincided with resonance, which continued up to death. Thus it would appear, that morbid resonance developed in the progress of pneumonia may be associated with bronchial respiration and crepitus, as in the cases above cited; or with total absence of all respiratory sound, as shown by Dr. Graves; and the evidence before us would seem to warrant the conclusion, that upon this distinction may be based the differential diagnosis between resonance due to air included in the tissue of a hepatized lung, and that depending upon aëriform accumulation in the pleura. Grisolle does not mention percussion resonance at all in connection with pneumonia; he only speaks of comparative dulness, and occasional absence of dulness, in the pneumonia of children. Dr. Stokes says: 'But of all these signs, the most remarkable is tympanitic clearness over the diseased lung, a phenomenon evidently proceeding from an effusion of air by secretion into the serous cavity.' Thus it would appear that at the date of his great work (1837), this eminent physician, than whom nobody in our day has contributed more to the physical diagnosis of thoracic disease, was still unaware, or at least unconvinced of the existence of the second form of pneumonic resonance, namely, that not depending upon pneumothorax.

"Dr. Williams includes amongst the essential and characteristic physical signs of pneumonia, 'dulness of sound on percussion,' to the presence of which, in the advanced stages of the disease, he makes absolutely no exception, and nowhere mentions tympanitic resonance as of occasional occurrence.

"Dr. Fuller, in his work, *Diseases of the Chest* (1862), does not allude to tympanitic dulness in pneumonia; and in the two most recent works on the Practice of Medicine, namely, those of Trousseau and of Hughes Bennett, the subject is not adverted to. As far as I have been able to make out, Dr. Hudson's cases, previously mentioned, were the first recorded examples of tympanitic resonance in pneumonia *not due to the presence of air in the cavity of the pleura;* and that Cases II. and VII. in his category, which so closely resembled my second case in all essential particulars, were of this nature."

I am convinced that of the varying conditions under which the phenomenon is observed one, at least, must always exist—viz., the expansion of the parietes consequent upon tumefaction or displacement of the lung. I add displacement, because the same sound, or one closely resembling it, is to be observed for a short time in cases of pleuritic effusion, and still more frequently, in my experience, in cases of copious liquid effusion into the pericardium. Tympanitic resonance is constantly observed in our wards in cases of pericarditis in young persons—a fact which was first recorded by Dr. Graves. I have also met with several examples of augmentation of the volume of the lung, and consequent expansion of the side, in case of plastic pneumonia, in which the tympanitic resonance on percussion was most remarkable.

This phenomenon is not merely interesting as a scientific curiosity; it is also worthy the attention of the student as an occasional aid to diagnosis. Let one instance suffice to show this.

Some time since I was called to see a case in consultation with a professional friend, which he believed to be one of hepatitis. He had, accordingly, applied a large number of leeches over the right hypochondrium, had given calomel freely, &c. &c. The patient presented the icteroid coloration of the surface described by Hasse and Zehetmayer in an unusually high degree, in fact he was subjaundiced; moreover, as my friend remarked, the right side was even clearer than the left, and no pneumonic crepitus was audible in any part. I, however, pointed out the peculiar tympanitic character of the clearness; the absence of crepitus was accounted for by the fact, that, the entire lung being engaged, there was no expansion of air cells to produce it: and I appealed to the future progress of the case for a confirmation of my diagnosis, predicting that on the commencement of resolution the existing clearness would be replaced by equally marked dulness.

At our meeting on the following day this had occurred, the entire side was dull on percussion, the other signs of pneumonia were present, and my friend acknowledged the value of a sign which previously he had not thought worthy his attention.

V.

In reference to the observations made at page 104, it may be said the analogy of action of putrid substances injected into the veins of animals, and that of the histories of poisoning by putrid ingesta, seem to support the view which would regard the primary diarrhœa of typhoid as an effort at elimination, made by the liver and intestinal glands. Thus by a reference to the experiments of injection of putrid pus, &c., into the veins of animals, performed by Magendie, Gaspard, Cruveilhier, &c., it will be seen that when the animal recovered it was after copious discharges of a vitiated character from the bowels. To these discharges the last-named writer attributes the recovery, and adds, that it is a fundamental fact of pathology that the intestinal canal is chiefly affected in diseases caused by miasmata. These experimenters also explain the fact not unfrequently observed, that of a number of individuals exposed to the same source of miasm, some will suffer an attack of typhoid fever, while others will be affected by diarrhœa or dysentery only; just as in the experiments of Gaspard, the recovery of the animal after putrid injection, was attended with profuse and offensive discharges—seemingly the mode of relieving the blood from the presence of the poison.

Again, if we refer to the published cases of poisoning from putrid ingesta, we see that besides those of irritant poisoning, in which the rapid rejection of the substances was followed by recovery, there is another class in which, after an interval allowing of the absorption of the poison into the circulation, a different set of symptoms followed; as in the following from Dr. Christison's work on poisons: "A family of five persons took for dinner broth made of beef, which, owing to its black colour, the master of the family had previously said to his wife he thought bad and unfit for use. In the course of some hours two boys were attacked with sickness and vomiting, but appear to have got soon well, probably from the early discharge of the poison. Next morning a washer woman, who had dined with the family, was seized with violent pain in the bowels, diarrhœa, racking pains, and weakness in the limbs, and did not recover for ten days. On the evening of the second day the master of the house was similarly affected and was ill for a fortnight. And a day later, his wife was also seized with a similar disorder preceded by soreness of the throat and tongue, and difficulty of swallowing, and ending fatally in fourteen days."

It is worthy of notice that the gravity of these cases was in proportion to the interval allowed for absorption of the poison—and in an inverse proportion to the severity of the primary symptoms of irritant poisoning.

The following observations were made by Dr. Todd upon a case of typhoid which terminated fatally by hemorrhage from the bowels (referred to at page 105). Subjoined is a similar case with the observations of Professor M'Dowell:—

I repeat that were it not for our experience of the constant accompaniment of a state of prostration, with a few ulcers of the small intestine, it would be impossible to believe that so grave an effect would follow such a cause. It is true that in this patient the ulcers were not few, but they were found in but a small portion of the intestine—namely, in a space three feet in length, leaving twenty-seven feet of the highest part of the bowel intact. I have, however, seen a state of as great, if not greater, prostration, where there were not more than four ulcers. What seems most essential to the production of this state of prostration is, that the sloughing and ulcerative process should be quick, and that it should be perforative in its tendency: that is, that it should eat quickly through the tunics of the bowel, as was the case with Gavin, in whom we found that at several points the coats of the bowel had been so eaten through as to leave only a little lymph, and a thin film of peritoneum, as their floors.

It seems to me that the production of this state is due, not so much to imperfect appropriation of food, as to the absorption of a matter from the ulcerated surfaces, which circulating with the blood, exercises a poisonous and depressing influence on the system, a matter of the nature of, if not identical with, pus, which is absorbed by the lacteals, and perhaps, also by the blood-vessels, but probably chiefly by the former, by which route it quickly reaches the lungs, without passing through the liver, where it may contribute to the increase of the bronchial congestion and irritation which so constantly accompany this typhoid typhus. This view I have often broached to you already at the bedside of patients suffering in this way.

I show you here a preparation which was put up for me some time ago by Dr. Lionel Beale. It exhibits a few well-marked deeply perforating ulcers of the ileum, having much the appearance, from the thick, swollen, and red margins, that the process of sloughing and ulceration was a quick one. In this case (the patient was a young woman) the fever ran its course in about three weeks, the diarrhœa was almost done, and the chief symptoms were a tympanitic abdomen, stupor (in fact, coma), bronchial congestion, and extreme prostration. A short time ago you may remember a woman of the name of Locke, who went off very quickly likewise with similar symptoms, the stupor being so great, that I was afraid that a few drops of laudanum, administered with starch to check diarrhœa, had narcotized her. There was in this case, in addition to the stupor, bronchial congestion, and prostration, but the diarrhœa was very slight and readily controllable.

Now, that the absorption of pus is capable of producing these depressing effects we have many proofs.

First in puerperal fever. In some cases the absorption seems to take place rapidly, and in large quantity, and under such circumstances the patient succumbs in a few hours from rapid prostration and pulmonary congestion, with more or less stupor. In other cases the absorption seems more gradual, the typhoid condition it induced more slowly, but very completely, and after a time purulent deposits are found in the joints and muscles, or elsewhere.

Secondly, in cases of erysipelas, in which this suppurative process is rapid, we have typhoid and comatose symptoms, which are out of proportion to the extent of lesion: in such cases doubtless pus finds admission into the circulation.

Thirdly, we sometimes have unequivocal evidence of the absorption of pus, as well as to the source whence it comes, as with respect to the secondary deposits.

I remember attending a case in private practice, where the pus showed itself in the anterior chamber of the eye. This case presented all the symptoms of typhus fever; and for a day or two I viewed it as such. One day I was much surprised at observing pus in the anterior chamber, which increased in quantity very rapidly, and pus was afterwards found in the elbow and shoulder joints. When we came to examine this patient we found an ulcer in the heart, at the base of one of the mitral valves. Some years ago we had a case in the hospital of a woman who was suffering from chronic bronchitis; she suddenly became typhoid, and I looked upon it as a case of most aggravated character. She died in a few days, and we found an abscess in the septum of the heart, which had burst, and thus the pus had entered the very fountain of the circulation, producing symptoms nearly resembling those which come on in a case of low typhoid fever.

There seems then sufficient grounds for explaining the prostration and fatal termination in Galvin's case without ascribing any ill effects to either what had been done for him, or to what had been left undone. The sloughing and ulcerative process undoubtedly interferes to a certain extent with the functions of the bowels, but it also furnishes a source of formation of a poisonous matter, which we know by experience of analogous cases, when taken into the system, creates symptoms of the same character as those of these fatal instances of typhoid fever.—*Clinical Lectures*, by Beale, p. 106.

Dr. M'Dowell exhibited a specimen of acute ulceration of the intestine occurring in typhoid fever. The patient was a remarkably fine looking young man, twenty-six years of age, who was admitted into the Hardwicke Hospital, January 10, 1857. He was a person in comfortable circumstances, well fed and well clothed, so that he was not exposed to those external influences, to which bad types of fever are so often referred—at least, among the poorer classes. We learned that the symptoms of continued fever had existed for twelve days, but so slightly that the patient had not been obliged to keep his bed until three days previously. On admission, and for two days after, the ordinary symptoms of continued fever were present. There was slight cough; the pulse was 112; the skin was warm, but not inordinately so; the tongue was disposed to be dry; the surface was free from eruption; the mind was a little confused, and the patient raved some at night, but slept a good deal. It is particularly to be observed, that neither diarrhœa or abdominal tenderness existed.

On the third day after admission (fifteenth day of fever) a few scattered papules were observed over the abdomen. The eyes were clear and bright, hearing unaffected; no sordes on the teeth. In these particulars, and in the absence of any marked signs of depression, the case was quite unlike one of ordinary typhus. On the evening of the same day, excited delirium, closely resembling that of delirium tremens, set in, which continued, but with diminished violence, during the next day. On the evening of that day, profuse bleeding occurred from the bowels; a large quantity of dark fluid blood was voided at the chair, and subsequently so much came away as to soak through the clothes and bedding to the floor. The following morning the man was greatly sunk, and his voice weak; his face was pale, and his pulse small and rapid; he was quite conscious, and the abdomen free from tenderness. During the night a profuse and sudden hemorrhage again occurred; he gradually sunk, and died on the nineteenth day of fever. On examination of the body, all the characteristic appearances of typhoid ulceration were observed in the cæcum, and lower thirty inches of the small intestines. The incipient condition of that affection was seen in the cæcum, in the form of small grayish elevations, on some of which a minute point of ulceration was visible. In the small intestine were numerous large oval ulcers, with elevated and thickened edges. Almost the entire of the mucous membrane of the lower end of the ileum was destroyed, but higher up the ulcers were confined to that portion of the intestine opposite the attachment of the mesentery. No spots showing incipient disease of the Peyerian glands, were found in the small intestine. At the highest point to which the disease had extended, the ulcers had attained a considerable size. The rapidity of the ulcerative process, seemed to be indicated by this fact. In many places not only the mucous, but even the muscular coat had been destroyed; and at one point the serous membrane had almost been penetrated. The mesenteric glands were considerably enlarged; and in some of them a whitish deposit (not pus) was found. The intestines did not contain any blood, nor were there any small coagula in any of the ulcers to denote the special source of the fatal hemorrhage. The tissue of the heart was firm, and looked healthy; the left ventricle was closely contracted; the lungs presented numerous bright petechial spots on their surface, and a section showed numerous small dark spots of extravasated blood in their parenchyma.

I have somewhat minutely detailed the preceding case, as its importance seemed to deserve more than a mere passing notice. The latency of the abdominal lesion, and the suddenness and profuseness of the hemorrhage, constitute its more remarkable features. The attack of nervous delirium on the fifteenth day, depended probably on the rapid development of the intestinal lesion, whilst, on the other hand, the intensity of the nervous symptoms would have masked any evidence of enteritis which might otherwise have existed. At the same time, a depraved blood, partly, perhaps, the result of absorption of the vitiated fluids of the intestinal ulcers might produce a similar train of symptoms, and of such an altered condition of the blood, the ecchymoses, on and in the lungs, are to be regarded as evidences.

The appearances in the intestines presented the full pathological history of typhous ulceration, the enlargement of a Peyerian gland, and elevation of the mucous membrane, by submucous deposit: ulceration of the summit of the mound thus formed in order to eliminate the morbid material. Lastly, sloughing of the mucous membrane over the whole extent of the diseased gland, and in some places the ulcerative process rapidly destroying even the muscular coats, and by opening the bloodvessels giving rise to one of the most serious accidents of such a lesion.—*Proceedings of the Pathological Society of Dublin.*

The following is the case of acute atrophy of the liver, by Dr. Gordon, referred to at page 109:—

During the past spring, a great number of applicants have been refused admission into the Hardwicke Fever Hospital for want of accommodation, and the fever has been of a very low type; in many instances, it was attended by complications which were also of a more or less depressing nature. I shall first endeavour to describe what these complications were, and then make some remarks on the nature of the fever itself. Of all the complications of fever which we had to treat, jaundice was, perhaps, the most alarming; it sometimes occurred at the commencement of the fever, and proceeded *pari passu* with it, while in other cases the fever was of several days' duration before the discoloration of the skin appeared; from time to time, cases have occurred in the Hardwicke Hospital so similar to those described by Rush and Louis as yellow fever, that there could be no doubt of their being the same disease. Such cases, however, were invariably sporadic, they were characterized by universal jaundice with extensive lividity of the surface, black vomiting, hemorrhages from the different mucous surfaces, and occasionally incipient gangrene of the nose, or of some one or more of the extremities. Within the last three months we had had none exactly of so severe a character, but the following case is sufficiently grave to show the alarming nature of the complication.

James Mahon, aged 18, was admitted into the Hardwicke Hospital, May 22d, with the ordinary symptoms of simple fever; he had been five days ill, complaining principally of rigors, headache, and palpitation; he was a clerk in an office, and had been generally very delicate. There was at first little worthy of observation in the case, for three days his pulse remained at 90, his skin was moderately cool, the bowels acted regularly, his tongue was white; thirst moderate; his only complaint was slight headache. On the 25th he was observed to be jaundiced, the colour was of an orange yellow; he lay on his back, and appeared greatly prostrated; the pulse was 76; he had some dark-coloured bilious vomiting; bowels not moved, urine high coloured and scanty. He was ordered a strong mercurial purgative, followed by the usual black draught.

26th. The bowels have been copiously moved without any amelioration of the symptoms; he appears more stupid, is very unwilling to answer questions, or be disturbed; extremities rather cold; pulse 60, weak; tongue dry and black, no vomiting, but tenderness on pressure at epigastrium, none over hypochondriac regions; abdomen tympanitic; the dejecta were untinged with bile, the surface of the body is more deeply jaundiced; respiration natural. A blister was applied to the back of the head, mercurial ointment was sedulously rubbed in, and the blistered surface was dressed with it, a turpentine enema was administered, and he was ordered a drachm of the oil of turpentine every third hour.

27th. Became greatly excited during the night, screaming frequently, and endeavouring to leave the bed. The tendency to coma has now returned; pulse 66, not so weak; he is perspiring freely; urine very high coloured, but secreted more freely; bowels free; evacuations clay coloured; extremities warm; abdomen less tympanitic.

28th. He is decidedly mercurialized, but does not seem in any respect improved; he is more deeply jaundiced, and the coma is more profound; pulse 48; report in other respects as yesterday. In the evening his respiration suddenly became very rapid and laborious, perspiration became most profuse, extremities cold, and he died in a few hours.

Post-mortem examination.—All the viscera were tinged of a deep yellow colour, the surface of the brain was rather dry, the fluid in the arachnoid sac was rather less than usual, the vessels of the pia mater were minutely injected, both pleuræ were extensively covered with small purpuric spots; the left lung presented a perfect example of true pulmonary apoplexy; the hemorrhage was very extensive, almost the entire of the lung was engaged; the apex and a small portion of the anterior part alone escaping, but only to show more accurately the defined margin of the diseased part. The parenchymatous structure of the lung was completely broken down, and in the words of Andral, the greater portion of the lung was reduced to a soft fluctuating mass, in which there could only be distinguished some *debris* of pulmonary parenchyma, and a quantity of effused blood, partly coagulated and partly fluid. The liver presented a more decidedly yellow colour than any of the other viscera; it was softer than natural, and broke down under the knife; some portions were much darker than others, in those parts all outline of lobules was effaced, the gall bladder was pale and contracted, and contained only about a dessert-spoonful of very thick, black, ropy bile, which, on pressure, flowed into the duodenum; the spleen was natural.

The gastro-intestinal mucous membrane presented throughout, patches of slight vascularity; the colour contained several gray indurated masses, similar to what are frequently found in the faces of jaundiced patients.—Dr. Gordon, *Dublin Hospital Gazette*, vol. i. 1854.

I commend to the careful perusal of the student, the two subjoined extracts which contain admirable descriptions of the morbid process set up in Peyer's glands in typhoid fever, from its commencement to its termination, by perforation of the intestine:—

The appearances which are most marked in the mucous membrane of the intestines, are those of increased action; vascularity sometimes occurring in patches of greater or less extent, without any obvious dependence or inflammation of the mucous glands, and occasionally extending, under some form or other, through the whole tract, from the pylorus to the rectum. But this vascularity is more generally connected with inflammation of the mucous glands, which often appear, like the smallpox on the second or third day of the eruption, elevated and almost transparent, and covered with minute vessels, which dip into them from the lining membrane of the intestines. They scarcely seem to go into a state of true suppuration, but become distended with a yellow cheesy matter, and slough off; or sometimes ulceration takes place upon their points externally, without any collection of yellow matter being perceptible. The same process, or nearly so, takes place both in the solitary and in the congregate glands; except that in the latter the appearance becomes much more formidable, and the mischief more extensive. The masses or clusters of congregate glands are chiefly placed along that part of the intestine which is furthest from the insertion of the mesentery; and when the parts are irritated from disease, three, four, or five considerable branches of vessels are seen passing on the mucous membrane, from the mesentery on each side towards the cluster of congregate glands: these divide and subdivide before they reach the glands, and running in part over the surface of the cluster, till their distribution is lost to the eye, enter apparently into the thickened mass of glandular structure beneath. The glands themselves seem first to enlarge, becoming distinctly visible to the eye, and after some time form a thick, flat mass of a lighter colour than the surrounding intestine; this sometimes increases to the thickness of a half-crown piece, and occasionally even spreads on the top, so that the surface overhangs the base nearly the sixth part of an inch; sometimes a dark coloured matter like grumous blood is deposited amongst the glands, so that the whole mass instead of being lighter than the intestine is of a brown colour, elevated evenly above the surface; but in either case the mucous membrane is at first only raised, and not broken. In a little time fissures are formed with ulceration on this mass, and ulcers, more or less deep, occupy the surface of the whole. Where the irritation is little the ulceration is often mild, and merely superficial; but when anything has occurred to irritate the ulcer it becomes deep and ragged, with an uneven bottom, caused apparently by the projecting remnants of the enlarged glands, or it is filled by a dense slough, stained of a yellow colour by the bile and feces. As the inflammation subsides the depth of the ulcer diminishes; and the greater part of the glandular structure being apparently removed by ulceration and sloughing, the edges fall down, and the ulcer becomes shallow, sometimes leaving the muscular fibres nicely displayed, or often exposing the internal surface of the peritoneum for the space of a quarter or half an inch square. This excavation is filled up by a process of granulation, which may be seen very beautifully, by suspending the intestine cut open before a lamp or bright sunshine, and examining it with a common lens; the granulations are then seen, sometimes arranged in broken lines in the direction of the muscular fibres, at other times arranged in radiated lines around a central point; and when the whole is healed a scar remains visible for some time, not unlike a superficial scar from the smallpox, and generally interspersed with slight elevations of a grayish colour. This scar appears to be covered with a true mucous membrane, the surface being quite continuous with the membrane lining the rest of the canal; indeed, when inspecting the ulcer in the process of healing, we perceive the vessels of the mucous membrane running over the surface to be repaired. The whole process of the ulceration and the healing, is quite analogous to those painful and irritating sores which frequently take place within the lips or on the mucous membrane lining the cheeks, where obstruction in the follicles, enlargement, ulceration, sloughing, and perfect repair, are all most distinctly and easily traced. The space occupied by the ulcers in the intestines is usually about two feet at the lower end of the ileum, and frequently the valve of the colon on the side next to the ileum, is the part where the disease is furthest advanced. A few ulcers are likewise often found in the cæcum, and some are occasionally dispersed along the colon, depending on a process very similar to that which I have described as taking place in the small intestines; but the glandular distribution being in this part more simple, the ulcers usually commence by small rounded elevations, and not in spreading masses.

The peritoneal covering of the intestines at the back of the ulcer is generally discoloured and vascular; but seldom appears actually inflamed, and the distribution of the vessels is somewhat different from that of the vessels which may be seen through the peritoneum on the mucous membrane, and is, perhaps, chiefly derived from vessels belonging to the muscular structure; for, instead of forming numerous branches, they arrange themselves in paralled lines with vessels crossing nearly at right angles. Occasionally, however, the mischief is not confined to the mucous or even the muscular covering, but the peritoneum becomes decidedly inflamed; in which case the symptoms are always greatly aggravated, and the tenderness of the abdomen is much more marked, and after death a sero-purulent effusion is found, and shreds of coagulable matter glue the convolutions together. In a few rare cases the ulceration finds its way completely through the peritoneum, and a portion of the contents of the intestine actually passes into the cavity of the abdomen, when general inflammation is excited, and death follows.

With these appearances of the intestines we usually find some considerable derangement in the structure of the mesenteric glands; they are almost always enlarged and vascular, often exceeding the size of a pigeon's egg, and appearing quite covered with turgid vessels. They are in general most affected immediately opposite to the ulcers of the intestines, and occasionally go into a state of complete suppuration, so that I have seen them apparently on the point of discharging themselves through the peritoneum, into the cavity of the abdomen; but I believe that not unfrequently the pus, even after it has been formed, is absorbed, and quietly subsides.—Dr. Bright, *Medical Reports*, vol. i. p. 180.

As demonstrator of morbid anatomy of St. Thomas's Hospital, it has fallen to my lot to examine a considerable number of cases of typhoid fever. These I have lately analyzed, and have been particularly struck by the large proportion of them in which death was caused by intestinal perforation. Some of the results at which I have arrived, in reference more especially to the latter complication, I propose to lay before the Society.

The exact number of fatal cases of which I have preserved records is 52. Of these 28 were of males, and 24 of females; *one* occurred between 5 and 10 years of age; *eight* between 10 and 15; *nineteen* between 15 and 20; *ten* between 20 and 25; *eight* between 25 and 30; *five* between 30 and 35, and *one* at 42.

In *eight* instances the patients died before ulceration had been thoroughly established in the diseased Peyerian patches, and evidently from the uncomplicated effects of the typhoid poison. In *nine* cases death occurred late in the disease, and when many of the ulcers had become more or less perfectly cicatrized from perforation, from relapse, or from pneumonia, pleurisy, abcess, or debility dependent upon other causes. In the remaining *thirty-five* examples, the fatal termination took place at various periods between the above limits, either as in the first class, from the direct influence of the disease itself, or from this associated, as in the second class, with diarrhœa, intestinal hemorrhage, bed-sores, pulmonary mischief, or perforation and peritonitis.

The small intestine was the seat of specific disease in every instance; the colon in 27 cases. The mesenteric and other glands connected with the bowels were, I believe (although their condition is not always recorded), invariably affected to a greater or less extent, being large, soft, and congested, presenting at times a fibrinous deposit, or even actual suppuration. In addition to the complications mentioned in a former paragraph, I observed in *one* case, cicatrices in the stomach; in another (that of a boy of *ten* years of age) half a dozen recent shallow ulcers in the same organ; in a third, a sloughy ulcer in the groin, connected with diseased lymphatic glands; and in a girl of 17, who had been suffering from gonorrhœa, superficial gangrene of the genital organs, together with a phagedænic condition of the intestinal ulcers. I have never yet met with laryngeal disease; nor have I recognized any visible affection of the brain beyond mere congestion. I may add, that my cases confirm the statement that there is no necessary connection between the intensity of the general symptoms of the disease and the extent of the intestinal mischief.

The remarkable tendency which typhoid ulcers have to produce perforation of the bowels, is shown by the fact that this lesion had occurred in no less than 15 of the above 52 cases, or in a proportion of 1 in rather less than 3.5. It happened *eleven* times in males, *four* in females; a disproportion dependent, in great measure, doubtless, on the different nature of their respective avocations, which, in the milder cases of the disorder, may be carried on up to the very instant of perforation. *One* individual died therefrom between 10 and 15 years of age; *seven* between 15 and 20; *one* between 20 and 25; *two* between 25 and 30; and *one* between 30 and 35; a result which seems on the whole to indicate that age exerts no special influence in its production. The periods of the disease at which it took place will perhaps be better judged of by the condition of the ulcers at the time of death, than by the history: in *nine*, it occurred during the process of separation of the sloughs; in *five*, after the sloughs had become completely detached; and in *one*, when cicatrization, excepting only in the ulcer perforated, was progressing. The number of ulcers present, and the degree of the general symptoms seem, neither separately, nor in combination, to have exerted much influence in producing the lesion in question, for in at least *two* instances (Cases 3 and 7), the patients, though complaining, continued to work up to the very moment of its occurrence; in several others, the febrile attack was more severe, though still mild, and in several also (Cases 4, 5, 6, 8, and 13), the accident occurred during symptoms of extreme intensity. Further, it appears that in *five* cases the ulceration was really very slight, and that in the remaining *ten*, it was more or less severe; in some instances, indeed, to an extreme degree.

This is not the place to detail symptoms, but I may briefly state, that in the cases in which the antecedent illness had been slight, and in which the patient was conscious, the access of peritonitis was characteristically indicated: and that in the remaining cases the symptoms resulting from perforation were more or less indefinite, in exact accordance with the degree of mental and bodily prostration of the patient; and that, therefore, in some instances, as might be supposed, the occurrence of the lesion was revealed *post-mortem* only.

The interval between perforation and death rarely exceeded a day or two, but varied between a few hours (Case 10) and two weeks (Case 3).

In every case the perforation had occurred in the ileum, and generally in quite the lower part. In one instance, the orifice was found two inches from the valve; in another at a distance of two yards; and in the remainder, at various points intermediate between these extremes. The actual perforation, was, I believe, without exception, due to laceration; at all events, I may venture to say that in every case in which speedy death ensued, the perforation was more or less linear in form, and the edges, consequently, though somewhat irregular, tending to a parallel direction. Nevertheless, this character was masked in various ways; as by adhesions binding the lips of the orifice to one another, or to some neighbouring loop of intestine, or by the occasional extreme tenuity and softness of the floor of the perforated ulcer. The size of the orifice was generally sufficient to admit a No. 4 or 6 catheter, but now and then little more than a mere pinhole.

The inflammation of the peritoneum following perforation was indicated by patchy congestion of its parietal layer, and of that covering the stomach, intestines, and pelvic organs, together with a certain amount of soft lymph smeared, with more or less uniformity, over the serous surface, and producing generally slight adhesions between contiguous viscera. There was present, also, a variable amount of gas and of fecal matter; the latter incorporated usually with lymph and pus, and, if sufficiently abundant, accumulated in the pelvis, or some other dependent portion of the abdominal cavity.

There are still two points to which I wish to draw attention, the importtance of which has been impressed upon me by more than one case that has come under my observation. They are points indicating a tendency towards cure, which might, perhaps, scarcely have been suspected in such cases as those under consideration, but which taken in conjunction with the actual progressive improvement temporarily manifested in some instances, hold forth a glimmering hope, which should not be lost sight of in the treatment which we may adopt. The first is, that in consequence of the disposition of parts in the abdomen, there is a great tendency for the peritonitis to become circumscribed. The great omentum, with the parts to which it is attached, forms a natural septum between the upper and lower halves; and we sometimes find inflammation strictly limited, through its instrumentality, to the lower half, in cases of typhoid perforation; to the upper, when the liver or stomach is the primary seat of the disease. The closure of the foramen of Winslow, again, occasionally forms a bar to the spread of inflammation from the sac of the great omentum, and conversely: while adhesions among the small intestines, and between them and neighbouring parts not infrequently limit inflammation and its results to one or other lumbar region, to the pelvic cavity, or even to some more unlikely situation than these. The second is, that there is a great tendency for the perforation itself to become closed by adhesions, and this within a very short period of its occurrence; so that in a considerable proportion (*six*) of the cases which I have examined, I have failed, notwithstanding the presence of fetid gas and of actual fecal matter in the peritoneum, to recognize any communication between the intestine and abdomen, until I have torn the very slight, though distinct, adhesions by which the margins of the orifice had become connected with some adjoining organs.

It is quite clear, therefore, that notwithstanding the almost utter hopelessness of these cases, there is still a curative effort on the part of nature: —firstly, to close the orifice by lymph, and thus to prevent further extravasations; secondly, to limit the material poured out, by adhesions between surrounding organs, to that portion of the abdominal cavity which is contiguous to the seat of mischief, and thus to circumscribe the more serious results of the accident. Of course, if the opening be large, or the effused fecal matters abundant, the probability is that the patient will die so speedily, that little or no indication will be presented of either of the above curative processes. But if the perforation be small, and a mere weeping of the bowel have occurred, there seems no reason whatever (so far as the actual condition of parts is concerned) why the patient might not recover without further complication. And even if a yet larger amount of matter have escaped, and this have become circumscribed by adhesion, there is still no reason derived from local considerations, why he should not ultimately do well; even though the progress of the the case would necessarily be retarded by the formation of a peritoneal abscess, which must discharge itself either through the intestines or abdominal parietes, or by some still less favourable route.

The case, No. 3, in the appended table, though it ultimately proved fatal, affords some justification on other than pathological grounds for the remarks just offered; and I may add, that some two or three years ago, I had under my charge, a case since published in the *Medical Times and Gazette*, of a little girl who recovered after paracentesis from an attack of circumscribed suppurative peritonitis. This, I believed at the time to have been due to perforation of the bowel, occurring in the course of a slight attack of typhoid fever, but the mildness of her antecedent symptoms, and her final recovery, combined to render the case inconclusive.—*Analysis of Cases of Perforation of the Bowel in the Course of Typhoid Fever*, by Dr. Bristowe, *Transactions of the Pathological Society of London*, vol. xi.

VI.

On the disputed question of the theory of uræmic poisoning, the observations of Dr. Oppler, of Berlin, deserve our attentive study. After detailing a number of experiments upon animals, this author says:—

I must now inquire to what conclusions the above detailed experiments lead. In the first place, I think they show undoubtedly that the decomposition of urea into carbonate of ammonia cannot be regarded as the cause of uræmic phenomena; for neither could I produce these phenomena by the injection of carbonate of ammonia, nor could I in the blood of eight uræmic dogs detect a trace of that substance. In one of the experiments indeed (the third) a little was found; but as the blood had been taken from the corpse of the animal, the carbonate of ammonia had, no doubt, been generated after its death. In the second place, chemical examination of the blood and of the muscles has supplied me with two positive facts, which may serve as a foundation for a different explanation of the uræmic phenomena. The facts I allude to are—the enormous increase of extractive matters in the blood, and the large quantity of kreatine and leucin found in the muscles.

As I discovered, on examining the blood of animals, which had not survived the operation more than forty or fifty hours, that the extractive matters often reached the amount of 18 or 19 in 1000. I entertain no doubt that this very considerable increase does not depend upon a simple mechanical retention of matters, which in the normal condition should be excreted by the kidneys; but that it is in great part occasioned by decompositions going on in various organs, in consequence of a change in the composition of the blood.

In order to arrive at certainty upon this point, I examined muscular flesh, because the products of its decomposition are best known and are most readily submitted to chemical examination. The result was unequivocal; and the large quantity of kreatine (34 grains in two pounds) and leucin showed distinctly that abnormal decomposition has been taking place. That the large quantity of kreatine was not merely the result of the impeded secretion from the kidneys and of that normally contained in muscles, is clearly proved by a comparison with the experiments of Schlossberger and Neubauer, the former of whom found five grains of kreatine in two pounds of human muscle, while the latter found that the quantity of kreatine secreted by the kidneys in 24 hours was $7\frac{3}{4}$ grains.

As it accordingly appears undoubted that, in consequence of the arrest of the functions of the kidneys, products of decomposition are formed and accumulate to an abnormal extent in muscles, it may almost with certainty be assumed that in the central organ of the nervous system, which is also exposed to be affected by the abnormal constitution of the blood, similar changes in its chemical composition must take place, although the products of the decomposition of the organ are as yet so little known that we cannot submit them to accurate chemical demonstration, as in the case of the muscles. If, therefore, we assume that products of decomposition are formed and accumulated to an abnormal extent in the brain and spinal marrow, we have, I think, a perfectly satisfactory explanation of uræmic phenomena. The occurrence of important chemical changes is quite sufficient to explain the derangement in the function of the brain, and to account for the characteristic symptoms, such as giddiness, dull headache, unconsciousness, stupor.

I therefore consider it an error to explain the occurrence of uræmic phenomena by the supposition that one or other of the elements of the urine, or of its decomposition, is retained in the blood, and acts on the brain in a deleterious manner; but I believe that a chemical change in the central organs of the nervous system is to be considered the most probable cause of uræmia.

With regard to the vomiting, which is so characteristic an occurrence in uræmia, I do not consider, as many do, that it is occasioned by irritation of the mucous membrane of the stomach by carbonate of ammonia, derived from the decomposition of urea, because I have often examined the vomited matter, as well as the contents of the stomach from uræmic animals just dead, and I have only found pure undecomposed urea present in large quantity. On the other hand, the fact, observed by me, that after ligature of the ureters vomiting appears much later than after extirpation of the kidneys, appears to justify the opinion that uræmic vomiting depends upon a sympathetic irritation of the gastric mucuous membrane, the result of irritation of the renal nerves. After ligature of the ureters the vomiting does not set in until the urine—having accumulated in the urinary tubules—produces on all sides pressure and irritation; whereas, in extirpation of the kidneys the renal nerves suffer a direct irritation.

I must next consider what conclusions are to be drawn from the parallel experiments, 2, 10, 11, 12. Formerly, as urea could not be detected in normal blood, it was supposed that it was formed in the kidneys, until Prevost and Dumas, and subsequently Gmelin, Mitscherlich, and Tiedemann, showed that it might be found in great quantities in the blood of animals from which the kidneys had been removed. These experiments, as well as a more accurate chemical examination, which shows that urea may be detected, although in small quantity in healthy blood, have led to the generally-received opinion that all the urea of the urine is formed in the blood by the oxidation of nitrogenous matter, and that the kidneys only effect its separation. Although the quantity of urea existing in the blood which passed through the kidneys, was known to be very small in proportion to the amount found in the urine, this was not considered as any objection to the supposition; for the continued circulation of even a very small quantity of a substance constantly formed by the metamorphosis of the tissues, might at length account for the presence of a larger quantity in the urine. In the same way was got rid of, the objection founded upon the known fact, that after extirpation of the kidneys a much smaller quantity of urea was found in the blood and in the tissues than would, under normal circumstances, have been eliminated in the same time by the kidneys. Bernard and Barreswill thought that the conversion of urea into carbonate of ammonia accounted for the difference; while others expressed the opinion that when urea had accumulated to a certain extent in the blood, its farther formation was thereby interfered with. It, therefore, seemed to be proved that the elements of the urine, especially the urea, arrived ready formed at the kidneys; and Ludwig maintained that the whole process of

18

secretion of urine in the kidneys depended simply upon relations of diffusion; that there was secreted from the glomeruli a very diluted fluid, which contained all the elements of the urine, and that this, in consequence of absorption of water by the capillaries which surround the urinary tubules, was concentrated to the condition of urine. Hoppe, however, showed that this simple theory could not be maintained; for if Ludwig's view of the formation of urine was correct, then, if the concentrated urine of the animal were separated from the blood-serum of the same animal by a membranous partition, no water should pass through from the serum to the urine; but experiment showed that this did take place, and it was therefore evident that the secretion took place in some other way. Now, on considering my parallel experiments regarding the amount of urine in the blood after extirpation of the kidneys, and after ligature of the ureters, I think I have arrived at a result of the highest importance regarding the physiological secretion of the urine—namely, that a large part of the urea which we find in the urine does not arrive ready formed at the kidneys, but that it is formed there. In proof of this I refer to the results of experiments 2, 10, 11, 12. After extirpation of the kidneys (Exp. 10) 2 pounds of muscle afforded about 1 grain of nitrate of urea. After ligature of the ureters (exp. 11) 1½ pound of muscle gave more than $3\frac{5}{8}$ grains of pure urea, which corresponds to nearly 62 grains of the nitrate. After extirpation of the kidneys (Exp. 10) 3½ ounces of blood gave four-tenths of a grain of nitrate of urea. After ligature of the ureters, 6½ ounces of blood were found to contain 7 grains of nitrate of urea. How can this great difference be accounted for, except on the supposition that the great excess of urea after ligature of the ureters is the expression of the amount formed independently by the kidneys? I think, therefore, that these facts clearly show that the kidneys are not to be looked upon as the centre of mere process of diffusion; but that they are to be regarded as having, if not the exclusive, at least a very large independent share in the formation of those substances which are to be considered as the final products of changes which the nitrogenous, especially the albuminous, materials undergo during the process of metamorphosis of the tissues. In what part of the organs this formation has its special seat, cannot be determined with certainty. Probably the epithelium plays an important part, as is rendered likely by what we know of the importance of epithelium in other glandular organs; by pathological observations, which show that by destruction of the epithelium the secretion of the urine is interfered with; and, by the statements of Busch and Willich, who have found uric acid in the epithelium of the urinary organs of snails and birds.

That kreatine plays an important part in the formation of urea is proved by this—that after extirpation of the kidneys, 34 grains of kreatine were found in 2 pounds of flesh, while after ligature of the ureters 1¼ pound of flesh only contained $7\frac{3}{4}$ grains. I purpose, however, to undertake further researches on this subject.[1]

I cannot too earnestly impress on you, says Sir D. Corrigan (referred to at p. 107), the necessity of closely watching the condition of the bladder. Murphy's case afforded a good illustration in point. He was very ill in maculated fever, so violent that it was necessary to put a straight waistcoat on him. His delirium was furious; his tongue dry and brown; his pulse above 130, and skin covered with both maculæ and petechiæ; he had not

---

[1] *Edin. Medical Journal*, October, 1861.

slept, and his eyes were suffused; he passed feces in bed, and we were positively assured by the nurse that he had also passed urine copiously under him. This statement seemed confirmed at first sight on turning back the bed-clothes, for there was a strong urinous smell from him; the sheets were wet, and the urine was seen welling from the orifice of the urethra, and dribbling over the thigh. Notwithstanding all this, I introduced the catheter, and there passed off more than two quarts of urine. You know what the effect on the brain and system would have been had the bladder been allowed to remain in this state. It is scarcely possible to suppose anything more calculated to extinguish life in fever than the bladder continuing in such a state. Remember what you saw in Murphy's case—that urine flowing out upon the sheets may be only the surplus which no longer finds room in the bladder.

I saw not long since, in private practice, another case illustrating the same point. In this case the patient was a lady under the care of a homœopath. You know a homœopath would not use a catheter. It was on the fifteenth or sixteenth day of fever. I found her in epileptic convulsions, which had continued for some hours, foaming at the mouth, insensible, unable to swallow, and, to all appearance, dying. On examining the abdomen, I felt the bladder extending as high up as the umbilicus. On introducing the catheter it was scarcely possible to bear the intolerable ammoniacal smell of the urine, which must have been shut up for several days. It continued to flow until some large basinfuls were drawn off. This patient recovered, but she suffered much for the neglect. Subacute and then chronic cystitis followed, under which she continued to suffer for more than a year afterwards.

M. Blot, whose observations are alluded to at page 119, submits the following among other conclusions: (1) That every case of eclampsia seen by him has been accompanied by albuminuria; (2) That the albuminuria of pregnancy is free from danger when uncomplicated with cerebral congestions; (3) That the relation between the two is probably sanguineous congestion, occurring at the same time in the kidneys and cerebro-spinal centres.—(*London Medical Gazette*, 1850.)

In the following case of uræmic coma, by Dr. Henderson, the condition of the cerebral circulation would seem to have played a part:—

JANET THOMPSON, æt. 24, single, resides in Leith; admitted March 12th, 1844; a stout, plethoric young woman; states, that she was seized with rigors on the 7th, which were accompanied with pain in the head, limbs, and over lumbar region. Previous to the shivering had felt indisposed for two or three days, but only slightly; had been in no communication with fever patients, but in the house from whence she came several persons had been ill in fever, though this was some time previously.

March 13th.—Sweated a good deal last night, and is now perspiring freely; does not think that she has sweated so much before during her illness, pulse 80, soft, of tolerable volume; pain in head, which was so severe yesterday, has entirely gone; complains of some stiffness in the arm and shoulder; bowels opened from medicine taken last night; tongue moist, but furred; has not passed any urine since her admission yesterday; no distension over hypogastrium, nor dulness on percussion; feels heavy and drowsy; every now and then falling to sleep, and waking in a short time with a start, and feeling (as she expresses it) as if she were falling out of bed: some dizziness and indistinctness of vision, with a feeling of heaviness over the eyes; no confusion in the mental powers, as she intelligibly answers every question made respecting her present condition, but cannot exactly state some of her previous symptoms. ℞.—Nitratis Potassæ ʒij, Liq. Ammon. Acet. ʒiij, Aquæ ʒv, Sit. mist. cap. ʒj, 4ta q. q. horâ.

14th.—Pulse 80, very irregular in its beats; began to pass water yesterday in an hour and a half after first dose of medicine; at seven o'clock P. M. had made six or eight ounces of highly coloured urine—the whole quantity for twenty-four hours is thirty-four ounces; has sweated a good deal since yesterday, feels much lighter, and the uneasiness in head is gone.

15th.—Had slight shivering this morning; skin hot and dry; pulse 88; urine, 24 ozs. sp. gr. 1.024. There is here an omission for some days as to the state of her case, and the next report is dated 22d —Pulse 120; skin hot; has an oppressed look; no stool to-day, but bowels were opened freely yesterday; no headache.—Mist. c. nit. potass et liq. ammon. acet. rep.

24th.—Pulse 120, of fair strength; skin hot and dry; countenance has much of a sottish expression; pupils pretty large, and equal; no decided difference between the two sides of the face when features are at rest, but a very apparent obliquity of the mouth and protrusion of the tongue to the right side when desired to put it out, the left angle of the mouth being drawn out at the same time. On speaking, left side alone moves. She is not insensible, but her intelligence is obscure. Moves her right arm freely enough but cannot squeeze with it so thoroughly as with the left. There is a bright red patch, and a series of others downwards, on the epigastrium and abdomen. A pale-coloured spot occupies the centre of the largest, and a similar one exists at carpal extremity of the left thumb. There is a superficial slough on left heel, and an ecchymosed spot on right instep; tongue dry; urine passed in small quantity.—Abradat capilitium. ℞—Pulv. nit. potassæ ʒiss; Aquæ ʒx; sit. mist. cap. ʒj, 2da q. q. hora.

25th.—Pulse 116, of good strength; the hemiplegia remains as before; tongue dry; had several stools; urine abundant, but not kept. On the left foot the upper surface of little toe is occupied by a small bulla of blood, another between that toe and the next. The spots on the epigastrium are not so large; that on the right thumb is converted into a straw-coloured bulla. Intelligence, pupils, &c., as before.—Mist. cont.

26th.—Pulse 104, of good strength; right cheek as before; right pupil smaller than left, but not contracted; tongue dry but clean; pus in the bulla of the thumb; the largest bulla on left foot has emptied itself of blood and the part is healthy below; the vibices are improving; passes a sufficient quantity of urine.—Habeat. Pulv. Jalap. comp. ʒj : mist. cont.

27th.—Was observed in the morning in a state of stupor, being incapable of observing or replying; was ordered by Dr. Craigie's clerk to be cupped about an hour ago; pulse, at present, 150, very feeble; is incapable of comprehending or replying to questions. The right pupil is considerably dilated; no effect is produced on the eyelid on thrusting the hand towards the eye; left pupil is smaller, not so contracted, and she winks on thrusting the finger towards the eye. Even irritation of the conjunctiva produces but little effect on the right side. No motion of the right arm is produced by irritating it, and it is quite in a state of resolution; pricking it, however, produces an expression of pain. Above the inner condyle of the right femur, and below the inner aspect of the left knee, parts which are said to have been lying in contact, there are spots of corresponding size of two inches long, by an inch and a quarter broad, consisting of red discoloration, containing a deep purple one nearly two-thirds of the size of the whole spot; around the redness there is a halo of a lemon colour; a scaly eruption of sudamina on the belly; a large bulla, containing apparently blood, has formed on the outside of the left foot; urine passed in bed, quantity of which cannot be accurately ascertained.—Habeat. sp. communis ℥iij, cap. ʒj in aqua calida omni secunda hora.

28th.—Died last night at eleven o'clock.

Remarks.—Infection seems to have given rise to her fever. On the sixth day of her illness there was some suppression of urine, which showed its effects in the cerebral symptoms, as noticed in the report. The next day (after the administration of pretty large diuretics), when the secretion of the kidneys was restored, the drowsiness and indistinctness of vision were removed. In the report of the 24th inst., and the seventeenth day from the commencement of her first attack, it is very evident that a serious affection of the brain existed; and as there was partial suppression of the urine, the diuretic medicine was again given; and although the kidneys were brought into action, yet the lesion, which had evidently taken place in the head, was in no degree removed. On the morning of the 27th the preludes to dissolution were indisputably present; the urine was voided in decreased quantity; the sensific nervous power in the affected side in a great manner lost, the intellect obscured, and other indications foretold the near approach of a fatal issue. The vibices and bulla spoken of in this case were, perhaps, in a great degree, dependent upon the circulation of urea, because this salt, by uniting with the elementary principles of water, is transformed into the carbonate of ammonia, which has the power of dissolving the fibrin of the blood, and thus rendering hemorrhagic effusions, transudations, &c., liable to occur. My friend, Dr. Michael Taylor, analyzed the blood taken by cupping, and discovered crystals of urea in considerable abundance. The postmortem examination showed an abnormal quantity of serous exudation in the ventricular cavities of the brain. When it is considered how great is the ambiguity which mystifies the fundamental pathology of fever, it becomes exceedingly interesting when the proximate cause of death is so fully ununderstood, as in the case above. Yet, on the other hand, there were instances occurred, in which a fatal termination ensued without any explicable cause being observable to elucidate the mode in which death had been produced.—*London Medical Gazette*, vol. xxxix.

VII.

After giving an analysis of the fatal cases of fever in which the head was examined, Dr. Reid observes (as referred to at page 128):—

" From this view of the *post-mortem* appearances observed in the brain after death from fever, and the symptoms with which they were attended during life; and from the contrast we have instituted between those cases where an abnormal quantity of serum was found within the cranium, and those cases where the usual quantity only was observed, we think we are justified in concluding that they afford no distinct evidence, that the serous effusion was in all cases, if in any, the cause of death. We have seen one case in which one ounce of serum was effused into the lateral ventricles, and yet nothing different from the usual confusion of thought was observed. We have also seen a cerebral derangement as strongly marked in those cases where no increased effusion of serum was found within the cranium after death, as in those where this was observed. Besides, it must be remembered that it is not unusual to find increased serous effusion within the cranium of old people, or when the patient has been emaciated by previous disease, in quantities equal to what we have described as occurring so frequently in fever. And this last statement naturally leads us to inquire into the probable effects of age, and the duration of the disease, upon the amount of this serous effusion. I find that the average age of the cases in which an increased effusion of serum was found within the cranium was $42\frac{21}{25}$ years; while the average age of those in which no increased effusion was found within the cranium was $26\frac{16}{19}$ years. The average age of the four fever patients in which seven drachms or upwards of serum were found within the lateral ventricle was $57\frac{1}{4}$ years; the youngest being 48, and the oldest 68 years of age. With regard to the average duration of the disease in the two classes of cases we do not find so striking a difference. The average duration of the disease in the cases in which an increased effusion of serum was found within the cranium (as calculated from twenty-four cases) was $12\frac{22}{24}$ days; while the average duration of those cases in which no increased effusion of serum was observed within the cranium was $11\frac{11}{13}$ days, or only a difference of nearly one day. . . . . It has already been stated that in most of the cases in which the brain was examined, we observed the bloodvessels well loaded with blood, as indicated by the numerous red points which presented themselves on the cut surfaces of the brain. In judging of the degree of vascular congestion in the brain we ought to remember that part of the increased quantity of blood in the vessels of the brain may be merely apparent, and arise from the fluid state of the blood; for it is obvious that if the blood remained fluid, the pressure of the knife used in slicing the brain will force the fluid blood through the open mouths of the vessels upon the cut surfaces. In judging of the probable causes of the congested state of the bloodvessels of the brain and of derangement of its functions, we must take into account the state of the respiratory organs; for it is apparent, that if there be any impediment to the passage of the blood through the lungs, this may influence materially the circulation within the cranium, even in those cases where the respiratory function is only secondarily affected through derangement of the central organs of the nervous system; for it is equally obvious that when from this cause there is any impediment to the circulation through the lungs it will react upon the central organs of the nervous system, and increase the primary derangement.

" It is probable, however, that the increased effusion of serum within the cranium, and the greater plenitude of bloodvessels of the brain, occur too frequently to be fully accounted for by the age of the patient, the previous emaciation of the body, and the derangement of the respiratory functions, and that it may, in a few cases, be owing entirely to, and in others much aided by, causes connected with the nature of the disease itself, and by which that disturbance of the cerebral functions so generally observed in fever is induced. The facts which we have stated ought, however, to render us very cautious in attributing the phenomena observed in any individual case of fever, to increased plenitude of the bloodvessels, or to the effusion of serum."—*Physiological and Pathological Researches*, p. 460. *Report on the Epidemic Fever of Edinburgh.*

Dr. Fritz arranges the symptoms met with in typhoid fever referable to the spinal marrow, under the heads of those of the spine proper, and those of the medulla oblongata. The first he divides into.

I. Derangement of sensibility.

(1). Hyperæsthesia, (*a*) cutaneous, (*b*) spinal, (*c*) muscular.

(2). Spontaneous pains comprehending, (*a*) rachialgia, or pain affecting different portions of the vertebral column; (*b*) radiations of pain from the spine into different parts, as the neck, shoulders, chest, arms, stomach, different parts of the abdomen, &c.; (*c*) pains in the inferior extremities, deep-seated, pungent, lancinating, terebrating, &c.—these pains being generally accompanied by muscular hyperæsthesia; (*d*) thoracic pain, occasioning difficulty of respiration; (*e*) perversions of sensibility, consisting in abnormal sensations, as tingling, pricking; (*f*) analgesia, including abolition of muscular sensibility.

II. Derangements of the motor functions of the spine, including paralysis, convulsions, and reflected movements.

Under the second head he details, at great length, a case in which the following symptoms, referable to the medulla oblongata, presented themselves, being preceded, in the first instance, by extreme cutaneous hyperæsthesia, rachialgia, and pain on moving the head, with retraction of the head upon the neck.

Panting respiration—an extreme degree of anhelation, resembling a paroxysm of asthma, the head being carried back on the neck with each inspiration—the tongue unable to articulate—intermittent spasm of the pharynx, causing dysphagia; violent spasms of the larynx, causing hissing inspirations, momentary closing of the glottis.

The following is the case referred to at page 133, reported by Mr. Todhunter:—

Julia Donnelly, æt. 34, cook, admitted into hospital, May 12th, 1866; states that she works hard daily, and sleeps in a small uncomfortable bed-room in a house in which the sewerage is out of order. A few weeks since, when getting up in the dark, she struck her head violently, and was stunned for a time, and had headache on the following morning She states that on the day before admission she washed her head, and afterwards sat in the sun; in the course of the day she was attacked with headache and vomiting. When admitted, she complained of headache and continued vomiting. Pulse 116 and weak, tongue thickly furred.

13th. Heart weak; pulse rapid and weak; tongue thickly furred; some cutaneous hyperæsthesia; no distinct eruption; bowels confined. Slept well, but still complains of sick stomach and headache; complaints uttered in a low whining voice, interrupted by sudden cries, accompanied by pressure of the hands over the forehead; frowning and tossing of the arms; headache said to be confined to one spot of the forehead, over the left eyebrow; slight tenderness of the occiput and nape of the neck; tenderness on pressure of every part of the abdomen; respiration very rapid, although she does not complain of her breathing; no abnormal chest signs; skin rather hot; complains of weakness in sitting up in bed.

14th. Seems much prostrated; heart weak; both sounds audible however, and rhythm pretty nearly normal; pulse weak, 100 in the lying posture, 120 when sitting; respiration still very rapid; cutaneous hyperæsthesia increased, and extending to the limbs; she cries out when the knees are held in the hand; does not complain of pain on pressure over dorsal spines, but feels pain on pressure in the occipital and cervical regions; pupils contracted; no photophobia, but says the sight is bad; has darting headache, extending over more of the forehead than yesterday; the whole scalp being tender, and feeling "*tight;*" cutaneous hyperæsthesia greatest in the abdomen, on which, as on the chest, there are several rose spots; gargouillement in the ileo-cæcal region; epistaxis; tongue thickly furred, the fur being greenish in the centre; bowels confined; urine dark; thinks that what she vomited before coming in was greenish. Two leeches to be applied to the nostrils, and six to the temple.

   R.—Calomel,   gr. j.
     Pulv. Jacobi,   gr. ij.
     Extract Aconiti, gr. ¼
   Ft. pilula quartis horis sumend.
   Cardiac mixture; beef tea.

15th. Heart rather weak; pulse rather weaker, 100 lying, 116 sitting; temperature as registered by thermometer in axilla, 100°; tongue same as yesterday; cutaneous hyperæsthesia less; headache not so violent; complains of pain and buzzing in her left ear; the upper part of the pinna being sore to the touch; dorsal spines slightly tender; gargouillement in the ileo cæcal region; spots much better marked than yesterday, and in considerable number over abdomen and chest; no diarrhœa; slight increase of previous converging strabismus of left eye; eyes a little suffused; not so much screaming.

Repeat the pills and mixture.

16th. Heart and pulse perhaps a shade weaker; pulse 100 lying, 108 sitting; temperature 101°.7; tongue rather cleaner, greenish hue still evident; cutaneous hyperæsthesia almost gone; headache better, and pain in ear gone; pinna numb; gargouillement in ileo-cæcal region, and slight meteorism of abdomen, spots fading; complains of "tightness" of the scalp, which is sore to the touch; slept pretty well, but had unpleasant dreams; "thought she was standing in an ash-pit, and the top of her head was taken off," &c.; respiration much less rapid; strabismus more marked; sees things badly, "a mist like" over her eyes; pupils still contracted; a few greenish stools; urine dark.

Treatment.—Repeat pills and mixture; evaporating lotion to the forehead.[1]

---

[1] The urine examined by Mr. Collins at this period was found to contain a marked excess of phosphates:—

  Specific Gravity, 1019.
  Reaction Acid.
   Amount of Phosphates in 100 grs. (*by weight*) *of urine.*

Phosphate of Lime	588,155
Triple Phosphate,	2,561,720
Total,	3,146
Normal quantity being,	1,000

17th. Heart very weak; pulse not weaker than yesterday, 100 lying, 100 sitting; hyperæsthesia of trunk and limbs gone; no pain in the back; headache worse, and scalp very tender; gets relief by pressing her hands to her head; more jactitation, and tossing of head on pillow; temperature 100.6; eruption almost gone from chest, considerable gargouillement, and slight tenderness of ileo-cæcal region; less strabismus; sight much better. Complains a good deal of her head, and cries out a good deal; wishes her head could be "taken off;" has buzzing in her ears, and "a sound like the wind in trees;" less prostration; respiration much more tranquil; no pain in the back; several greenish stools; gums slightly touched.
Pills omitted.

℞.—Tinct. Aconiti, ℨj.
   Mixt. Cardiac, ℨvj.
   St. ℨj ter. indie.

℞.—Liquor Opii (Battley), m. 30.
   Aquæ Camphor., ℨj.
   Ft. haustus nocte sumend

18th. Had several hours' sleep; has less headache; no cutaneous hyperæsthesia, except of scalp; heart is slightly weaker; pulse soft, 92 lying, 96 sitting; temperature 100; tongue cleaning; tenderness in ileo-cæcal region: spots almost faded from chest, a few still scattered over abdomen, which is very slightly tympanitic; no splenic dulness; sight better; pupils rather more contracted than yesterday; able to sit up without feeling giddy; feels stronger; some epistaxis last night; some retching this morning; no stool since yesterday; gums still a little sore.
Treatment same as yesterday.

19th. Slept pretty well; sight better; soreness of scalp almost gone; laughing very much at her dreams; says she never raved so much in her life as she did last night; dreamed she had several heads, &c. &c.
Heart still weak; pulse stronger, 88 lying, 92 sitting; temperature 98.5; tongue cleaner, and not *green*; no hyperæsthesia; bowels once moved.

Sunday, May 20th. Complains very much of headache; heart, perhaps a shade stronger, pulse stronger; 80 lying, 88 sitting; tongue cleaner; tenderness and tympany of abdomen.
Omit the Liq. Opii at night.

Monday, 21st. Slept well; still complains of headache; no soreness of scalp; sight quite natural; tenderness of ileo-cæcal region; heart sounds distinct, though still weak; pulse, 88 sitting, firm; temperature 98.5.

May 22d. Slept well; still complains of headache, which she says is worse than yesterday; when sitting up she feels giddy, and her head trembles; no abdominal tenderness; eruption quite gone; bowels regular.

23d. Pulse, 92 lying, 100 sitting; temperature, 98.2; tongue clean; still complains of headache, and giddiness on sitting up. A small abscess in left axilla. This has been coming on for the last few days, but only become painful yesterday.

24th. Heart a shade stronger; pulse, 88 lying, 104 sitting; temperature, 98.5; tongue pretty clean; still complains of headache, and some soreness of scalp over left ear; also of cough and occasional retching.

Friday, May 25th. Heart very weak; pulse, 88 lying, 104 sitting; headache worse.
Rept. mistura, sine Tinct. Aconiti. To have four ounces of wine.

May 26th. Slept well; feels better. Had no farther bad symptom.

The following memoranda referred to at page 140, will be read with interest, as illustrating the influence of pre-existing disease upon the type of the nervous symptoms in fever:—

I recollect the late Dr. ——— from an early age—I believe about twelve or thirteen years. As a boy, as a youth, and a man he was always delicate. I have heard his mother state that he was reared with difficulty—that he suffered grievously from the ailments of childhood, and at all times required the most anxious care. From the earliest period his intelligence was of the highest order, and he always evinced an extraordinary love for study. I recollect his suffering from various ailments during his earlier life, especially glandular swellings in the neck, lithic diathesis, causing at times painful micturition—a certain irritability of the heart, which rendered him liable to very distressing palpitations on running or making exertion—and several attacks of acute tonsillitis, which left some permanent enlargement of the amygdalæ. For some years before his death he suffered a good deal from dyspepsia, not causing the loss of appetite, but distressing pyrosis, with more or less of rumination.

As a rule, Dr. ——— was negligent of his health, and it was with difficulty that those around him could induce attention to the most ordinary and obvious precautions. Coming to the time more immediately preceding his fatal illness, I should observe that I was in less close contact with Dr. ——— than previously. Nearly all that year I lived in Paris, and when at home, as I stopped in the country, we were at some distance from each other, and consequently seldom met.

I believe it was during this year, and very shortly after his marriage, that he suffered from slight hæmoptysis, for which he consulted you. Of this attack I was not aware until you informed me of it about the time of his death.

However, about the spring of the year (1859), I heard that he was complaining of some affection of his sight. I forthwith went to him, and on cross-examination discovered a series of symptoms, of which he made light, but which appeared to me very serious.

He told me that very often—perhaps once in a month or three weeks—he found his sight confused, especially that of the left eye, which he felt himself obliged to close while reading or writing. This state of things would last for a day or two, and then pass off. On examining the eye, I found the pupil decidedly dilated, and when his back was turned to the light, I thought I observed some degree of ptosis. I further made out that at times he experienced uneasy sensations in the left arm and leg, though not the slightest loss of power. I learned from Mrs. ——— that his sleep at night was uneasy, and that he was latterly profoundly lethargic and unwilling to rise in the morning. The tongue was fairly clean, pulse regular, and digestion rather better than usual.

I immediately brought him to Dr. Hutton, who, having examined him most carefully, expressed an opinion to the effect that the symptoms were probably due to a cerebral affection of a strumous character. He advised an issue in the arm, and regular doses of the proto-iodide of mercury. These measures were carried into effect, and with decided benefit at the time.

Shortly after (May, 1859) I went abroad again, and did not return to Dublin until the middle of September. I then found him looking poorly, worn and haggard, complaining of fatigue on the least exertion; but, as regards the special symptoms detailed above, he assured me that they were much lessened, although not entirely gone. I saw him again on the 19th of September, and to my eye he was obviously ailing. On taxing him therewith he parried the questions, and assured me that he felt well; yet, his tongue was loaded, his pulse about ninety, and he was prostrate. He agreed to take some gray powder and James' powder that night.

The following days he was still, to my judgment, ill; but persistently denied it, although his wife assured me he had bad nights and occasional rigors. On the evening of Friday, the 23d September, he suffered greatly from headache. I was absent from town, but he had some leeches applied to the nape of the neck, and took some aperient medicine which acted smartly. On Saturday he maintained that he was perfectly well. Sunday morning, September 25, not being satisfied with his state, I stripped him, and examined his skin carefully. On the abdomen I found a few rose-coloured spots, which I recognized as those of typhoid fever. He would not even then go to bed, and with difficulty I obtained his consent to bring you to see him. You may recollect that it was on the afternoon of that Sunday you first visited him, in his study. Here we enter upon the history of the terrible fever of which he died on the 8th of October. I shall not attempt to give a daily account of its progress. I made no notes of the case, and indeed was too much occupied and overwhelmed to retain very accurate recollections of it. Doubtless you remember the case better than I do.

The first four days and nights passed very quietly by. On the 30th of September he began to suffer from facial neuralgia, which yielded on the application of a blister to the nape. That night, however, he got no sleep, and next day his pulse was dicrotous and somewhat irregular, and he was ordered brandy in small quantities. All that day, namely, Saturday the 1st of October, he was excited and hysterical. That evening he got a suppository containing a quarter of a grain of morphia. During the earlier portion of the night he slept quietly. I recollect his awakening just at 2 o'clock, as I was preparing to lie down on a sofa in his room. He awoke excited and hysterical, and began to shout in a frightful manner. From this state he passed into coma, about 3 o'clock that morning While insensible he had slight convulsions, principally confined to the left side. He was roused to sensibility by blistering of the scalp, together with the internal use of musk and camphor, with small doses of brandy, which his faltering pulse indicated.

After this period of coma, I think he never perfectly recovered control over the left arm and hand. October 2d, Sunday, he had no sleep, nor on the following night. October 3d, he was again greatly excited, when you and Dr. Stokes saw him, and it was agreed to give him three drops of tincture of opium in camphor mixture. Shortly after he relapsed into a comatose state, from which he was only roused by strong coffee. The following day he vomited unceasingly. There was no nausea, but whatever he swallowed he rapidly gulped up in a fashion which reminded me of the same symptom in some cases of hydrocephalus. Profuse sweating also set in—I think the most profuse I ever saw—another link—as I am disposed to interpret it—in the chain of symptoms directly referable to the nervous system.

About this period he began to sink decidedly, passing into a condition of low muttering delirium, which lasted until his death, on the 8th of October, and from which he only roused occasionally for a few moments when spoken to. Finally, the abdomen became more and more tympanitic—the evacuations were passed unconsciously, and the pulse gradually failed. During the six or eight hours immediately preceding death he again became perfectly insensible.

"Our experience," says the Reviewer, "fully bears out the statements of the Author. We have repeatedly seen cases in which opium alone failed to procure rest, yet in which the combination had the happiest effect. We had an opportunity of witnessing a most remarkable instance, which lately occurred in the practice of Dr. Lees, of this city, who has kindly furnished us with notes of the case. The remedy appeared to have an almost magical effect. The patient, a gentleman of nervous temperament and high literary attainments, and suffering under his first attack of typhus, had an imperfect crisis on the fourteenth day.

"His pulse had been 140; he was incoherent, looked wildly about him, and towards morning became extremely low, and said he was dying. He then broke out into profuse perspiration, had great tremor of the hands, and appeared in extreme terror. He got some stimulants and soon fell asleep, awaking in four hours much refreshed, with his pulse having fallen to 112. On the sixteenth day he again became greatly excited. He struggled violently, his features twitching and his pupils greatly dilated. His state now became terrible. The subsultus amounted to almost convulsion; his pulse 132 and miserable; opium and musk were given without effect. The head was shaved, but he continued as violent as ever, thrusting downwards in the bed, sobbing and screaming, and with the subsultus like tetantic shocks. Under these circumstances the following was ordered by Dr. Lees:—

℞.—Tartrat. Antimonii gr. vj.
Liquor Opii sedativi ʒj.
Misturæ Camphoræ ℥viij.

"Of this mixture he got one ounce in a single draught. He fell asleep, and slept calmly for three hours when he awoke in great terror; all the bad symptoms returned. Another ounce was administered, and soon after a third, with the same happy effect. Thus he continued to the evening of the next day, comparatively rational and taking nourishment, when he began to look wild, got restless, and tossed himself about; his tongue was dry and glazed. Another dose was administered, which did not produce sleep; but the patient lay quiet, passed much urine, and from this time began to recover. The gentleman is now in perfect health; and we hesitate not to say that, in our experience of typhus fever occurring in the upper ranks of society, we never witnessed a recovery so distinctly attributable to medicine."—From a Review of Dr. Graves's *Clinical Medicine*, referred to at p. 147.

The observations of Dr. Oppler, referred to at page 222, have been already given in Appendix vi.

The following case from Dr. Graves' *Clinical Medicine*, illustrates the eliminative action of tartar emetic alluded to at page 227. Few writers on fever have contributed more valuable observations on the cerebro-spinal derangements of typhus than Dr. Graves. Fettered by no theory or prejudice, he recognized the various conditions entering into different forms of nervous derangement, ignoring none of them, and treating each with rare discrimination. Thus, he once observed to me, "I recognize the case in which opium alone, or opium combined with tartar emetic, or tartar emetic alone, should be given, and I can tell in which case either of them will succeed."

"Maculated typhus with determination to the head, when properly treated, terminates not unfrequently about the tenth, eleventh, or twelfth day; sometimes it is protracted to the thirteenth or fourteenth, but most usually it ends fatally about the eleventh or twelfth. In neglected cases, the cerebral symptoms frequently assume a fearful violence on the seventh, eighth, or ninth day; and in such instances it must be expected that the best and most appropriate plan of treatment will fail in rescuing the patient from impending dissolution. If, however, we can find out a remedy which, in many cases apparently desperate, succeeds in rescuing the patient from the jaws of death, we must be satisfied. A case of this description has occurred since our last meeting. It has excited the attention of all who witnessed it, as well from the violence of the symptoms, and apparently hopeless state of the patient, as from the rapidity with which the exhibition of the remedies employed, was followed by a striking and decided alteration in the symptoms. Any one who saw him yesterday, would scarcely recognize him as the same individual to-day. This man, named Fogarty, was admitted about the seventh or eighth day of his fever, according to the account of his friends. Of course, in such cases, we cannot give implicit credence to those loose statements; for the lower class of persons in this country never calculate the time during which the patient remains out of bed struggling against the disease, a period which, in a people inured to suffering and privation, frequently lasts three, four, or even six days. Well, this man aged five-and-twenty, and of rather robust constitution, was admitted on the 20th of December, being then about eight or nine days ill. Previously to admission he had taken purgative medicines, had his head shaved and six leeches applied behind his ears or to his temples: I forget which. Now all these measures, although, perhaps, insufficient, were extremely proper, and must have produced more or less benefit. When we examined him on the 21st, we found him in a state of high excitement, as manifested by continued mental wandering, incessant talking and raving, and frequent attempts to get out of bed. He had illusions of the senses of sight and hearing, consisting of terrific ocular spectra and alarming sounds, which threw him into a state of intense agitation; his eye was red and watchful, and he never slept. Here, then, was a very threatening array of symptoms—perfect insomnia, ocular spectra, illusions of the sense of hearing, a fiery eye, and incessant mental wandering. To this was added great derangement of the whole nervous system: his body was agitated from head to foot by continual tremors, and he had violent and persistent subsultus; his respiration was interrupted; suspirious and irregular, amounting at one time to forty in the minute, afterwards not exceeding twenty-five; the acts of inspiration and expiration were extremely unequal, and occasionally accompanied by blowing and whistling. In a former lecture, I made some observations on this form of respiration, which I termed cerebral, from having first observed it in persons subject to apoplectic attacks, either before or during the paroxysms. It is frequently observed in bad cases of fever, and is a symptom of the greatest importance. He also lay constantly on his back; his pulse 120, soft and very weak, so that the canal of the artery could be obliterated by very slight pressure; his pupils were somewhat dilated; tongue parched and brown in the centre, red at the tip and edges; skin covered with maculæ; abdomen soft and full. Those who have witnessed the case, will acknowledge that the picture I have drawn is not too highly coloured, but on the contrary falls far short of the reality; and, no doubt, you all expected that if we did not succeed at once in arresting the progress of his symptoms, the case must have proved rapidly fatal. Observe the position in which we were placed; in the commencement of the fever, certain appropriate, but inadequate remedies had been employed, and under a treatment proper, but insufficient, the disease had progressed. It was an example of one of the worst forms of fever, characterized by intense cerebral excitement, and accompanied by total want of sleep, persistent delirium, and excessive disturbance of the nervous functions. All these symptoms had come on gradually, and arrived at their acme at a period when the low and debilitated state of the patient precluded the use of depletive measures to such an extent as to exert any efficient control over the most dangerous symptoms. The application of a few leeches would be extremely hazardous, and blistering would have been wholly useless and nugatory, for before the blister could rise the man would be dead. For these reasons we concluded, that the only remedy we could have recourse to with any prospect of success, was tartar emetic. We therefore ordered a draught, composed of two drachms of mint-water, two of common water, and a quarter of a grain of tartar emetic, to be given every hour, until it produced some decided effect on the constitution. You will recollect here that the scale was vibrating between life and death—that it was necessary that our plan of operation should be at once prompt and prudent, decisive and cautious. One of the pupils promised to stay by him the whole day and watch the effects of the remedy, and I determined to visit and examine him personally in the afternoon. In the course of four hours he took four doses of the tartar emetic: the first and second—in fact, almost every dose vomited him, but not immediately. He retained each dose for a considerable time, and then threw up. After the fourth dose it began to act on his bowels, and then the medicine was suspended for some time, and a small quantity of porter administered. When I saw him at eight o'clock in the evening, he had been freely purged, and had discharged a considerable quantity of bilious yellow fluid from his bowels. He had also enjoyed about an hour's sleep; his respiration was now more uniform and natural; his raving greatly diminished; the subsultus and tremors were nearly gone, and the man appeared quite tranquil. I then ordered him a wineglassful of porter, with two drops of black drop, to be repeated every second hour for three or four turns successively. I saw that the cerebral symptoms were evidently diminished, and that there was a tendency to returning tranquillity and repose, and I wished to follow up and assist the operations of nature.

"To-day this man is in a most favourable state. His skin is covered with a most profuse warm perspiration; he has slept well; belly soft and natural, respiration slow and regular, and pulse diminished in frequency; he is calm, rational, composed, and I think I am not too sanguine in anticipating for him a speedy and certain recovery."—*Clinical Medicine*, vol. i. p. 897.

VIII.

The following cases are referred to at page 167 :—

A boy, aged ten years, was admitted into the Meath Hospital, labouring under symptoms of typhoid fever. He gradually lapsed into a state of stupor, with low muttering delirium; a few dark-coloured petechiæ appeared on the surface, which became generally dark and muddy in colour. On examination, after death, this was found not to be a case of typhoid fever, but of purulent infection. The boy had suffered for years from disease of the inner ear, and the disease had at length caused a minute opening in the petrous portion of the temporal bone which communicated with the lateral sinus, and had thus produced the septicæmia and typhoid state which simulated true typhoid fever.

Several years since I saw in consultation a lady who, many weeks before, had presented the usual symptoms of early pregnancy, more especially morning sickness. Gradually, however, the sickness assumed more the characters of that arising from gastritis, and was treated accordingly by her medical attendant, who began to entertain doubts of the existence of pregnancy, which evidently was not progressing, if it had ever existed. I was struck with the sallow, muddy colour of the surface, which presented a dusky, mottled look, without, however, any distinct eruption. There was much prostration and emaciation, which, however, might reasonably be ascribed to the long-continued daily vomiting. I found a circumscribed globular swelling in the region of the cæcum, with a remarkable doughy feel, and I detected some crepitation over the ileo-cæcal valve.

It appeared to me that the patient was sinking into a typhoid condition, and I suggested the exhibition of hydrochloric acid and bark. I did not again see the case; but I was favoured with a report of the examination after death.

The cæcum was found distended, with a mass of feces of pultaceous consistence; its mucous membrane was the seat of a large, evidently chronic ulcer, with jagged, thickened edges; a shrivelled fœtus, evidently long dead, was found in the uterus. In this case I have no doubt the typhoid state was due to septicæmia, caused by absorption from the surface of the ulcer. I had not long since, under my care, in the Meath Hospital, a case of typhoid fever in a young woman—the cæcum being distended and doughy to the feel, in which a highly septicæmic condition coexisted. The case proved rapidly fatal; but unfortunately no examination of the body was permitted.

The cases of exposure to moisture by immersion, related in the Second Lecture, and the case by Dr. Maguire (Appendix ii.), are examples of septicæmia produced by this cause. I have seen several instances in which the septic type of the fever was apparently due to the emanations from decomposing animal or vegetable substances, and the writings of the older physicians contain numerous examples of the fact.

The theory of the difference of species is by no means new. The clinical differences have been observed by all our best writers on fever, and their observations have been confirmed, illustrated, and explained by recent investigations. As soon as Louis, with his characteristic precision, determined the anatomical character of typhoid, it became manifest that his description did not apply to the fever prevalent at the time in London, Dublin, or Edinburgh. Gradually, however, its existence was recognized in each of the three cities, as by Dr. Bright, who gave an admirable description of it in London; by Dr. John Reid, who recognized its exceptional occurrence in Edinburgh, or rather in the neighbourhood; and by Dr. Stokes, who has fully described it, as it occurred in Dublin,[1] during a few years previous to the outbreak of typhus in 1836.

Meantime, numerous observers, who had opportunities of studying the typhoid form under Louis, were struck with the marked differences presented by the typhus of Dublin or Edinburgh, and suggested that the two diseases might be of different species. I may refer to the papers of Dr. H. Kennedy, Dr. Lombard, and Dr. Staberoh in the early numbers of the *Dublin Medical Journal*, and to Dr. Gerhard's able memoir on the typhus epidemic in Philadelphia in 1836, in which the characteristic differences between this disease and the endemic typhoid fever of the country are admirably portrayed.[2] But it is unquestionably to Dr. A. P. Stewart that we owe the most complete and extended series of observations of both diseases on a large scale. I need not repeat his conclusions, as I have referred to them in the lecture on the diagnosis of fevers.

---

[1] Clinical Lectures, *London Medical and Surgical Journal*, vol. v.
[2] *American Journal of Medical Sciences*, 1836-7.

The symptomatology, pathology, and morbid anatomy of the two fevers being so manifestly different, the question soon arose, can the exciting cause be the same?—and since that of typhus is proved to be contagion, while the French physicians, Andral, Louis, and Chomel, are all disbelievers in the infectious power of typhoid, may it not be connected with the existence of malaria in our large cities? Physicians of eminence had, from time to time, been struck with the occasional coincidence of enteric fever with defective sewerage. Thus Dr. Christison observes:—

"In great towns cases are met with during the intervals between epidemics, and in a station of life where epidemic fever in epidemic seasons of the worst kind is seldom witnessed. A fever of this description, tedious in its course, characterized by much nervous and muscular depression, without any particular local disturbance, and, especially, without the marked disorder of the functions of the brain which distinguishes most cases of epidemic typhus and synochus, was so prevalent among the better ranks in certain streets of Edinburgh some years ago, at a time when fever was not prevalent among the working classes, that a general impression arose among professional people of the existence of some unusual local miasma. A great variety of parallel facts might be referred to—all leading to the general conclusion, that a disease if not identical with, at all events closely resembling, synochus and typhus as described above, may arise without the possibility of tracing it to communication with the sick. A statement of this kind acquires great weight in the instance of such a visitation of disease as that just alluded to, which prevailed among people in easy circumstances in a great town."[1]

Very similar is the testimony of Dr. Cheyne: "For several years the fever appeared in families only in solitary instances, or if more than one were affected they were seized nearly at the same time, but it did not extend so as to lead us to think that it propagated itself. We are unable to assign the cause of the disease further than that we observed in several houses, in which our patients lay, that fetor which is discoverable when a sewer is choked, and, in some instances, upon inquiry it was found that the sewer leading from the house had been improperly constructed and neglected."[2]

---

[1] Library of Practical Medicine; Art. Fever.
[2] Cyclopedia of Practical Medicine; Art. Epidemic Gastric Fever.

Histories of the outbreak of endemic enteric fever, in various places, apparently arising from this cause, rapidly accumulated. A remarkable instance occurred in Birmingham, a town previously peculiarly exempt from fever. The circumstances of the outbreak are thus described by Dr. Ogier Ward:—

"The river Rea, that separates Birmingham from its suburb Badesley, and serves as a cloaca maxima to both, carries its filthy stream onward, partly to turn a mill, and partly to fill a mill pond. During the drought which prevailed last year the water was very low in the main stream and mill pond, and the mills not being regularly worked became quite stagnant and offensive. The back stream also became dry and showed its mud banks, that were only occasionally wetted by a flush of the washings of the town after a shower, or by the small surplus accumulated during the cessation of the working of the mills. The exhalations from the half dried mud and putrid water were so disagreeable at night as to nauseate the more delicate inhabitants of the adjoining streets, and soon produced disease in the form of typhoid fever of an infectious (?) character." He goes on to state that about fifty cases—some fatal—occurred in the immediate vicinity of the stream, and "still lower down the stream, where the water was as black as ink, there were thirteen pauper cases in one yard, and many others, both pauper and private, along the same line." That this fever was owing to the state of the stream is proved by the disease being confined to the locality, the small number affected in so large a population as Birmingham, the season of the year, and the exemption of this town from the causes which aid contagion.—*Provincial Medical Transactions*, vol. 6.

A number of similar instances have been collected by Dr. Murchison, who has added many valuable facts resulting from his own investigations, and which conclusively prove the causal relation between civic miasma and enteric or typhoid fever. Year by year the conviction is strengthened in the minds of the profession, that in typhus and typhoid fevers we have two distinct species arising from different causes. There are still however among us those whose opinions we must respect, who think differently, and who point to the many symptoms common to the two diseases, to the relapse, as they term it, of the one into the other, and to their occasional coexistence as proofs of their identity. Of this theory my friend Dr. Henry Kennedy, who was one of the first to recognize the differences between the typhoid of Paris and our typhus, is now one of the most zealous and determined advocates, and during the last few years has published three papers in its support. One of these was read before the Medico-Chirurgical Society of London, and published in the *Edinburgh Medical Journal* for September, 1860. The second was published in the 34th vol. of the *Dublin Quarterly Journal;* and the last was read before the Medical Society of the King and Queen's College of Physicians during the last session. Having read each of these memoirs carefully, I find the clearest and most succinct statement of Dr. Kennedy's views in the second of them, in the following passage:—

. "The conclusion, then, to which, after the fullest consideration of this question, I have arrived, is the same as that of two years since, but with still stronger convictions on the point. *I believe that the two fevers known as typhus and typhoid are the result of a single poison; and that no other hypothesis can explain so well all the difficulties of the case.* I consider further that those who hold for a plurality of poisons are bound to explain the facts already given in this paper. They should tell us why the symptoms of these two affections so often run the one into the other; why the same type of fever, whether typhus or typhoid, presents such marked contrasts; why typhoid may assume the character of putrid ataxic, or inflammatory fevers, febricula, meningitis, &c., and still be typhoid all the time; and this, be it observed, is described by those who believe in the two distinct poisons. They will also have to answer the argument taken from analogy, and tell us if scarlatina affords the most marked contrasts, why fevers should not do the same; also, how it has happened that symptoms which one writer considers essential to the natural history of typhoid, are ignored and made little of by another? And, in the last place, an explanation must be given of what has occurred in Dublin this year—that is, the union of typhus and typhoid in the same subject. Now, one and all these points may be satisfactorily explained on the idea of the existence of but one poison. I confess, however, it appears to me impossible to explain them on the theory of two."

Let us examine these difficulties of belief seriatim.

First, we are called upon to explain why the two fevers—assuming them for the moment to be two—have so many symptoms in common, and each present such marked contrasts, by which expression I presume is meant want of conformity of each to its own type.

To the first part of this question, I answer, first, that pyrexia is common to both, and many of the symptoms referred to are common to all forms of pyrexia; and secondly, that the great similarity of these fevers, in many particulars, is no more than might be looked for in the case of two morbid poisons, each of which affects the entire organism, while each has its own special affinities. When Dr. Kennedy speaks of the symptoms of the one running into the other, he refers to the diarrhœa, epistaxis, and intestinal hemorrhage of typhoid, the delirium, and other nervous symptoms of typhus, met with also in typhoid, and the eruptions of the two fevers.

It is true that diarrhœa, which is all but constant in typhoid, does occur in some cases of typhus. Moreover, in some epidemics, as in that described by Dr. Da Costa, referred to at page 107, it seems to be not the exception, but the rule. But diarrhœa is only a symptom arising from one of several conditions. Is the condition producing it the same in typhus as in typhoid? By no means, on the contrary, the testimony of numerous observers is, that the specific affection of Peyer's patches is invariably absent. Such was the case in one of the examples adduced by Dr. Kennedy himself.—(Case of Bellew, *Dub. Quarterly Journ.*, v. 34.)

The same observation applies to intestinal hemorrhage. "This," says Dr. Kennedy, "is looked on as a symptom of typhoid, but I find it more frequent in typhus."

We answer, that typhus being a blood disease, and frequently complicated with septicæmia, hemorrhage from a mucous surface is to be regarded as likely to occur; but this does not prove that it is connected with, or symptomatic of, a special local lesion, as in the case of typhoid. On this point Dr. Kennedy's testimony may be quoted against himself: "In another place," says Dr. Kennedy, "I have put on record some thirty cases of well marked typhus—some of them examined after death—in which there had been extensive hemorrhage, and yet not a trace of ulceration was found; and when the patient survived the bleeding a week, I was unable to say from what part of the canal the blood had come. It must, in fact, have been an exudation."

I need not say that the intestinal hemorrhage of typhoid has a very different signification, and that it too frequently indicates the existence of a serious local lesion.

Of epistaxis, little need be said but that the experience of Dr.

Kennedy as to its greater frequency in typhus, is opposed to that of the profession. With regard to delirium, and other nervous symptoms, all observers agree that the attraction of the poison of typhus for the nervous centres is more marked than that of typhoid, and the primary nervous symptoms more constant; but that those secondary lesions which supervene at a later stage, and are due to reactive inflammatory irritation, to septicæmia, or to uræmic poisoning, are common to both. The more accurately we discriminate between the different forms of nervous derangement the more satisfied we shall be that, in regard to these, there is no confusion of type, no running of the symptoms of one form of fever into the other. As to the eruptions, I must consider Dr. Kennedy's objections vitiated by the mixing up of petechiæ with the exanthem of typhus and with the rose-spots of typhoid, and by the want of clearness in his definition of the characters of the fever in which the alleged mingling and confusion of eruptions occurred. If—as, I believe, is the case—this commingling occurred in the cases of mixed type, in which both poisons were present, the presence of both eruptions is accounted for, and is consistent with our theory.

But Dr. Kennedy overlooks the important fact that petechiæ, properly so termed, are not peculiar to fever of either form. In fact, in this instance he does not distinguish between the phenomena of fever and its epiphenomena, between its essential and its non-essential characters. And moreover he confounds typhus and typhoid fever with the typhoid state. To this we must attribute such an objection as the following: "Why typhoid may assume the characters of putrid ataxic, or inflammatory fevers, febricula, meningitis, &c., and still be typhoid all the time; and this, be it observed, is described by those who believe in the two distinct poisons."

Dr. Kennedy next proceeds to demand an answer to "the argument taken from analogy"—and to the question, "If scarlatina affords the most marked contrasts, why fever should not do the same." We reply that we fully admit the close analogy between the poison of scarlatina and that of typhus in their modes of origin and diffusion (by ochlesis and contagion)—in the effects of intense poisoning—it being sometimes a matter of difficulty to decide which poison has produced the fearfully rapid result—in the definite course of the eruption and in the special affinities of the poison.

We even recognize the analogy in the varying amount of eruption and in the occasional occurrence of petechiæ.

But will Dr. Kennedy assert that there exists another form of scarlatina presenting a contrast to typhus in these particulars, and an equally complete analogy to typhoid; or that there are any two forms of this exanthema presenting characters as widely differing from each other as do the invasion, course, duration, eruptions, and anatomical lesions of typhus and typhoid?

If, however, by the term contrasts is meant that scarlatina, usually analogous to typhus, occasionally presents also analogies to typhoid, we freely admit the statement, and find an explanation in the fact, that the poison is frequently met in combination with miasmatic poison, to the influence of which the occurrence of the intestinal affection is to be ascribed. Dr. Kennedy refers to a remarkable case of this *typhoid* scarlatina, by Dr. Anderson of Glasgow, who has had immense experience of fever; but he omits to mention that Dr. Anderson's conclusions are the opposite of his own. The whole passage is so important that I shall quote it at length. "J. F., aged 27, a domestic servant, died on the eighteenth day of a fever, characterized by the presence of a *very copious and livid typhus eruption* over the body and extremities. On inspection of the body we found the mucous membrane of the duodenum covered with enlarged solitary glands, softened and ulcerated so that in some places the peritoneum alone remained of the coats of the bowels. Peyer's glands were enlarged throughout the greater part of the jejunum and ileum, a patch in the lower part of the latter being about five inches long, and an excavated ulcer existed in its vicinity. The small and large intestines were besprinkled with numerous, enlarged, solitary follicles"

"The generalizing faculty would readily conclude from this case that typhus and enteric fevers are identical; but listen to two more:"—

"Jane S. died of asthenic confluent *smallpox* at the period of maturation. On opening the body we found on the mucous surface of the ileum a number of enlarged Peyer's glands, two of which were ulcerated."

"Mary S., aged 16, was admitted on the fourth day of fever; there was copious *scarlatinous* efflorescence over the body; she had sore throat, and a tongue white and florid at the tip and edges. Two days after she died, having passed a large quantity of blood by stool. On inspection, we found the whole surface of the ileum of a deep red on its mucous surface, and thickly besprinkled with enlarged solitary follicles. Peyer's glands were also much enlarged and ulcerated, and many swollen follicles were scattered over the colon."

"Now can we suppose it possible that smallpox, scarlatina, and typhus are *all* identical with enteric fever, because these cases prove that the special lesions of that disease may coexist with each of them? Thus by a *reductio ad absurdum* the position falls."[1]

"But lastly," says Dr. Kennedy, "an explanation must be given of what has occurred in Dublin this year—that is, the union of typhus and typhoid in the same subject." Here again Dr. Anderson draws an opposite conclusion from the same facts. In the passage following that just quoted he says, "I have notes of the case of a young man, aged 22, who was a patient in the hospital here, with every symptom of enteric fever, including the characteristic eruption. After a month's stay, he went out convalescent, but while at his work five days thereafter, was suddenly taken ill again and readmitted, having the *typhus* eruption out on him, though the symptoms of this second fever were still partly intestinal. He slowly recovered. This case seems to me to place the non-identity of the fevers beyond a doubt; and as we have already had proof that measles and scarlatina may coexist, why should we hesitate to admit that typhus and enteric fever may also occur combined? Besides the late Dr. Todd stated that he was convinced he had observed the coexistence of the two eruptions.

Dr. Anderson concludes his argument by remarking that "the coexistence of typhus and typhoid is no greater proof of their identity than a man having the misfortune to break *both* his arm and his leg by a fall, would be a proof that these injuries were in fact the same."

But Dr. Kennedy asserts that "all the above difficulties may be satisfactorily explained on the idea of the existence of but one poison."

As the explanation is not given we cannot judge if it is adequate, but we may remark that it appears to involve certain postulates which are far from being proved.

The first of these is the dogma of Hunter, that no two morbid poisons can set up their action together in the system. That they

---

[1] Lectures on Fever, p. 112.

may, has been proved by the observations of Dr. Robert Williams, Dr. Murchison, and other British and foreign physicians, as also that the one poison may remain latent during the operation of the other, and appear upon its subsidence. Thus, I some time since attended the daughter of a physician in this city in scarlatina, the sore throat and eruption being well marked. On the decline of the disease, and before the young lady had left her bed, she again sickened with equally well-marked measles.

Now, in this case we must believe either that two poisons were present together in the blood; or that measles and scarlatina—which, like typhus and typhoid, sometimes coexist in the form of a hybrid disease—arise from one poison.

Secondly—it would seem to follow from the above that the perfect or typical case of fever is that which presents the exanthem of typhus with the rose-spots and enteric lesion of typhoid; that each form is but a part of the disease, and consequently that it does not protect from a future attack of the other—in other words that each of the groups of phenomena, which we call typhus and typhoid fevers, is the complement of the other, and that in fact, the fever poison never sets up its complete operation in the blood unless both groups are present together, or in succession—that is to say, scarcely once in a hundred instances. Now, to the advocate of the opposite theory, it appears that when, with rare exceptions, all the most recent and accurate observations agree with those made in the London Fever Hospital—(where we are told during fourteen years, and in many hundred bodies dissected, no single exception has been met with)—"that when lenticular rose-spots appear in successive crops, the abdominal lesions of enteric fever are invariably present, while in exanthematous typhus they are invariably absent"[1]—the presumption is strongly in favour of these hybrid cases being the exception, not the rule; of their being due to the presence of *two* poisons in the system rather than of the ninety-nine per cent. being cases of the imperfect development of *one*.

Thirdly—it is assumed that the two forms of fever are intercommunicable. We have, however, no satisfactory evidence of a fact which should be as easily proved as the contagiousness of typhus. The cases adduced by Huss and others prove no more than the coexistence of the two diseases in a locality, a fact which admits of a simple and satisfactory explanation.

[1] Murchison, p. 581.

Lastly—the theory of unity of poison necessarily assumes the fact of a common origin. Either different forms of fever must be derived from a single source, or different sources must contain the same poison.

Dr. Kennedy's argument not only overlooks the *possible* coexistence of two poisons in the body, but also the *probable* coexistence of two distinct sources of poisons in the dwellings of the class from which our hospital patients are derived. He thus—not very clearly—expresses his views on this point:—

"Not that I have the slightest faith in bad sewerage causing typhoid more than typhus."

This sentence may mean that Dr. Kennedy doubts the power of fecal miasm to produce any form of fever; or that he considers it may equally cause both. Either reading leads us to infer that Dr. Kennedy considers the etiology of fevers, so far as their exciting causes, as of little moment in regard to their natural history and classification. We cannot agree with this view, and we consider that the science of pyretology has been much advanced by the large and daily increasing mass of observations made on this subject: observations which fully prove the law, that the factors of fever, or those substances which act as poisons when—whether generated in, or received into, the living body—they are *retained in the blood*, have a special affinity for the surfaces from which they are normally excreted. That in accordance with this law the poison generated by the accumulated and confined emanations from the lungs and skin of crowded collections of human beings (ochlesis), has an attraction for those surfaces which is manifested in typhus and in scarlatina; while that arising from the decomposing excreta of the intestines has a similar attraction for that surface, and is eliminated by it. Dr. Kennedy does not dispute the action of the first of these agencies; but he seems to deny or at least ignore the influence of the other, on what grounds he does not inform us.

Now, we contend for this influence for two reasons. One is the all but entire want of evidence of the existence of a contagious property in typhoid fever. Not that we would be understood to assert that it cannot be propagated by contagion—for example, in the manner suggested by Dr. Budd—but that in the vast majority of cases it is not, we have the strong testimony of so accurate and experienced an observer as Andral, who says:—

"In Paris, either in the hospitals or out of them, we never recognized in this disease (dothinenteritis) the slightest appearance of a contagious character. In the hospitals we do not see it transmitted from the individual who brings it from without to those who are lying in the beds next his own; neither do we see that the patients who lie in a bed previously occupied by a person who has recovered from, or who has died of a dothinenteritis, are attacked by it; neither are the physicians or medical students who come there attacked with it, more particularly those who have had to come in contact with patients labouring under the disease. Out of the hospital what circumstances are more favourable to contagion than those generally found combined in the case of medical students who attend their companions when affected with typhoid fever? Shut up in a room which in general is very small, they pay them the most assiduous and devoted attention night and day; if the affection were contagious almost all of them would contract it, and yet we do not remember to have seen the disease even once arise in this way in a healthy individual."[1]

It is true that many cases occur from time to time in a house or family; but, as Dr. Murchison remarks, at such intervals of time as do not admit of the idea of communication from one patient to another. Thus six cases were admitted into the London fever hospital, from the one house, one in June, 1849, one in October, 1851, one in February, 1854, and so on. A similar example of successive cases admitted into hospital from a small hamlet occurred to me some years ago, which I may mention. Thirty cases of typhoid occurred in twelve houses in the course of four months without my being able, by strict and repeated inquires, to obtain evidence of contagion in a single case. In some instances the attack was nearly simultaneous in several members of the family, in others three months elapsed between them. In others the patients first affected were in hospital when other members of the family sickened. A number of facts, which it is unnecessary for me to detail, proved that the fever in this instance arose from malaria, which was removed, along with the endemic, by six weeks of almost constant rain.

But not to dwell on the negative proof, it may be said that upon no subject in practical medicine is there a larger or more constantly increasing mass of evidence than as to the power of fecal miasm to generate typhoid fever, and to the fact that it does so.

[1] Clinique Médicale, by Spillan, p. 728.

We may consider this evidence both as it is presented on the large scale in our great cities, and as offered in single instances of fever originating in rooms, houses, or limited localities.

Of the first of these an able writer in the *British and Foreign Medico-Chirurgical Review* observes: "The position of civic populations with respect to the miasmata derived from animal excreta is this—a daily and hourly accumulation of the fecal evacuations of the population, and of the domestic animals they employ, takes place with the most unerring continuousness, unvaried by any circumstance whatever. On the other hand, the removal of these excreta is uncertain, and depends upon various circumstances, as the proper construction of water-closets, and of sewers, a supply of water, due diligence on the part of scavengers, &c. Recent inquiries have shown that all cities, towns, and villages are more or less defective as to their arrangements for the prompt removal of these noxious agents, whilst in numberless instances there are scarcely any arrangements whatever, and they are left to be completed as chance or necessity may dictate. The consequence is that great numbers of people are exposed to these fecal miasmata, and especially in the summer months, when decomposition is accelerated by the augmented temperature of the season."

Such, this writer goes on to show, are the conditions of London and of Paris; and to the gradual growth of a similar condition in Edinburgh, Dr. Murchison attributes the marked increase of typhoid in that city of late years.

For instances of the generation of typhoid in smaller localities I may refer to Dr. Murchison's work. One of many which have come under my notice in this city, I may mention.

I was called to attend one of the students of this hospital in exquisitely marked typhoid fever. I was told he had contracted the disease in our fever wards. Not believing that exposure to typhus can produce typhoid, I expressed my doubt of the fact. No other cause could, however, be discovered at the time, and in particular I was assured that there was no defect in the sewerage of the house in which the patient resided. Within a year after this occurrence the proprietor of the house called upon me and stated the following fact, which, he thought, might interest me. It appeared that our student occupied a small sitting-room, a few steps below the hall, the floor of which rested on the ground, having no room underneath; here he passed much of his time reading. For some reason the owner of the house wished to increase the height of his room by sinking the floor, when, on removing the boards, a large quantity of ordure, in an advanced stage of decomposition, was found collected under them, having leaked from a cesspool belonging to the adjoining house. It was conjectured that the leakage was of long date, as no complaint of the smell had been made by the gentleman, who had for many months occupied the room.

We maintain the true explanation of the mixed types of fever is to be found in the unquestionable fact, that in few of the zymotic diseases is one poison alone present in the blood. I have alluded (in Lecture II.) to the action of malaria as a predisponent to other zymotic diseases; but there can be little doubt that this poison not only predisposes the system to such disease, but, moreover, modifies and renders it more complex when it is set up.

This is the case with epidemic cholera, and with scarlatina and diphtheria; and if in these, why not in typhus?

In proof of the supervention of the one upon the other, either during its course or in succession, we point to the analogous combination of scarlatina with measles, and of typhoid with scarlatina. In his valuable paper on the coexistence of several morbid poisons Dr. Murchison, remarking on the coexistence of scarlatina and typhoid in the London fever hospital, says: "I have notes of nine such cases, and in four, at least, of these the eruptions of the two diseases were present at one and the same time."

"With regard to one of the cases my notes are very imperfect, and I am unable to state at what period of the pythogenic fever the scarlet fever supervened.

"In the second and third cases the scarlet fever appeared in the third week of convalescence, and five weeks after admission; and in one of these cases the scarlet fever was followed by enlargement of the submaxillary glands, general dropsy, and albuminuria; and during the persistence of these symptoms, about the thirtieth day from the supervention of the scarlet fever, well-marked variola showed itself. In a fourth case the scarlet fever supervened nine days after admission, and on the twenty-first day of the primary fever. It was followed by glandular swellings, and discharge from the ears, and proved fatal. No mention is made of rose-spots after the appearance of the scarlet rash, but diarrhœa, which had been a prominent symptom before, still continued. In a fifth case, scarlet fever appeared six days after admission, and on the sixteenth day of the primary fever. Rose-spots were noted three days before the appearance of the scarlet rash, and it is not impossible but that they existed afterwards. In all of these cases, as well as in those about to be mentioned, the usual symptoms of scarlatina, in addition to the rash, were present. In the four following cases the eruptions of pythogenic and scarlet fever existed simultaneously."[1]

If the fact is admitted that two morbid poisons may, and not unfrequently do coexist in the human body, one of these being malarial and, as such, having a special affinity for the intestinal glands, all the exceptional cases given by Huss and others are accounted for, and all Dr. Kennedy's difficulties are explained.[2]

I may be allowed to observe that I cannot agree in Dr. Kennedy's estimate of the value of the researches of London physicians, or think that on this subject they can be fairly said "to have ignored anything but what has occurred in their own city."

It is true that Dr. Cheyne[3] distinctly recognized and described the *clinical* characters of the three forms of fever prevailing in Dublin 50 years since; but it is by observers in London, Scotland, and America, that their natural history has been investigated, and their classification determined, with an accuracy and a precision which must eventually carry conviction to every unprejudiced mind.

[1] *British and Foreign Medico-Chirurgical Review*, vol. xxiv.

[2] The following is the case by Huss, so frequently referred to by Dr. Kennedy:—

"A man had died, it was stated, of typhus fever, the brother and his wife went to live in the house of the deceased, and used his clothes without previous airing and cleaning. They were soon taken ill and brought to the hospital, where they both died. The husband had violent delirium and a profuse petechial eruption; the post-mortem examination showing no change of the intestinal glands. The wife had milder cerebral symptoms, and a very scarce crop of eruption, but on examination swollen mesenterical glands, and swollen and ulcerated Peyer plaques were found in abundance."

[3] Dublin Hospital Reports. Vols. 1 and 2.

INDEX.

Abdominal lesions in fever, 101
    post mortem, 108
    symptoms, how derived, 101
Abdomen, pain in, during fever, 106
Access of fever, characteristics of, 107
Access, insidious, its influence on prognosis, 193
Access of typhus, date of the, 108
Acid, sulphuric, in typhoid, 218
Age, its influence on the result of fever, 189
Air, pure, in fever, importance of, 213
Albumen in the urine of fever, 115
Alvine discharges, characteristics of, in fever, and their import, 104
Ammonia, exhalation of, in fever, 112
    carbonate, its exhibition during treatment, 220
Anatomical lesions in typhus contrasted with those of typhoid, 181
    lesions in typhus contrasted with those of typhoid, by Dr. A. P. Stewart, 182
Anæsthesia, cutaneous, its value as a prognostic, 199
Anorexia in fever, 102
Antidote to fever poison, the? 207
Antimony and opium in fever, 227
    exhibition of, towards the crisis, 240
Arachnitis in fever, 159
Armstrong, Dr., on fever without reactive stage, 40
Arresting fever? 207
Arteriotomy in fever, 230
Aspect of the typhus patient, 124
Battley's sedative in typhoid, 218
Bernard, Claude, on influence of the blood and nervous system, 39
Blisters in fever, 220
    the inefficacy of, as a guide to prognosis, 202
    occasional ill effects of, 241
Blood contamination, secondary, 72
    Coleman and Reid's experiments illustrating the utility of V. S. in blood stasis, 99
    letting, in stasis, 99
        in arresting fever? 208
        modus operandi of, 210

Blood—
    letting, question of, in the treatment of fever, 231
    poisoning, signs of, in relapsing synocha, 174
    stasis, effects of, its treatment, 96
        cases illustrating, 96–97
    secondary contamination of, 153
    in the urine, as a symptom, 115
Boiling water as a powerful counter-irritant, 220
Brain, acute congestion of, symptoms, 128
    anœmia and hyperœmia of, difficulty of distinguishing, 141
    congestion of, its connection with nervous anxiety and overwork, 137
Brandy in fever, 220
Breathing, cerebral, diagnosis of, 87
Bright's disease, a predisponent, 51
    with fever, 118
Broncho-pneumonia, signs of, 88
Bronchitis, suffocative, in typhus, treatment of, 219
Bullæ, 78
    indicative of urœmia, 121
Calomel in fever, 211
Cantharides in fever, 220
Capillary circulation, its dependence on the left ventricle, 86
Cardiac and pulmonary lesions, 80
    phenomena, their importance, 86
Catarrhal typhus, 220
Cerebritis, case illustrating, 141
Cerebral lesions, their latency in fever, 138
    their occasional connection with insolation, 139
    their dependence on previous disease, 139
    symptoms from fever poison and from inflammation contrasted, 129
    respiration, definition of, Graves, 87
Cerebro-spinal arachnitis, 132
    lesions, 123
    group of symptoms indicative of, 133
    summary of, 151
    functions, derangement of, 123
Chambers, Dr., on the reception of fever poison, 211

Chloroform in fever, 228
Cholera, resemblance of, to typhoid, 133
Circulatory and respiratory systems in fever, rules for studying the, 100
Circulation and respiration, derangements of the, 80
Circulation, signs derivable from the, in forming prognosis, 197
Clinical study of fever, in what it consists, 43
Cold, combined with alcohol, a predisponent, 51
  exposure to, a predisponent, 51
  and moisture, effects of, on the economy, Lehman's experiment, 51
  Coleman and Reid's experiments, 99
Colour and expression in typhus and typhoid contrasted, 76
  the, in epidemic relapsing fever, 77
  changes in, to what due, 78
Condition of the patient at the time of seizure, its effect on prognosis, 191
Congestion, pulmonic, in fever, 88
Consciousness, normal condition of, in typhus, 125
Contagion of fever, imported, 61
  of typhus, typhoid, and epidemic fever, distinct, 67
Convalescence in relapsing synocha, characteristic of, 171
  in typhoid, dangers during, 203
Convulsion, its connection with crisis, 147
  with blood contamination, 155
Corrigan's, Sir D. J., example of the coexistence of different morbid poisons, 65
  on the urine of fever, 116
  on management, with reference to sleep, 215
Crisis, what, 42
  duties of the physician during the period of, 235
  modes of, their influence on treatment, 239
  nervous symptoms as modes of, 146
  prognostic signs observable at, 202
  of relapsing synocha, 170
  of typhus, characteristics of, 172
  of typhoid, 174
Critical days, 234
Critical and post-critical periods of the three fevers contrasted, 177
Crowding-over, its influence as a predisposing cause of fever, 49
  as an exciting cause, 60
  on prognosis, 192
Cullen on the effect of light in the treatment of typhus, 124
Cure, the, of fever? 204
Cutaneous transpiration, how modified, 78
Days, critical, 234
Day of the fever, what? importance of the question, 234

Digestive system, the signs derivable from, in forming prognosis, 198
Diagnosis of fever as an essential disease, 162
Diarrhœa in fever, its import as a symptom, 104
Delirium, characteristics and varieties of, in typhus, 124
  importance of, in diagnosis, 200
  following crisis, treatment of, 240
  tremens and typhus, difficulty of diagnosis, 164
Discolorations, yellow, in typhus, Hasse on, 154
Diseases, other, aids to their differential diagnosis from fevers, 163
  their differential diagnosis from fever, difficulties of, 163
  liable to be confounded with fever, 163
  pre-existing, their influence on prognosis, 190
Diurnal periods, effects of, 236
Dothinenteritis, tenderness of, not to be confounded with hyperæsthesia, 107
Dysæsthesia, muscular and cutaneous, 129.
Dysentery in relapsing synocha, 181
Dyspnœa depending on derangement of pulmo-cardiac circulation, 96
Emetics in arresting fever, use of, 208
  modus operandi of, in fever, 210
Emotions, depressing, influence of, in fever, 52, 150, 191
Epileptiform convulsions near to or after crisis, to what due, 149
Epistaxis in fever, 106
Epithems, turpentine, in fever, 220
Eruption of typhus and typhoid contrasted, 76
  difficulties in distinguishing, 77
  coexisting, 77
Etiology of fever, 45
  summary of, 55, 66
Exanthemata and fever, differential diagnosis, 166
Exciting cause of fever, 45, 57
Exhalations of decomposing organic matters a cause of fever, 49
Family peculiarities, their influence on predisposition, 54
  on prognosis, 190
Famine fever, its characteristics, 66
  a predisponent, 50
Fatigue a predisponent, 50
  effects of, on the progress of fever, 50
Fatal termination, signs portending, 201
Fever, access, sudden, of, 168
  arrest of, 208
  cause of, essential, 45
    predisposing, 45
  colour and expression in, 76
  contagiousness of, 61
  deportation of, 61

INDEX. 313

Fever—
    diagnosis of, 162
    epidemic relapsing, colour in, 77
    investigation of a case of, mode of, 70
    and pneumonia, relation between, 91
    poison, 37
        primary action of, 38
    predisponent, its essential, 47
    predisposition to, definition of, 48
    process, first step in, 38
    reaction, stage of, 41
        absent, 39
        Armstrong on, 40
        post-mortem appearance, 41
    relation of, to condition of persons, places, and periods, 57
    relapsing, cause of relapse, 174
    study of, its importance, 34
        its difficulties, 34
    studying, mode of, 35
    symptoms of first stage, or that of depression, 39
    theory of, 36
        Virchow's, 39
Fevers, classification of, question of, 183
    initiatory paroxysms of, 167
    laws regulating the diffusion of, 185
    management and treatment of, 206
Food, rules for administering, in fever, 214
Gargouillement, its import as a symptom, 107
    in relapsing synocha, 181
    in typhus, absence of, 107
Glands of Peyer and Brunner, progress of the morbid process in, in typhoid, 110
Gooch on the faculty of observation, 79
Graves, the late Dr., on cerebral symptoms of fever, 129
Hæmaturia as a symptom, 116
Headache, in fever, 128
    its import in prognosis, 195
Heart, fibrinous coagula in, 96
    condition of, in fever, Dr. Stokes on, 81
    congestion of right cavities of, how caused, 94
    effect of "bleeding" on, 210
    importance of the study of, in fever, 80
    softening of, auscultatory signs of, 82
    sounds of, when diagnostic of cerebral irritation, 130
Hemorrhage, intestinal, in fever, 105
    treatment of, 218
Henderson, Dr., on enlargement of the spleen in fever, 106
    on convulsions in relapsing fever, 223
Hoffman on narcotics in fever, 202
Humoralists and Solidists, their controversy, 38
Hunter's, Dr. (1773), examples of fever, the result of ochlesis, 59

Hyoscyamus, with antimony and camphor, in fever, 227
Hyperæsthesia, cutaneous, 107
    its value as a prognostic, 199
Hysteria, connection with crisis, its, 146
    import of, as a prognostic, 196
    observations on, of Cheyne and Graves, 150
Idiosyncrasy, effects of, on fever, 54
Ileo-cæcal region, tenderness over, indicating, 107
Incubation, prolonged, its influence on prognosis, 193
Intellect, condition of, in fever, 124
    signs derivable from condition of, in forming prognosis, 200
Intestine, affection of, in typhoid, its treatment, 218
Intestinal follicles, morbid change in, 110
    perforation, symptoms of, 108
Invasion of fever, mode of, its influence on prognosis, 192
James' powder in fever, 211
Jaundice in fever, 103
Judgment, in typhus, 124
Kennedy, Dr. Henry, on general paralysis from fever poison, 148
Kidney, condition of the, in fever, 221
    disease of, chronic, in fever, 118
    due performance of its function impeded in fever, 113
Laws regulating pathological phenomena, utility of knowing, 72
    the diffusion of fevers, 185
    the fatality of fever, 185
Leeches in fever, 230
Lesions, abdominal, in fever, after death, 107
    cardiac and pulmonary, 37
    cerebro-spinal, 123
    characteristic of typhoid, 110
Life, mode of, its influence on prognosis, 190
Light, effects of, in typhus, 124
Liver, pathology of, in fever, 109
Locality, its influence on the result of fever, 190
Lungs, anatomical lesions of, in fatal cases of fever, 88
    inflammatory congestion of, during fever, 93
Malaria, a predisponent, 49
    effects of, on the blood, 50
Management and special treatment of fevers, rules for the, 242
Medulla oblongata, symptoms referable to, 151
Memory, condition of, in typhus, 125
Meningitis low, and typhoid with cerebro-meningitis, differential diagnosis, 165
Menstruation, suppression of, in fever, its effects, 154
    function of, causing susceptibility, 48

814   INDEX.

Mercury in fever, 232
Metamorphosis of tissue, regressive, effects of cold on, 51
    effects of alcohol on, 51
    causes why the products of, are present in the blood, 153
    products of, in the blood, modifying symptoms, 153
Meteorism, its importance in fever, 106
Miasm, animal, fever generated from, character of, 62
    civic or fecal, character of fever from, 64
    a cause of fever, 63
Moan, peculiar, a characteristic of uræmic poisoning, 156
Morbid poisons, different, the existence of, 65
Mortality in fever, rate of, varying, 188
    influence of age on, 189
    of typhus and typhoid, rate of, 179
Motor functions, how affected in fever, 126
Murchison, Dr., on the anatomical lesions of typhoid and typhus, 182
Muscular tremor, 155
Narcotics in fever, 202
Nausea and vomiting in fever, 105
Nephritis in fever, 117
Nervous lesions, predisponents, 137
    symptoms consequent on imperfect crisis, 146
        following crisis, 147
        in typhoid, date of, 173
    system, predisponents acting through the, 52
Nutriment in fever, 214
Occupation, its influence on prognosis, 180
Ochlesis, definition of, 49
    a cause of fever, 49, 59
    capable of generating typhus, 60
    fever generated by the contagiousness of, 60
    a predisponent, 49
Ochrey stools of typhoid, 105
Odour from surface of typhus and typhoid patients, 78
Opium combined with wine and brandy, when indicated, 229
    contra-indicated, when, 226
    in typhoid, 219
Paralysis following crisis, 148
Parkes, Dr., on the characters of urine in fever, 114
    Virchow's theory of fever, 39
Pathological changes of the, each form of fever, contrasted, 175
    phenomena the result of fever poison, 71
    causes of, in typhus and typhoid, 176
    mode of studying the, 73
Perforation of the bowel, symptoms of, 108

Perforation of the bowel—
    treatment to prevent, 219
    treatment of, 219
Peritonitis, low, after typhoid, treatment, 242
    tubercular and typhoid, differential diagnosis, 165
    symptoms indicating, 108
Petechiæ of relapsing fever and typhus, how they differ, 78
Peyer, glands, pathology of, in fever, 110
Phenomena referable to the surface in fever, 78
    sensible, value of, in diagnosis, 79
Phthisis and typhus, differential diagnosis, 165
Pleuritis in typhus, 238
Pneumonia, congestive, signs of, 89
    and fever, relation between, 91
    hypostatic, treatment of, 220
    and pulmonic congestion, their pathology contrasted, 88
    and typhus, differential diagnosis, 163
Poison of fever, action of, 37
    concentrated dose of, its effects, 45
    conformity of, to the laws of other morbid poisons, 45
    efforts at the elimination of, 42
    order of its operations, 43
    Simon, Mr., on, 46
Poisoning, uræmic, symptoms of, 155
    treatment of, 223
    theory of (Dr. Oppler), 222
    various sources of, 117
Post-mortem appearances of the cerebrum and its appendages in fever, 128
    Dr. Reid's observations on, 128
Potassa, nitrate of, in fever, 211
Predisponent of fever, the essential, 46
Predisponents acting through the nervous system, 52
    existing previously in the blood, 46
    influence of, continuing during the course of the disease, 53
    influencing the results of fever, 53
    to typhus, 183
Predisposition, degrees of, under different circumstances, 48
    local or special, 55
    Mr. Simon on, 46
    peculiar, or idiosyncrasy, 54
Prognosis, rational, 187
    age of patient, influence on, 189
    caution in forming a, 188
    rules for forming a, 187
    in typhus and typhoid, contrasted, 203
    in typhoid, to be guarded, 204
    unfavourable, indications of, 195
    conformity to type, 195
    daily observations of symptoms, 196
    phenomena of crisis, 202
    temperature, 204

INDEX.

Prognostic indications derivable from treatment, 201
Ptosis as a symptom, 156
Puerperal condition one of susceptibility, 49
Pulmonic congestion, in fever, 88
   and loaded right ventricle of heart, their connection, 95
Pulse, when indicating an unfavourable prognosis, 195
   radial, notation of, its importance, 86
Purgatives in fever, 211
Rash of typhus, 75
   date of appearance, 172
   of typhoid, 76
Reid's and Coleman's experiments, 99
Relapse in relapsing synocha, cause of, 174
Remora in the great veins, cause of, 80
Respiration and circulation, functional derangement of, 80
   derangements of, mode of studying the, 81
   cerebral, definition of, 86
   diagnosis of, 87
   condition of, affecting prognosis, 197
   mode of derangements of, 87
Rigor of crisis in relapsing fever, 171
   when an unfavourable symptom, 195
Rokitansky's atonic ulcer of intestine, 111
Season, effects of, on the mortality of typhoid, 188
Senega in fever, 220
Sensation, derangements of, in fever, 127
Senses, special, condition of the, in fever, 127
   signs derivable from, in forming prognosis, 199
Sensorium, condition of, in typhus, 124
Septicæmia confounded with typhus, 166–7
   in typhoid, 173
Sex, influence of, on the mortality of typhus, 189
Sleeplessness and delirium following crisis, 147
Sleep in fever, 127
   importance of procuring, 215
Solar rays, direct, their connections with cerebral lesions, 139
Solidists and Humoralists, their controversy, 38
Spleen, condition of, in relapsing fever, 106
   rupture of, in fever, 109
Stasis, blood, effects of, and treatment of, 96
   cases illustrating, 96
Stimulants, combined with bloodletting, use of, 99
   during the period of crisis, 236
Stokes, Dr., on arrested typhus, 92
   on softened condition of heart in fever, 82

Stomach, irritable, import of, in fever, 102
Strabismus a symptom, 156
Study, excessive, a predisponent, 55
Sudamina, 77
Surface, appearance of, in different forms of fever, 75
   value as a prognostic, 76
Symptoms of access of typhoid, 169
   of typhus, 168
Symptoms, first, of fever, 39
   of reactive stage of fever, 41
   order of, in typhoid and typhus, 176
Synocha, relapsing, character of, the convalescence from, 171
   mortality, rate of, in, 180
   prognosis in, 204
   reaction, state of, 169
Temperature, change and range of, in typhus, 73
   change and range of, in typhoid, 74
   changes, marked, significancy of, 204
   in relapsing fever, 74
   in typhus, typhoid, and relapsing fever, contrasted, 74
   rules for taking the, in fever, 74
   in stage of reaction of relapsing synocha, 170
Terminations, fatal, characteristics of, 179
Thermometric observations in fever, 74
Thirst in fever, 102
Tissue, retrograde metamorphosis of, its connection with susceptibility, 48
Tongue, in early stages of typhus, characteristics of, 101
   prognostic indications derived from state of, 198
Transpiration, cutaneous, modification of, 78
Treatment of fever, 206
   dietetic and medical, in typhus and typhoid, contrasted, 217
   general, as quasi antidotal, 213
   indications of, 226
   prognostic indications derivable from, 201
   special, what it depends on, 207
Tremor, muscular, in certain cerebral states, 151
   in uræmia, 155
Turpentine in catarrhal complication, 220
   in typhoid, 218
   as a purgative in fever, 233
Types of constitution, family, effects of, on the progress of fever, 54
Typhus, arrested, after Stokes, 91
   access of, symptoms of, 168
   and delirium tremens, differential diagnosis, 164
   duration, progress, and termination of, contrasted with relapsing synocha, 173
   and pneumonia, differential diagnosis, 163

Typhus—
  and typhoid, pathological phenomena contrasted, 176
    summary of differences, 183–4
  softening of the heart in, signs of, 82
    Stokes on, 81
  typhoid, and relapsing synocha. distinct diseases, 183
  typhoid, and relapsing fever, their mode of invasion contrasted, 70
  unfavourable signs in, 203
Typhoid, access of, symptoms of, 169
  anatomical lesions of, contrasted with those of typhus, 183
  convalescence from, treatment during, 241
  duration, progress, and termination, contrasted with relapsing synocha, 172
  and acute phthisis, differential diagnosis, 164
  favourable signs during, 202
  prone to attack the young, 184
  state, the, 167
  and tubercular peritonitis, differential diagnosis, 164
  unfavourable signs in, 203
Uræmia, symptoms and characteristics of, 119
Uræmic convulsions in fever, treatment, 224
  poisoning, symptoms of, 155
Urea, deficient excretion of, result of, 155
  excretion of, in typhoid, 174
Urinary functions, treatment of the, in fever, 221
  system, aids for investigation, 113
    its condition in fever, 121
    derangements of, in fever, 112

Urine, bloody, 116
  blood in the, characteristics and importance of, 115
  deviations of the, from its typical characteristics, 114
  febrile, typical characteristics of, 114
  normal, in fever, 190
  suppression of, in fever, treatment, 223
Ventricle of the heart, the left, its condition in fever, 81
  the left, its connection with the systemic capillary circulation, 86
  the left, its condition weakened, Stokes on, 81
  the right, condition of, how to study, 93
  the right, diminished power of, the influence of, on the pulmonic capillaries, 95
  the right, diminished contractility of, how manifested, 93
  the right, its loaded condition and pulmonary congestion, 95
Vibices, 78
Virchow's theory of fever, 39
Vomiting in fever, 105
  green in, 129
  import as regards prognosis, 196
Water, external and internal use of, in fever, 217
Wine in fever, 215
  contraindicated, when, 226–238
  effects of, on capillary circulation, 86
  exhibition of, question entered into, 237
  withdrawal of, when indicated, 240
Yeast, in typhoid, 218
Youth, susceptibility of, to fever, 98

ON

RAILWAY AND OTHER INJURIES

OF THE

NERVOUS SYSTEM.

ON

RAILWAY AND OTHER INJURIES

OF THE

NERVOUS SYSTEM.

BY

JOHN ERIC ERICHSEN,
FELLOW OF THE ROYAL COLLEGE OF SURGEONS,
PROFESSOR OF SURGERY AND OF CLINICAL SURGERY IN UNIVERSITY COLLEGE;
SURGEON TO UNIVERSITY COLLEGE HOSPITAL;
EXAMINER IN SURGERY AT THE UNIVERSITY OF LONDON;
AND FORMERLY SO AT THE
UNIVERSITY OF DURHAM, AND THE ROYAL COLLEGE OF PHYSICIANS.

"JE RACONTE, JE NE JUGE PAS."
MONTAIGNE.

PHILADELPHIA:
HENRY C. LEA.
1867.

PHILADELPHIA:
COLLINS, PRINTER, 705 JAYNE STREET.

NOTICE.

The following Lectures were addressed to the Students attending the University College Hospital in the Spring of this year.

They are published as they were delivered; hence their colloquial style.

My object has been to describe certain forms of Injury of the Nervous System that commonly result from Accidents on Railways, to which I have reason to believe the mind of the Profession has not been directed with that amount of attention which their frequency and the important questions involved in them, appear to demand.

JOHN E. ERICHSEN.

6 CAVENDISH PLACE, LONDON,
    June 15, 1866.

CONTENTS.

LECTURE THE FIRST.
INTRODUCTION.

                                                                                   PAGE

Introductory Remarks—Injuries of Spine—Importance of Subject—Conflict of Professional Opinion not confined to Medical Profession—Necessity of Precision of Statement—The "Railway Spine"—Opinion of Surgeons—Bacon's Opinion—Case of Count de Lordat—Conclusion 17

LECTURE THE SECOND.
EFFECTS OF SEVERE BLOWS ON THE SPINE.

Concussion of the Spine from Direct and Severe Injury—Opinions of Authors, Cooper, Mayo, Bell, Boyer, Abercrombie, Ollivier—Case 1, Recovery—Case 2, Partial Recovery—Case 3, Permanent Paralysis—Case 4, Death after Concussion of the Spine from Direct Injury to the Back—Effects of Severe Blows on the Spine—Fatal Result of Concussion of the Spine—Hemorrhage within Spinal Canal—Laceration of Membranes—Inflammation of Cord and Membranes—Cases—Complications of Injury of Cord—Rupture of Ligaments of Spine—Inflammatory Softening 26

LECTURE THE THIRD.
ON CONCUSSION OF THE SPINE FROM SLIGHT INJURY.

Concussion of Spine from Slight Injuries—Railway Injuries Peculiar but not Special—Effects of Slight Blows on Spine, Case 5—Concussion from Railway Injury, Case 6—Concussion from Railway Injury, Case 7—Concussion from Carriage Accident, Case 8—Concussion from Falls—Long durations of Symptoms 43

LECTURE THE FOURTH.
CONCUSSION OF THE SPINE FROM GENERAL SHOCK. TWISTS AND WRENCHES OF THE SPINE.

Concussion from General Shock—Case 9. Concussion from Shock to Feet—Case 10. Concussion from Railway Shock—Case 11. Concussion

viii   CONTENTS.

                                                                      PAGE
from Railway Shock—Case 12. Concussion from Railway Shock—
Twists, Sprains, and Wrenches of the Spine—Case 13. Wrench of
Spine—Case 14. Twist of the Spine.—Effects of Twists and Wrenches
of the Spine   55

LECTURE THE FIFTH.
SYMPTOMS AND PATHOLOGY OF CONCUSSION OF THE SPINE.

Period at which Symptoms begin to develop—Length of Time that often
elapses—Concussion not associated with other Injury—Nature of
Changes produced by Concussion—Early Symptoms of Railway Concussion—Detail of Symptoms of Railway Concussion—Symptoms of
Railway Concussion—Interval between Accident and Symptoms—
Pathology of Railway Concussion—Mr. Gore's Case   72

LECTURE THE SIXTH.
DIAGNOSIS, PROGNOSIS, AND TREATMENT.

Diagnosis from Cerebral Concussion—From Rheumatism—From Hysteria
—Prognosis of Concussion of the Spine—What is meant by Recovery
—Probability of Recovery—Period of Fatal Termination—Treatment
—Importance of Rest—Counter Irritation—Medical Treatment   .   91

INJURIES OF THE NERVOUS SYSTEM.

LECTURE THE FIRST.

Introductory Remarks—Injuries of Spine—Importance of Subject—Conflict of Professional Opinion not confined to Medical Profession—Necessity of Precision of Statement—The "Railway Spine"—Opinion of Surgeons—Bacon's Opinion—Case of Count de Lordat—Conclusion.

INTRODUCTORY REMARKS.

GENTLEMEN: It has justly been said by one of the greatest masters of the Art of Surgery that this or any other country has ever produced—Robert Liston—that no injury of the head is too trivial to be despised. The observation, true as it is with regard to the head, applies with even greater force to the spine; for if the brain is liable to secondary diseases in the one case, the spinal cord is at least equally, and probably more so, in the other.

My object in these Lectures will be to direct your attention to certain injuries and diseases of the spine arising from accidents, often of a trivial character—from shocks to the body generally, rather than from blows upon the back itself—and to endeavour to trace the train of progressive symptoms and ill effects that often follow such injuries.

These concussions of the spine and of the spinal cord not unfrequently occur in the ordinary accidents of civil life, but from none more frequently or with greater severity than in those which are sustained by passengers who have been subjected to the violent shock of a railway collision; and it is to this particular class of injuries that I am especially desirous of directing your attention. For not only have they, in consequence of the extension of railway traffic, become of late years of very frequent occurrence, but, from the absence often of evidence of outward and direct physical injury, the obscurity of their early symptoms, their very insidious character, the slowly progressive development of the secondary organic lesions, and functional disarrangements entailed by them, and the very uncertain nature of the ultimate issue of the case, they constitute a class of injuries that often tax the diagnostic skill of the surgeon to the very utmost. In his endeavours not only to unravel the complicated series of phenomena that they present, but also in the necessity that not unfrequently ensues of separating that which is real from those symptoms which are the consequences of the exaggerated importance that the patient attaches to his injuries, much practical skill and judgment are required.

The secondary effects of slight primary injuries to the nervous system do not appear, as yet, to have received that amount of concentrated attention on the part of surgeons that their frequency and their importance demands; and this is the more extraordinary, not only on account of the intrinsic interest attending their phenomena, but also from their having become of late years a most important branch of medico-legal investigation. There is no class of cases in which medical men are now so frequently called into the witness-box to give evidence in courts of law, as in the determination of the many intricate questions that often arise in actions for damages against railway companies for injuries alleged to have been sustained by passengers in collisions on their lines; and there is no class of cases in which more discrepancy of surgical opinion is elicited than in those now under consideration.

It is with the view and in the hope of clearing up some of the more obscure points connected with these injuries, that I bring this important subject before you; for I believe that, as these cases come to be more carefully studied, and consequently to be better understood, by surgeons, much of the obscurity that has hitherto surrounded them will be removed, and we shall less frequently see those painful contests of professional opinion which we have of late been so often constrained to witness in courts of law.

That discrepancy of opinion as to relations between apparent cause and alleged effect; as to the significance and value of particular symptoms, and as to the probable result in any given case, must always exist, there can be no doubt, more especially where the assigned cause of the evil appears to be trivial, where the secondary phenomena develop themselves so slowly and so insidiously that it is often difficult to establish a connecting link between them and the accident.

And for the existence of such discrepancy of opinion, and for the expression of it, if necessary, on oath,—as a matter of opinion, merely—a very undue amount of blame has been cast on members of the medical profession.

In well-marked and clearly-defined cases of injury, where the physical lesion is distinct—as in a fracture—or the general symptoms unmistakable, as in the loss of sight or hearing, or in the sudden and immediate induction of paralysis—no discrepancy of opinion can or ever does exist; and I have no hesitation in saying that in at least nineteen-twentieths of all the railway or other accidents that are referred to surgeons of experience for arbitration or opinion, there is no serious difference as to the real nature of the injury sustained, or as to its probable result on the patient, either locally or constitutionally, immediately or remotely. But in a certain small percentage of cases, in which, as has already been said, the relation between alleged cause and apparent effect may not always be easy to establish, in which the symptoms come on slowly and gradually, in which they may possibly be referable to other constitutional states, quite irrespective of and pre-existing to the alleged injury, and in which the ultimate result is necessarily most doubtful, being dependent on many modifying circumstances; in such cases, I say, discrepancy of professional opinion may legitimately, and indeed must necessarily, exist. There is no fixed standard by which these points can be measured. Each practitioner will be guided in his estimate of the importance of the present symptoms, and of the probable future of the patient, by his own individual experience or preconceived views on these and similar cases. But, in these respects such cases differ in no way from many others of common and daily occurrence in medical and surgical practice. We daily witness the same discrepancies of opinion in the estimate formed by professional men of obscure cases of any kind. In cases of alleged insanity, in the true nature and probable cause of many complicated nervous affections, in certain insidious and obscure forms of cardiac, pulmonary, and abdominal disease; in such cases as these we constantly find that "*quot homines tot sententiæ*" still holds good. Even in the more exact science of chemistry, how often do we not see men of the greatest experience differ as to the value of any given test, as to the importance of any given quantity of a mineral—as of arsenic, mercury, or antimony, found in an internal organ—as an evidence of poisoning.

Were public discrepancies of opinion confined to the members of the medical profession it would be a lamentable circumstance, and one which might justly be supposed to indicate a something deficient in the judgment, or wrong in the morale, of its members. But when we look around us, and inquire into the conduct of members of other professions, we shall find that in every case in which the question at issue cannot be referred to the rigid rules of exact science—whether it be one of engineering, of law, of politics, or of religion—the same conflict of opinion will and does, as a matter of necessity, exist, and the same subjects and the same phenomena will present themselves in very varying aspects to the minds of different individuals—conflict of opinion being the inevitable result.

Look at any great engineering question. Are engineers of the highest eminence not ever to be found ranged on opposite sides in the discussion of any point of practice that has become one of opinion, and that cannot be decided by a reference to those positive data on which their science is founded? Is there no discrepancy of opinion often manifested amongst gentlemen of unimpeachable integrity in their profession, as to the possible causes of that very accident, perhaps, which has occasioned the catastrophe that has led to your presence in the witness-box?

Is the law exempt from conflicts of opinion, independently of those that are of daily occurrence in its courts? Are there no such institutions as courts of appeal? Are decisions never reversed? Are the fifteen judges always of one mind upon every point that is submitted to them? Do we never see conflict of opinion spring up in the Lords and Commons, amongst the magnates of the legal profession, on questions that involve points of professional doctrine and practice?

Is the Church herself free from differences of the widest kind on questions that we are taught are of the most vital importance? Have we not for years past heard questions of doctrine, of practice, of ritualism, discussed with an amount of vehemence and zeal to which we can find no parallel in our own profession? Are not angry passions roused in quarters where they are little to be expected, and may we not at times be tempted to exclaim, "*Tantæne animis cœlestibus iræ?*"

These conflicts of opinion, gentlemen, are common to all the professions and to every walk of life. Religion and politics, law, medicine, and the applied sciences, all contain so much that is, and ever must be, matter of opinion, that men can never be brought to one dead level of uniformity of thought upon any one of these subjects; and out of the very conflicts of opinion that are the necessary consequence of the diversity of views that are naturally entertained, Truth is at last elicited.

Far be it from me to do otherwise than to speak with the utmost respect of a learned and liberal profession, when I say that slight discrepancies of opinion arising between medical men are often magnified by the ingenuity of advocates, so as to be made to assume a very different aspect to that which they were intended to present, and are exaggerated into proportions which those who propounded them never meant them to acquire. Medical men deal habitually with the material rather than the ideal, with facts rather than with words, and are frequently, perhaps, at times somewhat inexact in the expressions they use. Mere verbal differences, mere diversities in modes of expressing the same fact, are thus sometimes twisted into the semblance of material discrepancies of statement and opinion. How often have I not heard in courts of law attempts made to show that two surgeons of equal eminence did not agree in their opinions upon the case at issue, because one described a limb as being "paralytic," whilst the other perhaps said "there was a loss of nervous and muscular power in it"—when one said that the patient "dragged" a limb, the other that he "walked with a certain awkwardness of gait." The obvious professional moral to be deduced from this is, that it is impossible for you to be too precise in the wording of your expressions when giving evidence on an obscure and intricate question. However clear the fact may be to your own minds, if it be stated obscurely, or in terms that admit of a double interpretation, you may be sure that the subtle and practised skill of those astute masters of verbal fence who may be opposed to you, will not fail to take advantage of the opening you have inadvertently given them, to aim a fatal thrust at the value of your evidence. It is your province to give a distinct and clear description of the facts that you have observed, and an unbiassed and truthful opinion as to the inferences you draw from them. It is their business to elicit the Truth, and to place the cause of their client in the best possible light, by questioning the accuracy of your facts and by sifting the validity of the opinions you have deduced from them.

I purpose illustrating these lectures by cases drawn from my own practice, and by a reference to a few of the more interesting published cases that bear upon the subject. In doing so, I shall confine myself to the detail of a few selected instances. It would be as useless as it would be tedious to multiply them to any great extent, as they all present very analogous trains of symptoms and phenomena. I will not confine my illustrations to cases drawn from railway accidents only, but will show you that precisely the same effects may result from other and more ordinary injuries of civil life. It must, however, be obvious to you all, that in no ordinary accident can the shock be so great as in those that occur on railways. The rapidity of the movement, the momentum of the person injured, the suddenness of its arrest, the helplessness of the sufferers, and the natural perturbation of mind that must disturb the bravest, are all circumstances that of a necessity greatly increase the severity of the resulting injury to the nervous system, and that justly cause these cases to be considered as somewhat exceptional from ordinary accidents. This has actually led some surgeons to designate that peculiar affection of the spine that is met with in these cases as the *"railway spine."*

But yet, though the intense shock to the system that results from these accidents naturally and necessarily gives to them a terrible interest and importance, do not for a moment suppose that these injuries are peculiar to and are solely occasioned by accidents that may occur on railways.

There never was a greater error. It is one of those singular mistakes that has arisen from men trusting too much to their own individual experience, and paying too little heed to the observations of their predecessors. It is an error begot in egotism and nurtured by indolence and self-complacency. It is easy for a man to say that such and such a thing cannot exist, because, " I, in my large experience at our hospital, never saw it," and not to trouble himself to learn, by the study of their works, that surgeons of equally large, or perhaps of far greater, experience in their generation, have seen and have described it.

Sir Astley Cooper, who certainly enjoyed a wider range of experience in surgical practice than has ever before or since fallen to the lot of any one man in this country, said that his experience, extensive as it had been, was only as a bucket of water out of the great ocean of surgical knowledge.

In the writings of Sir A. Cooper himself, in those of his predecessors and contemporaries, especially of Boyer, of Sir C. Bell, and, at a later period, of Ollivier and Abercrómbie, you will find many cases recorded that prove incontestably that precisely the same trains of phenomena that of late years have led to the absurd appellation of the "Railway Spine," had arisen from accidents, and had been described by surgeons of the first rank in this country and France, a quarter of a century or more before the first railway was opened, and that they were then generally recognized by surgeons as arising from the common accidents of civil life. The only difference being, that those accidents having increased in frequency and intensity since the introduction of railways, these injuries have become proportionally more frequent and more severe.

Bacon truly said, "They be the best physicians which, being learned, incline to the traditions of experience, or, being empirics, incline to the methods of learning." The same remark is applicable to surgeons, and that observation is as true at the present day as when it was made, nearly three hundred years ago.

Yes, truly, gentlemen, if you are "empirics," incline to the methods of learning. Do not trust wholly to your empiricism;" in other words, to your own individual experience; but learn what has been seen by others of equal, perhaps of greater, experience than yourselves; as accurate in observing, and as truthful in recording. The study of the works of such men is not a vain and futile learning, but one replete with valuable results. In reading their works, you feel that you come into direct communion with these great men—with the Boyers, the Bells, and the Coopers—and from them you will learn many a lesson of practical wisdom, the direct result of their own accurate observations.

But you may go further back than the writings of these great men, and you will find scattered here and there throughout medical literature occasional most interesting cases that bear upon this very point. You will find much in this literature that anticipates what are often erroneously supposed to be more recent discoveries, and many a man, thinking that he has struck out a new vein of truth, and finding that it has already, years ago, been explored and the ore extracted by his predecessors, may exclaim, "*Pereant ante nos qui nostra dixere.*"

If you take up the third volume of the "Medical Observations and Inquiries," you will find that in 1766, exactly one hundred years ago, a case is related by Dr. Maty of "a palsy occasioned by a fall, attended with uncommon symptoms," which is of so interesting a nature, and which bears so closely upon our subject, that I feel that I need no apology for giving you an abstract of it here, although, as it occurred between sixty and seventy years before the first railway was opened in this country, it might at first appear to have less relation to railway accidents than it really has, for it is identical in its course and symptoms with many of them.

This case, which is given at length, is briefly as follows:—

Count de Lordat, a French officer of great rank and merit, whilst on his way to join his regiment, in April, 1761, had the misfortune to be overturned in his carriage from a pretty high and steep bank. His head pitched against the top of the coach; his neck was twisted from left to right; his left shoulder, arm, and hand much bruised. As he felt at the time little inconvenience from his fall, he was able to walk to the next town, which was at a considerable distance. Thence he pursued his journey, and it was not till the sixth day that he was let blood on account of the injury to the shoulder and hand.

The Count went through the fatigues of the campaign, which was a very trying one. Towards the beginning of the winter (at least six months after the accident), he began to find an impediment to the utterance of certain words, and his left arm appeared to be weaker. He underwent some treatment, but without much advantage; made a second campaign, at the end of which he found the difficulty in speaking and in moving his left arm considerably increased. He was now obliged to leave the army and return to Paris, the palsy of the left arm increasing more and more. Many remedies were employed without effect. Involuntary convulsive movements took place all over the body. The left arm withered more and more, and the Count could hardly utter a few words.

This was in December, 1763, two years and a half after the accident.

He consulted various physicians, and underwent much treatment without benefit.

In October, 1764, three years and a-half after the fall, Dr. Maty saw him. "A more melancholy object," he says, "I never beheld. The patient, naturally a handsome, middle-sized, sanguine man, of a cheerful disposition and an active mind, appeared much emaciated, stooping, and dejected. He walked with a cane, but with much difficulty, and in a tottering manner."' His left hand and arm were wasted and paralyzed; his right was somewhat benumbed, and he could scarcely lift it up to his head. His saliva dribbled away; he could only utter monosyllables, "and these came out, after much struggling, in a violent expiration, and with a low tone, and indistinct articulation." Digestion was weak, urine natural. His senses and the powers of his mind were unimpaired. He occupied himself much in reading and writing on abstruse subjects. No local tumour or disease was discoverable in the neck or anywhere else. From this time his health gradually declined, and he finally died on the 5th March, 1765, nearly four years after the accident.

On examination after death, the pia mater of the brain was found "full of blood and lymph;" and towards the falx some marks of suppuration. The medulla oblongata is stated to have been greatly enlarged, being about one-third larger than the natural size. The membranes of the cord were greatly thickened and very tough. The cervical portion of the cord was hardened, so as to resist the pressure of the fingers.

"From these appearances," says Dr. Maty, "we were at no loss to fix the cause of the general palsy in the alterations of the medulla spinalis and oblongata." That the twisting of the neck in the fall had caused the membrane of the cord to be excessively stretched and irritated; that this cause extended by degrees to the spinal marrow, which, being thereby compressed, brought on the paralytic symptoms.

This case is of the utmost interest and importance; and though it occurred more than a century back, and was published exactly one hundred years ago, it presents in so marked a manner the ordinary features of a case of "concussion of the spine," arising from injury, that it may almost be considered a typical case of one of those accidents.

The points to which I would particularly beg to direct your attention in this case are these:—

1st. That there was no evidence of blow upon the spine—merely a twist of the neck in the fall.

2d. That no immediate inconvenience was felt, except from the bruise on the shoulder and hand.

3d. That the patient was able to walk a considerable distance, and to continue his journey after the occurrence of the accident.

4th. The symptoms of paralysis did not manifest themselves for several months after the injury.

5th. They were at first confined to the left arm and to the parts of speech.

6th. They very slowly but progressively increased, extending to the left leg and slightly to the right arm.

7th. This extension of paralysis was very gradual, occupying two or three years. The sphincter were not affected, and the urine was healthy.

8th. The general health gradually but slowly gave way, and death at last ensued, after a lapse of four years, by a gradual decay of the powers of life.

9th. After death, evidences of disease were found in the membranes of the cord, and the cord itself. The narrator of the case stating that the membranes were primarily, and the cord secondarily, affected.

You will find, as we proceed in the investigation of this subject, that the symptoms, their gradual development, and the after-death appearances presented by this case, are typical of the whole class of injuries of the spine grouped together under the one common term "Concussion," from whatever cause arising.

LECTURE THE SECOND.

EFFECTS OF SEVERE BLOWS ON THE SPINE.

Concussion of the Spine from Direct and Severe injury—Opinions of Authors, Cooper, Mayo, Bell, Boyer, Abercrombie, Ollivier—Case 1, Recovery—Case 2, Partial Recovery—Case 3, Permanent Paralysis—Case 4, Death after Concussion of the Spine from Direct Injury to the Back—Effects of Severe Blows on the Spine—Fatal Result of Concussion of the Spine—Hemorrhage within Spinal Canal—Laceration of Membranes—Inflammation of Cord and Membranes—Cases—Complications of Injury of Cord—Rupture of Ligaments of Spine—Inflammatory Softening.

It is not my intention in these Lectures to occupy your time with any remarks on those injuries of the spine that are attended by obvious and immediate signs of lesion to the vertebral column itself, such as fractures and dislocations of it, or direct wounds of the cord. The nature and the consequences, proximate and remote, of such injuries as these are so obvious and so well understood by all engaged in surgical practice, that their consideration need not detain us.

My object is to bring under your observation the effects, local and constitutional, immediate and remote, of certain forms of injury to which the spinal cord may be exposed without lesion of its protecting column or enveloping membranes. These injuries, from the obscurity of their primary symptoms, the very slow development of their secondary phenomena, and from the ultimate severity and long persistence of the evils they occasion, are of the greatest interest to the practical surgeon.

In considering these injuries, I shall adopt the following arrangement:—

1. The consideration of the effects of slight and apparently trivial injuries applied directly to the spine.

2. The effects that injuries of distant parts of the body, or that shocks of the system, unattended by any direct blow upon the back, have upon the spinal cord.

3. The effects produced by wrenches or twists of the spine.

Before, however, proceeding to the consideration of these questions, it will, I think, be important to inquire into the effects produced by those forms of concussion of the spinal cord which follow immediately and directly upon a *severe* degree of external violence applied to the vertebral column, as by so doing we shall be able to understand more clearly the phenomena resulting from the slighter form of injury.

It is by no means easy to give a clear and comprehensive definition of the term, "Concussion of the Spine." Without attempting to do so, it may be stated, in explanation of this phrase, that it is generally adopted by surgeons to indicate a certain state of the spinal cord occasioned by external violence; a state that is independent of, and usually, but not necessarily, uncomplicated with any obvious lesion of the vertebral column, such as its fracture or dislocation—a condition that is supposed to depend upon a shake or jar received by the cord, in consequence of which its intimate organic structure may be more or less deranged, and by which its functions are certainly greatly disturbed, various symptoms indicative of loss or modification of innervation being immediately or remotely induced.

In fact, it appears to me that surgeons and writers on diseases of the nervous system have included four distinct pathological conditions under this one term, "Concussion of the Spine," viz., 1. A jar or shake of the cord, disordering, to a greater or less degree, its functions, without any obvious lesion cognizable to the unaided eye. 2. Compression of the cord from extravasated blood. 3. Compression of the cord from inflammatory exudations within the spinal canal, whether of serum, lymph, or pus; and, 4. Chronic alterations of the structure of the cord itself as the result of impairment of nutrition consequent on the occurrence of one or other of the preceding pathological states, but chiefly of the third. These various conditions differ remarkably from one another in symptoms and effects, and have only this in common that they are not dependent upon an obvious external injury of the spine itself, as the laceration or compression of the cord by the fracture or dislocation of a vertebra.

Concussion or commotion of the spinal cord as a consequence of severe and direct blows upon the back is an injury that has long been recognized and described by those writers who have occupied themselves with the consequences of accidents applied to this part of the body.

Sir A. Cooper[1] relates two cases of concussion of the spine, one terminating at the end of ten weeks in complete, the other in incomplete recovery.

Mayo[2] relates two cases. In one at the end of six months there was no amelioration. In the other at the end of four months symptoms of inflammatory softening of the cord set in.

Sir Charles Bell[3] relates two most interesting cases of concussion of the spine, both occasioned by falls and blows upon the back, in one of which the symptoms were immediate, in the other only developing themselves slowly after an interval of some months.

Boyer[4] relates two cases. In one the patient struck his loins by falling into a deep ditch. He was affected by complex paraplegia, and speedily died. On examination no morbid appearances could be detected, neither fracture, dislocation, effusion, or any lesion of the cord or its membranes. In the other case, a man amusing

---

[1] Dislocations and Fractures of Joints, 8vo. ed., p. 526 *et seq.*
[2] Outlines of Pathology, Lond. 1836.
[3] Surgical Observations, London, 1816.
[4] Maladies Chirurgicales, vol. iii. p. 135.

himself with gymnastic exercises strained his back between the shoulders. He became paraplegic, and died in a few weeks. After death no lesion of any kind was found in the spine or cord.

Abercrombie, in his well-known and philosophical treatise on the Brain and Spinal Cord[1] has a short chapter on this injury, in which he relates several cases from his own observations and from the practice of others, in which the characteristic symptoms of concussion of the cord followed blows upon the spine.

Ollivier[2] has collected, from his own practice and that of others, thirteen cases of this injury. They are detailed with much minuteness. Several of these proved fatal, and of these the after-death appearances are given at length.

Concussion of the spine from a direct and severe injury of the back may terminate in four ways: 1. In complete recovery after a longer or shorter time. 2. In incomplete recovery. 3. In permanent disease of the cord and its membranes; and, 4. In death.

The probability of the termination in recovery does not depend so much on the actual severity of the immediate symptoms that may have been occasioned by the accident as on their persistence. If they continue beyond a certain time, changes will take place in the cord and its membranes which are incompatible with the proper exercise of its functions.

The following cases will illustrate these forms of spinal concussion from the infliction of severe and direct injury to the spine. The first case is an instance of complete recovery—after severe and uncomplicated concussion of the spine. The second case is one of partial recovery after incomplete paralysis from concussion. The third case is a remarkable instance of incurable paraplegia following concussion; and the fourth case is one of death following a direct blow on the spine.

Case 1.—A man, 42 years of age, a clerk, fell whilst getting down from the roof of an omnibus, striking his back heavily upon the ground. He tried to get up, but was unable to do so, and was carried to University College Hospital where he was admitted in February, 1857, under my care.

On examination it was found that he had a transverse bruise upon the back, in the dorso-lumbar region, probably from coming

---

[1] London, 1828, p. 375.
[2] Traité des Maladies de la Moelle Epinière. Paris, 1837.

in contact with the step of the vehicle in his fall. He suffered pain on pressure about the bruised part; but there was no irregularity to be detected in the line of the spinous processes or any other sign of fracture or of injury to the vertebræ. The ecchymosis extended over the two or three last dorsal and the first lumbar vertebra. His consciousness was in no way disturbed. He could not stand, his legs giving way under him. He complained of complete numbness in the left leg, but in the right there was a certain degree of sensibility associated with tingling, pricking sensations. When laid in bed he could not move the left lower extremity, but he could flex the right thigh upon the abdomen and draw up the knee, though he could not raise the foot. The catheter was passed and clear urine drawn off.

He was ordered complete rest in bed; five grains of calomel, to be followed by a purgative enema, and the use of the catheter, if necessary, every eighth hour.

Febrile reaction set in, which continued for three or four days. He was quite unable to empty the bladder; the urine was consequently drawn off by the catheter. There was no incontinence of flatus or of feces. No change in the state of the lower extremities.

At the end of a week he was decidedly better; he could raise the right foot from the bed, and the normal sensibility of that limb had in a great measure returned. He could draw up the left knee, and there was some sensation in the leg and in the dorsum of the foot. The retention of urine continued.

At the end of a fortnight motion and sensation had returned in the right lower extremity, but the left limb was still weak and partially numb, with formications and tinglings. He now began to pass his urine—which was acid—without the use of the catheter. During the whole of this period the only treatment that had been adopted was rest in bed, with an occasional aperient. He was now ordered to sit up, and had dry cupping to the lower part of the spine.

At the expiration of another week he was able to move about on his feet with a tottering, straddling gait, by the aid of a chair and stick. He now steadily improved both in appearance and in power of moving. At the end of the first month he could walk with but little assistance; he was still very weak in the left leg, which was partially numb; it felt as if asleep, and tingled.

Stimulating embrocations were ordered to the spine, and he was put on the bichloride of mercury, gr. $\frac{1}{12}$, in tincturæ cinchonæ co. ʒi, thrice a day. Under this treatment he steadily improved, and was able to leave the hospital at the end of the sixth week, walking with the aid of a stick. He was treated as an out-patient with strychnine and iron, and the local application of galvanism, for two or three weeks longer, and then dismissed cured.

This case is related as an instance of not very uncommon occurrence, in which, after a severe and direct blow upon the spine, paraplegic symptoms are suddenly developed, which again disappear completely in the course of a few weeks under the influence of rest and appropriate treatment. The only point of special interest in this case is, that although there was paralysis and complete retention, the urine continued acid throughout. It is probable that the pathological lesion in such a case as this, consists of some intra-vertebral extravasation of blood, the compression exercised by which occasions the symptoms, which disappear as it gradually becomes absorbed.

Case 2.—A painter, 30 years of age, was admitted into University College Hospital under my care, in June, 1865, under the following circumstances. He states that whilst painting a house he over-reached himself, and fell with the ladder to the ground, a height of about thirty feet, striking his back upon a gravel walk. His hand was cut in the fall but his head was uninjured. On admission he was found somewhat collapsed, cold, and with a feeble pulse. There was no evidence of fracture either of spine or pelvis, but the back was ecchymosed to some extent about the centre of the dorsal region. He could not stand, but when lying in bed could draw up the knees nearly to a right angle, although he was unable to raise the feet. He complained of numbness and tingling in both legs and feet, but could feel when pinched or pricked. The patient had perfect control over his sphincters, and the urine was acid.

He was treated by rest in bed, dry cupping to the spine, and occasional aperients. At the end of a month he had not improved, being as nearly as possible in the same state as on admission. He was now put on small doses of the bichloride of mercury in bark, and had counter-irritation applied to the spine. Some little amendment took place under this plan of treatment, and in August he was able to sit up, but could neither walk nor stand without support, and continued to complain of the numbness and tingling in his legs. Towards the end of the month he seemed to have acquired slight power over the legs, and could manage, by dragging them along, and leaning on a chair and stick or crutch, to move across the ward. He now very slowly improved, and by the end of September was able to leave the hospital. He was emaciated, cachectic-looking, and could barely manage to walk and drag his leg, by holding on to the furniture, or by pushing a chair before him. He continued through the winter mending but slowly. Towards the early part of the following year he was taken charge of by the Sisters of Mercy, who sent him to their establishment at Clewer. There he gradually regained a certain degree of health and strength. I saw him again on April 20th, exactly ten months after the accident; he was then in the following state:—

He describes himself as being languid, depressed, and as if going out of his mind. His memory has become very bad—at times all seems a blank to him. When he goes on an errand he often cannot recollect what it is about; is always obliged to write it down. His thoughts are confused; he often mixes up one thing with another. He is very nervous and easily frightened. He dreams much, and is told that he talks and cries out in his sleep.

He is "not the same man that he was," and thinks he never will be. He cannot do ordinary work as before the accident—only "odd jobs." He cannot walk more than a mile; cannot carry a pail of water without great exertion.

He is never free from an aching, throbbing pain in the back; most severe in the middle dorsal region. There the spine is very tender on pressure, and the tenderness extends to some distance on either side of it, more especially on the left. This pain is greatly increased by movement of any kind, especially by bending backwards. He stoops with great difficulty, and is obliged to go upon one knee in order to pick anything off the floor. He walks in a shuffling, unsteady manner, and always uses a stick. He complains of numbness and "pins and needles" in the right leg and foot. There is no difference in the size of the limbs.

He has suffered since the accident from muscæ volitantes and coloured spectra, "like the rainbow" before his eyes. Light does not distress him, but loud noises do. His hearing is very acute indeed.

No irritability of bladder; holds and passes his water well; urine is acid.

This case presents a good example of concussion of the spine followed by partial paralysis of sensation and motion of the lower

CASE 3.—CONCUSSION, PERMANENT PARALYSIS.

limbs without affection of the sphincters or alkalinity of urine, terminating in incomplete recovery.

It appears to me doubtful whether intra-vertebral hemorrhage took place in this case; but there can be little doubt that the spinal cord had sustained some serious organic lesion which interfered with complete recovery.

In some cases, however, the result is not so satisfactory even as in this; the symptoms that are immediately developed continuing for many years, even for the remainder of the patient's life, without change.

The following[1] is one of the most remarkable cases on record, of long persistent paralysis after a blow on the spine, the loss of sensation being so complete that the patient submitted to the amputation of both thighs without feeling the slightest pain. As this case has never, I believe, been published in this country, and is of so very remarkable a character, I have thought that it might not be out of place to give an abstract of it here.

Case 3.—A man, 22 years of age, in felling a tree, was struck on the back part of the head and between the shoulders by a large bough. This accident occurred in 1845. The force of the blow expended itself chiefly on the lower cervical spine and the shoulders. A complete paralysis of sensation and motion, of all the parts below this, was the immediate result. This condition continued without the slightest change. The vital and animal functions were naturally performed. Respiration, circulation, digestion, secretion, and assimilation were all normal. There was a sensible increase in the frequency and volume of the circulation, and respiration was noticed to be slightly increased in frequency above the normal standard. The weight of the body became greater after than it had been before the injury, and the lower limbs retained their natural heat and physical development.

The patient evidenced an unusual share of mental vigour after the injury, and possessed a resolution and determination that are described as truly surprising in his helpless condition. He threw himself into the midst of society for excitement, and was fond of travelling, lying on his back in his carriage.

In 1851, six years after the accident, he presented himself in the

---
[1] Eve's Surgical Cases, p. 90; and New York Journal of Medicine, 1853. By Wm. D. Purple, M. D., of Greene, New York.

County Medical Society (Greene, New York), and requested the amputation of his lower extremities, which he stated were a burdensome appendage to the rest of his body, causing him much labour in moving them, and stating that he wanted the room they occupied in his carriage for books and other articles. He insisted on the operation with his wonted resolution and energy. The surgeon whom he consulted at first refused to consent to amputation, not only objecting to so extensive a mutilation for such reasons as he gave, but fearing lest the vitality of the vegetative existence enjoyed by his limbs might be insufficient for a healthy healing process. The patient, still determined in his resolve to have the limbs cut off as a useless burden to the rest of the body, sought other advice, and at last had his wishes gratified.

Both limbs were amputated near the hip-joints, without the slightest pain or even the tremor of a muscle. The stumps healed readily, and no unfavourable symptoms occurred in the progress of perfect union by the first intention. In this mutilated state he was perfectly unable to move his pelvis in the slightest degree. He resumed his wandering life, and travelled over a great part of the States. He died in May, 1852, of disease of the liver, brought on by his excesses in drink, to which he had become greatly addicted since his accident. No post-mortem examination was made.

This case is a most remarkable one in several points of view, and in none more than in this, that a double amputation of so serious a character could be successfully practised on a person affected by complete paraplegia, and yet that the stumps healed by the first intention. Besides this remarkable fact, there are two special points of interest in this case which bear upon the subject that we are now considering, viz., that the weight of the body is stated to have increased after the accident, and that the limbs which were so completely paralyzed as to admit of amputation without the patient experiencing the slightest sensation of pain, had in no way wasted during the six years that they had been paralyzed, but retained "their normal physical development," as is expressly stated in the report of the case. We can have no stronger evidence than this to prove that mere disuse of a limb for a lengthened period of years even, is not necessarily followed by the wasting of it.

Case 4.—J. R., a clerk by occupation, was admitted under my care into University College Hospital, October 2d, 1862. He had been knocked down half an hour previously by a Hansom cab, the horse falling partly upon him, and striking him with its knee on the neck. He never lost consciousness, but being quite unable to move, was carried to the hospital; on his way he passed his urine and feces involuntarily.

On examination after admission, it was found that he had an abrasion and ecchymosis on the left side of the neck. There was no inequality or irregularity about the spinous processes, or any evidence of fracture of the spine, but the patient complained of severe pain at the seat of the bruise. There was complete paralysis of sensation and of motion in the lower extremities and the trunk as high as the shoulders—incontinence of feces, retention of urine. The breathing was wholly diaphragmatic. He was quite conscious, and gave a description of the accident. He had suffered from urethral stricture for thirty-three years, so that only a No. 5 catheter could be passed.

On the following day his state was much the same. He complained of great pain in the right arm and hand, which were bruised. He said he thought he was paralyzed, as he could not move his legs; but on being pressed to do so, after some difficulty he succeeded in raising both legs, and in crossing them. Sensation appeared to be completely lost. His most distressing sensation was a feeling of tightness as of a cord tied tightly round the abdomen below the umbilicus.

5th. He had slept well, and was able to move his legs with less difficulty. Pulse 64, strong; passes feces involuntarily. Urine drawn off, and is ammoniacal. He was placed on a water mattress, as his back was becoming excoriated. Ordered quinine and acids.

8th. Is able to move his head and neck from side to side. Has less pain. Urine more ammoniacal; feces pass involuntarily. Bedsores over sacrum have much extended.

10th. Difficulty of breathing came on, but was relieved by the 11th. On the 12th it returned, with mucous râles, and he died that night—ten days after the accident.

On examination after death the head and brain were found uninjured and healthy. On exposing the vertebral column, it was found that the sixth and seventh cervical vertebræ had been separated posteriorly. The vertebræ themselves, and their arches, were quite sound, but there was a fissure without any displacement, extending through the articulating processes on the left side. A large quantity of blood was extravasated into the spinal canal, lying between the bones and the dura mater. There was a considerable quantity of reddish serous fluid in the arachnoid. The pia mater of the cord had some blood-patches upon it on the lower cervical region. The cord itself was quite healthy.

In this case it will be observed that the paralysis was most extensive, as much so as is compatible with life. The loss of sensation appeared to be more complete than that of motion, the patient being able, by an effort of the will, to cross his legs, but he could not feel when they were pinched or pricked. The fracture of an articulation without displacement was an accidental and insignificant complication, the real injury consisting in the extravasation of blood within the vertebral canal, which, by compressing the cord, induced the paralysis, that ultimately proved fatal. Death being, doubtless, hastened by the effusion of a large quantity of serous fluid from the irritated arachnoid.

The primary symptoms of concussion of the cord immediately and directly produced by a severe blow upon the spine will necessarily vary in severity and extent according to the situation of the injury, the force with which it has been inflicted, and on the amount of organic lesion that the delicate structure of the cord has sustained by the shock or jar to which it has been subjected.

A severe blow upon the upper cervical region may produce instantaneous death.

A severe blow inflicted on the dorsal region may induce more or less complete paraplegia.

In some cases the paralysis of the lower limbs has been complete and instantaneous; has affected both sensation and motion, with loss of power over the sphincters.

In other cases there has only been paralysis of motion, sensation continuing perfect.

The reverse has been met with, but less frequently and less completely, there being loss of sensation and impairment, though not complete loss of power over motion.

One leg is frequently more severely affected than the other. Or the two legs may be unequally affected as to sensation and motion; both sensation and motion being impaired, but in varying degrees in the two limbs.

There may be complete loss of power over the sphincters both of the bladder and anus, with incontinence or retention of urine and feces; or the loss of power may be confined to the bladder only.

This is especially the case when there is paralysis of motion rather than of sensation in the lower limbs.

The state of the urine will vary. If there is no retention, it will continue acid. When there is retention the urine usually becomes alkaline, but sometimes even when there is complete retention it remains strongly acid; and Ollivier has noted the very remarkable circumstance in one case of retention that there was an enormous formation of uric acid, so that the catheter became loaded with it.

Priapism does not occur in concussion, as it does so often in cases of laceration and irritation of the cord.

The temperature of the paralyzed parts is generally notably lower than that of the healthy parts of the body, and in some cases an absence of normal perspiration has been observed.

The *secondary* symptoms of severe concussion of the spine are usually those of the development of inflammation in the meninges and in the cord itself. They consist in pain in some part or parts of the spine, greatly increased by pressure and on motion, consequent rigidity of the vertebral column, the patient moving it as a whole. The pain is greatly increased by all movements, but especially by those of rotation.

It frequently extends down the limbs or around the body, giving the sensation of a cord tied tightly.

If the case goes on to the development of acute inflammatory action in the cord and its membranes, spasms of a serious character come on—at first, usually of the nature of trismus—then general spasms of the body and limbs, usually followed by speedy death from the exhaustion produced by the repetition of these violent convulsive movements.

If the inflammatory action assume a chronic and subacute character, permanent alterations in the structure of the cord will ensue, which will lead to paralytic affections of an incurable nature, usually confined to the lower extremities, but sometimes influencing the brain, and associated with great and deep-seated derangement of the general health.

Concussion of the spine from a severe and direct blow upon the back may prove fatal at very different periods after the injury. The time at which death occurs will depend partly on the situation of the blow, and in a great measure on the lesions to which it has given rise.

Concussion of the spine may, and often has, proved fatal by the sudden induction of paralysis, without there having been found after death any lesion of the cord that could explain the fatal termination of the case.

Abercrombie says: "Concussion of the cord may be speedily fatal without producing any morbid appearance that can be detected on dissection." And he refers to the case related by Boyer, and four by Frank in confirmation of this remark.

But in other cases the fatal result may have been occasioned by direct and demonstrable lesion of the spine or cord.

There appear to be four forms of lesion that will lead to a fatal result in cases of spinal concussion.

1. Hemorrhage within the spinal canal.
2. Laceration of the membranes of the cord, and extravasation of the medullary substance.
3. Disintegration and perhaps inflammatory softening of the cord.
1. Hemorrhage within the spinal canal may occur—
   1st. Between the vertebræ and dura mater;
   2d. Between the membranes and the cord;
   3d. In both situations.

In these respects intravertebral extravasations resemble closely those that occur as the result of injury within the cranium. The three following cases are illustrations of these three forms of hemorachis.

Sir A. Cooper mentions one case, to which I shall have occasion hereafter to refer, in which, in consequence of a strain of the neck in a boy of 12 years of age, symptoms of paralysis slowly supervened, which proved fatal at the end of a twelvemonth.

On examination after death, "the theca vertebralis was found overflowing with blood, which was effused between it and the inclosing canals of bone." This extravasation extended from the first cervical to the first dorsal vertebra.

Müller[1] relates the case of a corporal of cuirassiers who fell from a hay-loft on to his back, striking it against a log of wood. He was found to be completely paralyzed in his lower limbs, but preserved his consciousness. He died on the second day. On examination it was found that there was a large quantity of blood extravasated between the spinal cord and its membranes. This extravasation extended from the sixth cervical to the ninth dorsal vertebra.

[1] Bull. des Sc. Médicales, 1826.

Ollivier[1] relates the case of a woman 49 years old, who threw herself out of a window in the fourth story, alighting on her back. There was complete paralysis of the lower limbs, with incontinence of urine. Her mental faculties were unimpaired. She died on the third day after the injury, and on examination it was found that there was a fracture, but without any displacement, of the tenth dorsal vertebra; at this spot blood was extravasated between the vertebra and the dura mater, and also into the sub-arachnoid cellular tissue.

2. Death may occur—in that form of severe concussion which we are at present considering—from laceration of the pia mater, and hernia of the cord. Of this form of fatal result, Ollivier records one case, that of a man, 46 years of age, who had fallen heavily on his back, striking the spine in the middle of the dorsal region. He had paraplegia, paralysis of the sphincters, violent pain in the spine at the seat of injury, and much constitutional disturbance. He died on the seventeenth day. On examination after death it was found that the pia mater of the cord had been ruptured at two places opposite to the seat of injury, giving exit to the medullary substance in two patches, each about the size of a halfpenny, about two to three lines in thickness, and of a reddish colour. These protrusions had passed out of two longitudinal slits in the meninges of the cord, each about one inch in length, situated at the medial and posterior part, and opposite to the fourth and fifth dorsal vertebræ. At the points opposite to these herniary protrusions, the spinal cord was much contracted, having lost a great part of its substance; but it preserved its normal consistence. The dura mater contained a large quantity of bloody serum.

3. The last condition of the cord that leads to a fatal termination in these cases of concussion arising from direct and severe injury is an inflammation, with, perhaps, suppuration of the meninges, with inflammatory softening and disintegration of its substance. This is, doubtless, of an acute and probably inflammatory character. The following cases will illustrate the morbid state.

Ollivier relates the case of a man, 28 years of age, who fell from the second story of a house, striking himself violently on his back, left hip, and thigh. His lower extremities became paralyzed completely, so far as motion was concerned; incompletely, as to sensa-

---

[1] Vol. i. p. 492.

tion. The sphincters were paralyzed. He died on the thirtieth day after the accident. On examination after death, it was found that the spinous process of the fourth cervical vertebra was detached but not displaced, and the twelfth dorsal vertebra was broken across but not displaced. The spinal cord was healthy in all parts except opposite this point, where it was soft, diffluent, of a yellowish-gray colour, and injected with capillary vessels.

A remarkable case is recorded by Sir C. Bell (op. cit., p. 145). It is that of a wagoner who was pitched off the shafts of his cart on to the ground, falling on his neck and shoulders. At this part there was evidence of bruising. He could not stand, and dragged his legs. He lay for nearly a week without complaint, and had during this time no sign of paralysis. But on the eighth day he was suddenly seized with convulsions over the whole of the body —which were relieved by bleeding. He became maniacal, but in the course of twelve hours the convulsions ceased and he became tractable. On the third day after this attack he complained of difficulty in using his arm, and on the fifth day he had total palsy of the lower extremities, regaining the use of his arm. He died about a week after this. On examination after death, it was found that a considerable space existed between the last cervical and the first dorsal vertebra. The intervertebral substance was completely destroyed, and an immense quantity of pus surrounded the bones. This purulent collection had dropped down through the whole length of the sheath of the cord to the *cauda equina*.

The following case offers a remarkable resemblance to the preceding one—being attended by nearly identical post-mortem appearances following the same kind of injury.

Dr. Mayes,[1] of Sumter District, South Carolina, relates the case of a negro who, while raccoon-hunting, fell a height of fifteen feet from a tree, striking his back at the lower cervical and upper dorsal region against the ground. He instantly became completely paraplegic, and died on the tenth day. On examination seven hours after death, it was found that the fifth and sixth cervical vertebræ were separated from each other posteriorly, but not fractured or dislocated. Here there was manifest injury to the medulla. As soon as the muscular coverings of the spine were cut through, the softened and disintegrated medulla gushed out "similar to the

---

[1] Southern Medical and Surgical Journal, 1847.

escape of matter from an abscess when opened by the lancet. The medulla spinalis was evidently at this point in a state of decomposition."

In this case it is evident that not only the meninges of the cord, but the ligamenta subflava were torn through, and the arches of the vertebræ separated to such an extent, that the softened and disorganized medulla found a ready exit through the gap thus made at the posterior part of the spinal column.

It is a point of much practical moment to observe that in this, as in several other of the cases of so-called "concussion of the spine," there is, in addition to the lesion of the cord, some serious injury inflicted on the ligamentous and bony structures that enter into the composition of the vertebral column, which, however, must be considered as accidental complications, as they do not occasion, or even aggravate, the mischief done to the medulla itself. Thus the ligaments, as in the case just related, may be torn through so as to allow of partial separation of contiguous vertebræ, or, as in Ollivier's or in Case 4, a vertebra may be fractured—but without any displacement of the broken fragments, or other sign by which it is possible during life to determine the exact amount of injury that has been inflicted on the parts external to the cord. In this respect injuries of the spine again closely resemble those of the head—their chief importance depending, not on the amount of injury to the containing, but on that inflicted upon the contained parts. In the spine just as in the head, it will sometimes be found after death from what appears to be, and in reality is, simple injury of the nervous centres, that the vertebral column in the one case, and the skull in the other, have suffered an amount of injury that was unsuspected during life; and which, though it may not in any way have determined to the fatality of the result, yet affords conclusive evidence of the violence to which the parts have been subjected, and the intensity of the disorganizing shock that they have suffered.

There is, however, this very essential difference between the spine and the head in these respects—that a simple fracture of the cranium may be of no moment except so far as the violence that has occasioned it may have influenced the brain. Whilst in the spine the case is not parallel; for as the vertebral column is the centre of support to the body, its influence in this respect will be lost when broken; even though the spinal cord may not have been injured by the edges of the fractured vertebræ, but simply violently and fatally concussed by the same force that broke the spine itself.

Boyer had previously noticed the very interesting practical fact, that when the interspinous ligaments were ruptured in consequence of forcible flexion of the spine forwards, no fatal consequences usually ensue, the integrity of the parts being restored by rest. But that when the ligamenta subflava are torn through, and the arches separated, paraplegia and death ensue. This he attributes to stretching of the spinal cord. Sir C. Bell, however, with great acuteness, has pointed out the error of this explanation, and states that "it is the progress of inflammation to the spinal marrow, and not the pressure or the extension of it, which makes these cases of subluxation and breach of the tube fatal" (p. 149). There can be no doubt that this explanation is the correct one, and that when once the spinal canal is forcibly torn open, fatal inflammation will spread to the meninges and to the medulla itself.

Perhaps the most marked case on record of inflammatory softening of the cord consequent upon concussion of it, unattended with any injury to the osseous or ligamentous structures of the spine, is that which occurred in the practice of Dr. Hunter, of Edinburgh, and is related by Abercrombie. It is that of a man thirty-six years of age, who fell from the top of a wagon, a height of ten feet, into a pile of small stones, striking his back between the shoulders. He was immediately rendered paraplegic. When admitted into the Edinburgh Infirmary at the end of a month he was greatly emaciated: there was paralysis of motion, but not of sensation, in the lower extremities, retention of urine, involuntary liquid motions, deep-seated pain on pressure in the region of the third, fourth, and fifth dorsal vertebræ. Three days after admission tetanic symptoms came on; then more general spasms of the limbs and body, of which he died in forty-eight hours. On examination after death there was no injury found to the spine itself. There was a high degree of vascularity of the pia mater of the cord in the dorsal region. There was most extensive softening of the body of the cord, affecting chiefly the anterior columns. "These were most remarkably softened throughout almost the whole course of the cord; in many places entirely diffluent; the posterior columns were also softened in many places, though in a much smaller degree" (p. 348).

This case epitomizes so succinctly and clearly the symptoms and after-death appearances occurring in cases of inflammatory softening after uncomplicated concussion of the cord from severe and direct violence, that it needs neither comment nor addition.

The consideration of these subjects in connection with concussion of the spine, as the result of severe and direct violence, will pave the way for what I shall have to say in the next lecture about concussion of the spine as the result of slight, indirect, and less obvious injuries.

LECTURE THE THIRD.

ON CONCUSSION OF THE SPINE FROM SLIGHT INJURY.

Concussion of the Spine from Slight Injuries—Railway Injuries Peculiar but not Special—Effects of Slight Blows on Spine, Case 5—Concussion from Railway Injury, Case 6—Concussion from Railway Injury, Case 7—Concussion from Carriage Accident, Case 8—Concussion from Falls—Long durations of Symptoms.

IN the last lecture I directed your attention to the symptoms, effects, and pathological condition presented by cases of concussion of the spine, proceeding from the infliction of severe injury directly upon the vertebral column so as immediately and injuriously to influence the organization and the action of the delicate nervous structures included between it.

My object in the present lectures is to direct your attention to a class of cases in which the injury inflicted upon the back is either very slight in degree, or in which the blow, if more severe, has fallen upon some other part of the body than the spine, and in which, consequently, its influence upon the cord has been of a less direct and often of a less instantaneous character.

These cases are extremely interesting to the surgeon, for not only is the relation between the injury sustained and the symptoms developed less obvious than in the former case, but in consequence of the length of time that often intervenes between the occurrence of the accident and the production of the more serious symptoms, it becomes no easy matter to connect the two in the relation of cause and effect.

Symptoms indicative of concussion of the spine have of late years not unfrequently occurred, in consequence of injuries sustained in railway collisions, and have been very forcibly brought under the observation of surgeons in consequence of their having been the fertile sources of litigation; actions for damages for injuries alleged to have been sustained in railway collisions having become of such very frequent occurrence as now to constitute a very important part of medico-legal inquiry.

The symptoms arising from these accidents have been very variously interpreted by surgeons, some practitioners ignoring them entirely, believing that they exist only in the imagination of the patient, or, if admitting their existence, attributing them to other conditions of the nervous system than any that could arise from the alleged accident. And when their connection with, and dependence upon, an injury have been incontestably proved, no little discrepancy of opinion has arisen as to the ultimate result of the case, the permanence of the symptoms, and the curability or not of the patient.

I will endeavour in these Lectures to clear up these important and very intricate questions; and in doing so I shall direct your attention most particularly to the following points:—

1. The effect that may be produced on the spinal cord by slight blows when inflicted on the back or distant part of the body.

2. The length of time that may intervene between the alleged injury and the development of the symptoms.

3. The diagnosis of the symptoms of "Concussion of the Spine," from those arising from other morbid states of the nervous system.

4. The grounds on which to form a prognosis as to the probable result.

I shall illustrate these various points by selected cases, not only of persons who have been injured on railways, but in the ordinary accidents of civil life.

I wish particularly to direct your attention to the fact that there is in reality no difference whatever between the symptoms arising from a concussion of the spine received in a railway collision and those from a fall or ordinary accident—except perhaps in severity—and that it is consequently an error to look upon a certain class of symptoms as special to railway accidents. I cannot, indeed, too strongly impress upon you the fact that there is in reality nothing special in railway injuries, except in the severity of the accident by which they are occasioned. They are peculiar in their severity, not different in their nature from injuries received in the other accidents of civil life. There is no more real difference between that concussion of the spine which results from a railway collision and that which is the consequence of a fall from a horse or a scaffold, than there is between a compound and comminuted fracture of the leg occasioned by the grinding of a railway carriage over the limb and that resulting from the passage of the wheel of a street cab across it. In either case the injury arising from the railway accident will be essentially of the same nature as that which is otherwise occasioned, but it will probably be infinitely more severe and destructive in its effects, owing to the greater violence by which it has been occasioned. I intend to draw my illustrations, to some extent at least, from ordinary accidents, as in these the question of compensation in money for injury sustained is not mooted, and hence an element which is usually alleged to have a disquieting effect on the nervous system of the sufferer is eliminated from our consideration.

The consideration of the effects that may be produced on the spinal cord by *slight* blows, whether applied to the back or to a distant part of the body, is not altogether a matter of modern surgical study arising from the prevalence of railway accidents, but had, long antecedent to the introduction of modern means of locomotion, arrested the attention of observant practitioners.

Abercrombie, writing in 1829, says that chronic inflammations of the cord and its membranes " may supervene upon very slight injuries of the spine;" and further on he says, "Every injury of the spine should be considered as deserving of minute attention. The more immediate effect of anxiety in such cases is inflammatory action, which may be of an acute or chronic kind; and we have seen that it may advance in a very insidious manner even after injuries that were of so slight a kind that they attracted at the time little or no attention" (p. 381).

Nothing can be clearer and more positive than this statement. These remarks of Abercrombie's are confirmed by Ollivier, by Bell, and by other writers on such injuries.

The following cases will illustrate this point.

The first two are cases of concussion of the spine resulting from railway accidents, in which there were at the time slight marks of external injury. The others are very similar cases occurring from other accidents than those received on railways.

Case 5.—Mr. R., 35 years of age, a farmer and miller, of very active habits, accustomed to field sports, and much engaged in business, habitually in the enjoyment of good health, was in a railway collision that took place on Nov. 4, 1864. He received a blow upon his face which cut his upper lip on the left side, and was much and severely shaken. He did not lose consciousness, and was able shortly to proceed on his journey. On leaving the station to proceed to his own home, it was observed by a friend who drove him that he did not appear to recollect the road, with which he was familiar, having been in the daily habit of driving over it for years.

On reaching home, feeling bruised, shaken, and confused, he took to his bed, but did not feel sufficiently ill to seek medical advice until November 9, five days after the accident, when he sent to Mr. Yorke, of Staunton, who continued to attend him. But notwithstanding every attention from that gentleman, he progressively, but slowly got worse.

I saw Mr. R. for the first time on the 18th February, 1866, fifteen months after the occurrence of the accident, when I found him in the following state. His face was pallid, much lined, indicative of habitual suffering. He looked much older than his alleged age (36 years). He was sitting with his back to the light, and had the venetian blinds drawn down so as to shade the room, the light being peculiarly distressing to him. His skin was cool. Tongue slightly furred, appetite moderate, digestion impaired. Pulse 104 to 106, weak and compressible. I understand from Mr. Yorke that it rarely fell below this, and often rose above it. He has not lost flesh, but all his friends say that he is quite an altered man.

He states that since the accident his memory has been bad—that he cannot recollect numbers—does not know the ages of his children, for instance—he cannot add up an ordinary sum correctly—he will add up the same set of figures if transposed differently. Before the accident he was considered to be a peculiarly good judge of the weights of beasts—since its occurrence he has lost all power of forming an opinion on this point. He has been quite unable to transact any business since the injury. Is troubled with frightful dreams. Starts and wakes up in terror, not knowing where he is. Has become irritable, and can neither bear light nor noise. He frowns habitually, so as to exclude the light from his eyes. He complains of stars, sparks, flashes of light and coloured spectra flaming and flashing before the eyes. He cannot read for more than two or three minutes at a time, the letters becoming confused, and the effort being painful to bear. On examining the state of the eyes, I find that vision is good in the right eye, but that this organ is over sensitive to light. Vision is nearly lost in the left eye, so much so that he cannot read large print with it.

His hearing is over sensitive with the right ear, dull on the left side. He cannot bear noises of any kind, more particularly if sudden; they are peculiarly distressing to him. Even that of his children at play annoys him.

He complains of a numb sensation accompanied by tingling, burning sensations on the right side, in the right arm and leg, more particularly in the little and ring-fingers, and along the course of the ulnar nerve. The rest of the right hand feels numb. He makes no complaint of the left arm or leg. These sensations are worst in the morning.

He cannot stand or walk without the support of a stick, or by resting his hand on a piece of furniture. He can do so in this way on the left leg, but if he attempts to do so on the right foot the limb immediatly bends, and sinks as it were under him. His gait is very peculiar. He separates the feet so as to make a straddling movement, and brings one foot very slowly before the other. He advances the right foot less than the left, and does not raise the sole as far from the ground. The foot seems to come down too quickly. He does not drag with the toes, but does not raise the heel sufficiently, and is apt to catch it in walking in inequalities on the ground. Flexion and extension are more perfectly and rapidly performed with the left than the right foot.

The attitude of his body in walking is very peculiar: the back is stiff, the head fixed, and he looks straight forward without turning it to the one side or the other.

He has great difficulty in going up or down stairs, cannot do so without holding on by the banisters. The difficulty is greatest in going down stairs, and if he attempts this without support he falls or rolls over to the right side.

There is no appreciable difference in the size of the two legs, but the right feels colder than the left. The patient complains of the coldness of both legs and feet.

The spine had lost its natural flexibility, so that the patient kept the body perfectly straight, fixed, and immovable. He could not bend the body in any direction without suffering severe pain. This was complained of equally whether the patient bent forwards, backwards, or sideways. It was most severe on any attempt being made to twist the spine. He sits in a rigid and upright attitude.

There was considerable pain at the occipito atloid articulation, as also at that between the axis and atlas. If an attempt was made to bend the head forcibly forwards, or to rotate it, the patient suffered so severely that it became necessary to desist. When directed to look round, the patient turned the whole body.

Owing to the rigidity of his spine he could not stoop so as to pick anything off the floor without going down on one knee.

On examining the spine by pressure and percussion, three tender spots were found; one in the upper cervical, the other in the middle dorsal, and the third in the lumbo-sacral region. There is pain both on superficial and on deep pressure at these spots. The pain is limited to the spine, and does not extend to the muscular structures on either side of it.

The power of retaining the urine is very materially diminished. He passes water four or five times in the night, and every second hour during the day. The urine is sub-acid.

The generative power, though impaired, is not lost. A remarkable circumstance has been noticed in this case by Mr. R.'s wife and his friends. It is that since the accident he is unable to judge correctly of the distance of objects in a *lateral* direction, though he appears to be able to do so when looking straight forward. Thus, when driving in the middle of a straight road he always imagines that the carriage is in danger of running into the ditch or hedge on the *near* side.

The opinion I gave was, that the patient had sustained an injury of the spinal cord, and that the base of the brain was also, to some extent, though probably secondarily, involved. That chronic subacute meningitis of the spine and base of the cranium had taken place. That it was not probable that he would ever completely recover, and that it was even doubtful whether, as the disease had up to the present time been progressive, it might not continue to be so, and terminate in incurable disorganization of the nervous centres. The patient was seen by Sir Charles Hardings and Mr. Carden, who took a similarly unfavourable view of his present state and probable future.

An action was brought at the spring assizes at Worcester, in 1866, against the company on whose line the patient had been injured. No surgical evidence was called for the company, the statement made by the plaintiff's medical advisers being accepted. The question of damages resolved itself, to a great extent, into one of loss of income and expense incurred. The jury awarded £5775.

Case 6.—Mr. J., 43 years of age, a wine-merchant, healthy and of active business habits, was in a railway collision on the 23d of August, 1864. He was suddenly dashed forwards and then rebounded violently backwards.

When he extricated himself from the ruins of the carriage in which he had been travelling (a third-class one), he believed himself to be unhurt—suffering from no immediate effect of the shock he had sustained. He assisted his fellow-passengers, many of whom were much injured, and was thus actively engaged for two hours.

On his return home the same evening, he was greatly excited and very restless; he felt chilly, and his arms and legs tingled. He could not sleep that night.

On the following day he felt ill and shaken; could not attend to his business, and was lame from some slight contusions on his legs. He continued much in this state for several days, and was seen by Mr. Everett, of Worcester (to whom I am indebted for the early history of this case), on the 1st of September, eight or nine days after the accident. He was then much disturbed in health; his pulse was feeble, he looked anxious and depressed; he complained of violent pains in the head, confusion of thought, and loud noises in the ears and head. He also complained, but slightly, of pain in the back.

These symptoms continued for some time without improvement. He found more and more difficulty in walking, and his right ankle often gave way. This appeared to Mr. Everett to be owing to some spasmodic action of the muscles of the leg rather than to any weakness of the joint itself.

He now began to show more serious symptoms in connection with the nervous system. His memory became worse and confusion of ideas greater; he often called people and things by wrong names; addressed his wife as "sir."

The pains in the head became more violent, and assumed a paroxysmal character. There was acute sensibility to sound in the right ear, deafness of the left. Vision of the right eye was rather dim.

This was his condition at the end of twelve weeks after the occurrence of the accident. The symptoms, though progressively assuming a more and more serious character, did not do so uninterruptedly, but, as Mr. Everett expresses it, were "undulatory,"—sometimes better, sometimes worse; but yet at the expiration of any given time of a few weeks' duration, decidedly and persistently worse than at an earlier period.

Three months after the accident he began to complain, for the first time, of contractions of the muscles of the right arm and hand. His fingers became flexed, so that force was required to straighten them. Shortly afterwards the left arm became similarly affected. These contractions assumed an intermitting and spasmodic character, and occurred several times daily.

The pain in the back, which was but slightly complained of at first, now became more and more severe. It was more acute over the sixth to the tenth dorsal vertebra, both inclusive. Spasms of the diaphragm now came on occasionally, and distressed him much.

His gait was peculiar; he seemed to be uncertain where to set his feet, and he kept his head steadily fixed.

On February 1, 1865, five months after the accident, he complained, for the first time, of pain in the neck, greatly increased on moving the head.

During the whole of this period his digestion had been fairly good. He had gained flesh since the accident. There had been no loss of power over the sphincters, and his urine was normal and acid.

I saw this patient, in consultation with Mr. Carden and Mr. Everett, of Worcester, on March 8th, 1865, and found that the symptoms above detailed continued, and had somewhat increased in intensity since the last report.

Loss of memory, confusion of thought and ideas, utter incapacity for business, disturbed sleep, pains and noises in the head, partial deafness of the left, morbid sensibility of the right ear, irritability of the eyes, rendering light very painful—though vision had become imperfect in the right eye. Numbness, tingling sensation, and formication in the right arm and leg, were the most prominent *subjective* symptoms.

He walked with a peculiar unsteady straddling gait; was obliged to feel with his right foot before planting it on the ground; did not raise the heel, but carried the foot flat, and let it fall suddenly; instead of putting it on the ground in the usual way; used a stick, or supported himself by the furniture.

He could stand for a moment on the left leg, but immediately fell over if he attempted to do so on the right.

His right arm and hand were numb; the little and ring-fingers contracted. He could not pick up a small object, as a pin, between his finger and thumb, nor could he write easily or legibly.

The spine was very tender at three points—in the upper cervical, in the middle dorsal, and in the lower lumbar regions. There was constant fixed aching pain in it in these situations. This pain was greatly increased on pressure; it was limited to the vertebral column, and did not extend beyond it.

Movement of any kind greatly increased the pain. If the head was raised by the hands and bent forward, or rotated, so as to influence the articulations between the occipital bone, the atlas, and the axis, the patient shrieked with the agony that was occasioned.

He could not bend the body either forwards, backwards, or sideways, the pain being so greatly increased in the dorsal and lumber regions by these movements. He consequently could not stoop.

The spine had entirely lost its normal flexibility. It was perfectly rigid, moved as a whole as if made of one bone. The patient could neither bend nor turn his head. Hence he could not look on the ground in walking to see where to place his feet; and when he wished to look round, he had to turn the whole body.

The pulse was feeble, about 98. Countenance pale, anxious, haggard. Tongue slightly coated. Digestive and other functions well performed. Urine clear and acid.

The case was tried at the spring Assizes at Worcester in 1865. The opinion expressed by Mr. Carden, Mr. Everett, and myself, amounted to this, that the patient was suffering from concussion of the spine, which had developed irritation or chronic inflammation of its membranes and of the cord, and that his recovery was very doubtful. The plaintiff recovered £6000 damages.

At this time (May, 1866), a year and nine months after the

accident, he is still an invalid, being so completely shattered in health that he has been obliged to winter in the southwest of England, and is quite unequal to attend to business of any kind.

Case 7.—The following case illustrates the fact that a train of symptoms of a most persistent nature, closely resembling those detailed in the preceding cases, may occur from other causes than railway accidents.

Captain N., 38 years of age, consulted me on October 27th, 1862. Looks careworn, pale, lined, and at least ten years older than his real age. He states that in November, 1854—eight years previously, he had been thrown out of a pony-chaise, which was accidentally upset. At the time he hurt his right knee and bruised the right arm, but sustained no blow or evidence of injury on the head or back. He was much bruised and shaken at the time, but did not suffer any serious ill effects for several months after the accident, although during the whole of this period he felt ailing, and that he was in some way suffering from the injury he had sustained.

About six months after the accident he began to be troubled with the following train of symptoms, which have continued ever since: Confusion of thought; his memory was impaired; he had giddiness, especially on moving the head suddenly; his sight became impaired; he suffered from muscæ volitantes; sparks and flashes of light; he could not continue to read beyond a few minutes, partly because the letters ran into each other, partly because he could not concentrate his thoughts so as to fix his attention.

He now began to suffer from a feeling of numbness and a sensation of "pins and needles" in both hands, but more particularly the left, and chiefly in those parts supplied by the ulnar nerve.

He complained of the same sensations in the left leg and foot. He walked with difficulty, and with the legs somewhat apart, using a stick, or else supporting himself by holding on to pieces of furniture in the room as he passed them. He can stand on the right leg, but the left one immediately gives way under him. He walks with great difficulty up and down stairs, obliged to put both feet on the same step. The spine is tender on pressure and percussion in the lower cervical region and between the shoulders. It is stiff; he cannot bend the back without pain, and cannot stoop without falling forward.

He has irritability of the bladder, passing water every second or third hour, and can only do so in a sitting position. He has completely lost all sexual power and desire. The urine is slightly acid. These symptoms have continued with varying intensity, since their commencement, about six months after the accident. He thinks they were most severe about a year after they began, and have somewhat improved since then. But he has never been free from them, or enjoyed a day's health, for the last seven and a half years, and never expects to do so.

This case closely resembles, in all its general features, and in many of its details, those that have just been related. It only differs in the symptoms being less intense, as would naturally be expected, from the accident that occasioned them being less severe than those which occur from railway collisions. The persistence of the symptoms for so lengthened a period as nearly eight years is significant of the long duration of the pernicious effects of these insidious injuries to the nervous system.

But the interminable duration of the most serious nervous phenomena, from comparatively slight injuries of the spine, receives additional illustration from the following case.

Case 8.—Miss B., 26 years of age, was brought to my house on the 11th April, 1866, by my friend Dr. Gibb. She looked moderately healthy, was of good constitution, with no discernible hereditary tendency to disease of any kind—was not anæmic. The digestive and uterine functions were well performed. She has had no disease except that from which she now suffers, no convulsions or fits in childhood.

When about eighteen months old, she fell out of her cot and injured her cervical spine. From that time to the present she has suffered from a continuous and remarkable train of nervous phenomena. These were aggravated about the period of puberty, and at the age of 17 were still further increased in consequence of her falling over a stile backwards. She has never had hysteria in any of its ordinary forms, paralysis, epilepsy, or convulsive attacks of any kind.

On examining the spine, I find it straight and the body well formed. There is a distinct projection backwards of the spinous processes of the fifth and sixth cervical vertebræ. She complains of a constant pressure and pain of a grating or grinding character in this region, as if the bones were in contact with one another.

There is no evidence of abscess or of any distinct mischief in or around the tender vertebræ, and nothing is to be observed with the laryngoscope at the anterior part of the cervical vertebræ or pharynx. From this point a peculiar sense of uneasiness spreads itself over the whole of the body and limbs, producing nervous sensations of the most distressing character. These sensations, which consist of tingling and painful feelings, prevent her sitting still or lying down quietly for any length of time. She is better when in movement. She cannot sleep for more than an hour or two at a time, and is conscious of her sufferings through her sleep.

Her power of movement has never been impaired, the distress being confined to sensation, and not producing any disturbance of motion.

She can walk well under certain circumstances, can stand, and in fact scarcely ever sits; but cannot turn suddenly without becoming giddy, with the fear of falling.

She can walk well so long as there is anything near her. Thus she can walk along a street guided by the area railings; but when she comes to an open space, as a square or crossing, she is lost, and requires to be guided or she would fall. She cannot bear the sensation of having a space around her, and would then fall unless supported.

She has unceasing loud noises in her head, which she compares to "gravel-stones" rolling through it. They are so loud that she thinks that other people must hear them.

Her hearing is good.

Her sight is strong, but she sees the circulation of the blood in her own eyes, the corpuscles spinning round in convolutions, and often coloured. No perversion of smell or taste. The hands and feet always cold, even in summer.

She has been from first to last under the care of at least thirty medical men. Has had every variety of treatment adopted—a seton kept open in the neck and the clitoris excised; but so far from benefiting has slowly but steadily become worse, and her general health is now beginning to give way.

This lady, who is remarkably intelligent, gave a lengthened and minute history of her ailments, of which the above is a sketch. She referred all her morbid sensations to the seat of excurvation in the cervical vertebræ. At this point there had evidently existed disease leading to organic changes to which the remarkable train of general phenomena presented by this case were doubtless referable. If I were to hazard an opinion, it would be that some thickening of the meninges of the cord had probably taken place, the effect of which was to interfere with the sensory portions of the cord, rather than with the motor.

LECTURE THE FOURTH.

CONCUSSION OF THE SPINE FROM GENERAL SHOCK.

Concussion from General Shock—Case 9. Concussion from Shock to Feet—Case 10. Concussion from Railway Shock—Case 11. Concussion from Railway Shock—Case 12. Concussion from Railway Shock—Twists, Sprains, and Wrenches of the Spine—Case 13. Wrench of Spine—Case 14. Twist of the Spine—Effects of Twists and Wrenches of the Spine.

THERE is another class of cases of an extremely insidious and protracted character to which I wish to direct your attention, *viz.*, those cases in which the patient has received no blow or injury upon the head or spine, but in which the whole system has received a severe shake or shock, in consequence of which disease is developed in the spinal cord, perhaps eventually extending to the membranes of the brain. These cases, although necessarily more frequent in railway than in other injuries, yet occasionally occur as a consequence of ordinary accidents. I will first relate a case of this kind, and then direct your attention to the details of several instances that have fallen under my notice of similar phenomena occurring after railway accidents.

Case 9.—On the 17th November, 1861, I saw, in consultation with Dr. Strong, of Croydon, Mrs. B., 32 years of age. She states that in November, 1860, whilst going down-stairs, she accidentally stepped upon the side of a pail, and slipped forwards, bumping down three or four stairs forcibly on her heels. She did not lose her footing, did not fall, and did not strike any part of the body or head. Of this she is quite certain. She felt nervous, faint, and shaken at the time, and was obliged to take some brandy. At the period of the occurrence of the accident, and up to that time, she had been a strong, healthy, and active woman. She was married, and the mother of two children. She had never suffered from any disease of the nervous system, or from any serious complaint.

Two days after the trifling accident that has just been described, she was attacked with neuralgic pains in the right side of the head —apparently hemicrania. For this she was treated in the usual way, and did not feel it necessary to lay up. About a fortnight after the accident, she felt numbness and tingling conjoined in the right arm, hand, and leg, and also on the right side of the head, where the neuralgia had previously existed. The numbness after a time extended to the right half of the tongue.

When I saw her three months after the accident the numbness and tingling existed unchanged in these parts, and the left hand and arm had also begun to be affected. She felt a numb sensation in the little and ring-fingers, and slightly in the middle finger.

Although there is this numb sensation in the hands, and in the right leg, she has no impairment of motion. She can pick up a pin, untie a knot, and otherwise use the right hand, which is the one most affected, in ordinary small occupations. She can stand; walk fairly well.

I saw the patient again on the 13th April, four months after the accident. Notwithstanding the treatment that had been adopted (iron and strychnine), she was weaker, looked anæmic, and was rather worse, so far as the paralytic symptoms were concerned. She could no longer pick up so small an object as a pin, but can pick up a piece of money—a shilling for instance. The right hand and leg are still the worst, but the left limbs are more affected than they were. In the left hand the numbness has now affected the little, ring, and middle fingers, with the tip of the forefinger.

From this time to the present there has been a very slow increase in the symptoms, notwithstanding a great variety of treatment to which the patient has been subjected by the many different medical men whom she has seen. On examining her, on April 10, 1866, about five and a half years after the accident, with Mr. Ayling, her present medical attendant, she tells me that she feels that she is gradually, though very slowly, getting worse. She has an anxious, anæmic look. She totters in walking, so that in going about the room she supports herself by the chairs and tables. She could not in any way walk a quarter of a mile. She can stand unsupported on the left leg, but she immediately falls over if she attempts to do so on the right. The right hand and foot are much colder than the left. The paralysis of the hands continues much the same, but a marked change has taken place in the right hand in consequence of the contraction of all the fingers, but more especially of the little and ring-fingers. They have become rigid, and the flexor tendons stand out strongly. She can, consequently, scarcely use this hand. On testing the irritability of the muscles in the opposite limbs by galvanism, the contraction was almost *nil* in those of the right arm and hand. Much stronger, though not normally strong, on the left side.

She complains of confusion of thought and loss of memory; the senses are unimpaired. Appetite is bad, and digestion imperfect. Urine is acid. Can hold her water well.

In this case a very trivial accident occasioned a jar communicated to the feet, and evidently transmitted to the nervous centres, leading to impairment of innervation, and eventually to progressive and incurable paralysis.

Case 10.—M. H. I., a surgeon, 43 years of age, naturally a stout healthy man, of active professional habits, consulted me on February 22d, 1865. He states that on 9th October, 1864, he was in a railway collision, by which he was thrown forwards, but without any great violence. He received no blow on the back, head, or other part of the body. He was much frightened and shaken, but did not lose consciousness.

Beyond a general sensation of illness, he did not suffer much for the first three or four weeks after the accident, but he was not able to attend to his business; could not collect his thoughts sufficiently for the purpose.

About a month after the accident he began to suffer from pain across the loins. He could not walk without great fatigue. He lost strength and flesh, and his pulse became habitually much more frequent than natural, being about 98 to 100.

At the present time, four and a half months after the accident, he continues much in the same state; is quite unfit for business, and has been obliged to relinquish practice; not owing to any mental incapacity, but entirely owing to his bodily infirmities. His mind is quite clear, and his senses perfect, though over-sensitive; loud and sudden noises and bright light being peculiarly distressing to him.

He complains chiefly of the spine. He suffers constant pain in the lower part of it, in the lower dorsal, and the lumbar regions.

He compares the sensation there experienced to that of a wedge or plug of wood driven into the spinal canal. It is a mixed sensation of pain and distension. The spine generally is tender, and the pain in it is greatly increased by manipulation, pressure, and percussion. It has lost its normal flexibility, moves as a whole, so that he cannot bend forwards or stoop. There is no pain in the cervical region, or on moving the head.

He complains of painful numbness and formications in the right, and occasionally down the left leg. The legs are stiff and weak, especially the right one. He cannot stand unsupported on this for a moment. He walks in a slow and awkward manner—straddling—not able to place the feet together. If told to stand on his toes, he immediately falls forwards. He has lost control over the limbs, and does not know exactly where to place the feet. He has a frequent desire to pass water, suffers greatly from flatus, and has completely lost all sexual desire and power. The pulse was at 98; appetite bad; digestion impaired.

I saw this patient again, at Brighton, towards the end of April, seven months after the accident, in consultation with Mr. Curtis, and found that his condition had in no way improved; indeed, that in some respects, so far especially as power of movement was concerned, it had progressively become worse.

In this case the injury produced by the shock had evidently occasioned mischief within the lower portion of the spinal canal, leading to partial paraplegia. I believe this mischief to have been of a chronic inflammatory nature; the tenderness of the spine, the feeling of distension, the pain in movement, and the habitually high pulse, point in this direction. This case was settled out of court for £2500.

Case 11.—Mr. C. W. E., about 50 years of age, naturally a stout, very healthy man, weighing nearly seventeen stone, a widower, of very active habits, mentally and bodily, was in a railway collision on February 3d, 1865. He was violently shaken to and fro, but received no bruise or any sign whatever of external injury. He was necessarily much alarmed at the time, but was able to proceed on his journey to London, a distance of seventy or eighty miles. On his arrival in town he felt shaken and confused, but went about some business, and did not lay up until a day or two afterwards. He was then obliged to seek medical advice, and felt himself unable to attend to his business. He slowly got worse, and more out of health. Was obliged to have change of air and scene, and gradually, but not uninterruptedly, continued to get worse, until I saw him on the 26th March, 1866, nearly fourteen months after the accident. During this long period he had been under the care of various medical men in different parts of the country, and had been most attentively and assiduously treated by Dr. Elkington, of Birmingham, and by several others, as Dr. Bell Fletcher, Dr. Gilchrist, Mr. Gamgee, Mr. Martin, &c. He had been most anxious to resume his business, which was of an important official character, and had made many attempts to do so, but invariably found himself quite unfit for it, and was most reluctantly constrained to relinquish it.

When I saw him at this time, he was in the following state:—

He has lost about twenty pounds in weight, is weak, unable to walk a quarter of a mile, or to attend to any business. His friends and family stated that he is, in all respects, "an altered man." His digestion is impaired, and his pulse is never below 96.

He complains of loss of memory, so that he is often obliged to break off in the midst of a sentence, not being able to complete it, or to recollect what he has commenced saying. His thoughts are confused, and he cannot concentrate his attention beyond a few minutes upon any one subject. If he attempts to read, he is obliged to lay aside the paper or book in a few minutes, as the letters become blurred and confused. If he tries to write, he often mis-spells the commonest words; but he has no difficulty about figures. He is troubled with horrible dreams, and wakes up frightened and confused.

His head is habitually hot, and often flushed. He complains of a dull confused sensation within it, and of loud noises which are constant.

The hearing of the right ear is very dull. He cannot hear the tick of an ordinary watch at a distance of six inches from it. The hearing of the left ear is normal, he can hear the tick at a distance of about twenty inches. Noises, especially of a loud, sudden, or clattering character, distress him greatly. He cannot bear the noise of his own children at play.

The vision of the left eye has been weak from childhood. That of the right, which has always been good, has become seriously impaired since the accident. He suffers from muscæ volitantes, and sees a fixed line or bar, vertical in direction, across the field of vision. He complains also of flashes, stars, and coloured rings.

Light, even of ordinary day, is especially distressing to him. In fact, the eye is so irritable that he has an abhorrence of light. He habitually sits in a darkened room, and cannot bear to look at artificial light—as of gas, candles, or fire. This intolerance of light gives a peculiarly frowning expression to his countenance. He knits and depresses his brows in order to shade his eyes.

The senses of smell and taste seem to be somewhat perverted. He often thinks that he smells fetid odours which are not appreciable to others, and has lost his sense of taste to a great degree. He complains of a degree of numbness, and of "pins and needles" in the left arm and leg, also of pains in the left leg, and a feeling of tightness or constriction. All these symptoms are worst on first rising in the morning.

He walks with great difficulty, and seldom without the aid of a stick; whilst going about a room he supports himself by taking hold of the articles of furniture that come in his way. He does not bring his feet together—straddles in his gait—draws the left leg slowly behind the right—moves it stiffly and keeps the foot flat in walking, so that the heel catches the ground and the limb appears to drag. He has much difficulty in going up and down stairs, cannot do so without support.

He can stand on the right leg, but if he attempts to do so on the left, it immediately bends and gives way under him, so that he would fall.

The spine is tender on pressure and on percussion of these points, *viz.*, at lower cervical, in middle dorsal, and in lumbar regions. The pain in these situations is increased on moving the body in any direction, but especially the antero-posterior. There is a degree of unnatural rigidity, of want of flexibility, about the spine, so that he cannot bend the body—he cannot stoop without falling forwards.

On testing the irritability of the muscles by galvanism, it was found to be very markedly less in the left than in the right leg.

The genito-urinary organs are not affected. The urine is acid, and the bladder neither atonic nor unduly irritable.

The opinion that I gave in this case was to the effect that the patient had suffered from concussion of the spine—that secondary inflammatory action of a chronic character had been set up in the meninges of the cord—that there was partial paralysis of the left leg, probably dependent on structural disease of the cord itself—and that the presence of cerebral symptoms indicated the existence of an irritability of the brain and its membranes. The patient brought an action for damages at the Gloucester Spring Assizes, April, 1866, against the company on whose line he had been injured, and, notwithstanding powerful adverse medical testimony, recovered £3500 damages.

Case 12.—The following case presents some very remarkable and unusual nervous phenomena, resulting from railway shock, which I will briefly relate to you.

"*March* 1, 1865. Mr. D. is a man of healthy constitution and active habits, aged 33. He was travelling in an 'express' (third class, with divided compartment), and was seated with his back to the engine. When near Doncaster, the train going at about thirty miles an hour, ran into an engine standing on the line. He was thrown violently against the opposite side of the carriage, and then fell on the floor.

"*Immediate effects.*—There was a swelling the size of an egg over the sacrum, severe pain in the lower part of the spine, which, on arriving at Edinburgh the same day, had extended up the whole back and into the head, producing giddiness and dimness of sight. These, with tingling feelings in the limbs (particularly the left), great pain in the back, and tenderness to the touch, sickness in the mornings, and lameness, continued for the first fortnight.

"The *treatment* adopted consisted of blisters and hot fomentations to the spine.

"The patient seemed to improve, and the pain to move more between the shoulders after these applications.

"28th. He was seen by an eminent surgeon, who ordered him to go about as much as possible, but to avoid cold. The result of this advice was that he found the whole of the symptoms much increased with prostration and lameness.

"*April* 20th. Left for London, breaking journey for a week in Lancashire, greatly fatigued by journey. A discharge came on from the urethra, lameness much increased, could not advance the left leg in front of the right, and great prostration."

I saw him, in consultation with Mr. Hewer, May 1, 1865, when I received the above account from the patient. He was then suffering from many of the "subjective" phenomena which are common to persons who have incurred a serious shock to the system.

But in addition to these, he presented the following somewhat peculiar and exceptional symptoms:—

1. An extreme difficulty in articulation, in the nature of a stammer or stutter of the most intense kind, so that it was extremely difficult to hold a continuous conversation with him. Although he had, previously to the accident, some impediment in his speech, this has been aggravated to the degree that has just been mentioned, so as to constitute the most intense stutter that I have ever heard in an adult.

2. A very peculiar condition of the spine and the muscles of the back.

The spine is rigid—has lost its natural flexibility to antero-posterior as well as to lateral movement.

There is an extreme degree of sensibility of the skin of the back, from the nape of the neck down to the loins. This sensibility extends for about four inches on either side of the spine. It is most intense between the shoulders.

This sensibility is both superficial and deep. The superficial or cutaneous sensibility is so marked, that on touching the skin lightly or on drawing the finger down it, the patient starts forwards as if he had been touched with a red-hot iron. There is also deep pain on pressure along the whole length of the spine, and on twisting or bending it in any direction.

Whenever the back is touched at these sensitive parts, the muscles are thrown into violent contraction so as to become rigid, and to be raised in strong relief, their outlines becoming clearly defined.

3. The patient's gait is most peculiar. He does not carry one leg before the other alternately in the ordinary manner of walking, but shuffles sideways, carrying the right leg in advance, and bringing up the left one after it by a series of short steps. He can alternate the action of the legs, but he cannot bring one leg in front of the other without twisting the whole body and turning, as on a pivot, on the leg that supports him. He cannot bend the thigh on the abdomen.

I saw this patient several times during the summer and autumn. In the early part of December, his condition was as nearly as possible the same as that which has been described in May, no change whatever in pain or in gait having taken place. There was not at this time, nor had there ever been, any signs of paralysis, but he complained of the sensation of a tight cord round the waist.

In addition to Mr. Hewer and myself, this patient was seen at different times by Sir W. Fergusson, Drs. Reynolds and Walshe. We all agreed that the patient was suffering from "concussion of the spine," and that his ultimate recovery was uncertain. Mr. D. brought an action against the railway company, which was tried at Guildhall, in December, 1865, and recovered £4750 damages.

Since the trial he has been continuously under my care, and I have seen him at intervals of about a month. He has been treated by perfect rest, lying on a Prone couch; by warm salt-water douches to the spine, for which purpose he has resided at Brighton, and by full doses of the bromide of potassium. Under this treatment he has considerably improved (May, 1866). The extreme sensibility of the back is materially lessened, and he can walk much better than he did. He also stammers less vehemently, but he still has considerable rigidity about the spine, can only walk with the aid of a stick, and retains that peculiar careworn, anxious, and aged look that is so very characteristic of those who have suffered from these injuries.

I shall now direct your attention to another very peculiar and interesting class of cases, those in which the spine has been violently twisted or strained, but not concussed or jarred.

TWISTS, SPRAINS, or WRENCHES OF THE SPINE, without fracture or dislocation of the vertebræ, may occur in a variety of ways.

Boyer relates a fatal case of this kind, occurring from an injury received in practising gymnastics. Sir A. Cooper gives an instance, to which I shall refer, of a fatal wrench of the spine from a rope catching a boy round the neck whilst swinging.

In two cases which I shall relate, the injury also arose from violence applied to the cervical spine; in one from a railway accident, in the other from a fall from a horse.

These wrenches of the spine are, from obvious reasons, most liable to occur in the more mobile parts of the vertebral column, as the neck and loins; less frequently in the dorsal region.

In railway collisions, when a person is violently and suddenly jolted from one side of the carriage to the other, the head is frequently forcibly thrown forwards and backwards, moving as it were by its own weight, the patient having momentarily lost control over the muscular structures of the neck. In such cases the patient complains of a severe straining, aching pain in the articulations between the head and the spine, and in the cervical spine itself. This pain closely resembles that met with in any joint after a severe wrench of its ligamentous structures, but is peculiarly distressing in the spine, owing to the extent to which fibrous tissue and ligament enter into the composition of the column. It is greatly increased by motion of any kind, and however slight, to and fro, and especially by rotation. The pains are greatly increased on pressure and on lifting up the head, so as to put the tissues on the stretch. In consequence of this, the patient keeps the neck and head immovable, rigid, looking straight forwards—neither turning to the right nor to the left. He cannot raise his head off a pillow without the assistance of his hand, or that of another person.

The lumbar spine is often strained in railway collisions, with or without similar injury to the cervical portion of the column, in consequence of the body being forcibly swayed backwards and forwards during the oscillation of the carriage on the receipt of a powerful shock. In such cases the same kind of pain is complained of. There is the same rigidly inflexible condition of the spine, with tenderness on external pressure, and great aggravation of suffering on any movement being impressed upon it, more particularly if the patient bends backwards. The patient is unable to stoop; in attempting to do so, he always goes down on one of his knees.

These strains of the ligamentous structures of the spinal column are not unfrequently associated with some of the most serious affections of the spinal cord that are met with in surgical practice as a consequence of injury.

They may of themselves prove most serious, or even fatal. Thus, in Case 13, we have an instance of loosening of the cervical portion of the spinal column to such an extent that the patient could not hold the head upright without artificial support.

In Case 14, we have an example of inflammatory swelling developing around the sprained part to such an extent as to compress the cord and spinal nerves, and thus lead to paralysis. And lastly, in Sir A. Cooper's case, we have an instance of a sprain of the spine terminating in death, and a description of the post-mortem appearances presented by this accident.

The *prognosis* will depend partly on the extent of the stretching of the muscular and ligamentous structures, partly on whether there is any inflammatory action excited in them which may extend to the interior of the spinal canal.

As a general rule, where muscular, tendinous and ligamentous structures have been violently stretched, as in an ordinary sprain, however severe, they recover themselves in the course of a few weeks, or at most within three or six months. If a joint, as the shoulder or ankle, continues to be weak and preternaturally mobile, in consequence of elongation of the ligaments, or weakness or atrophy of the muscles, beyond this period, it will, in all probability, never be so strong as it was before the accident.

The same holds good with the spine; and a vertebral column, which, as in Case 13, has been so weakened as to require artificial support, after a lapse of eleven months, in order to enable it to maintain the weight of the head, will not, in all probability, ever regain its normal strength and power of support.

One great prospective danger in strains of the spine is the possibility of the inflammation developed in the fibrous structures of the column extending to the meninges of the cord. This I have several times seen occur, and I believe that in Cases 6 and 11 this happened. We see that this is particularly apt to happen when the strain or twist occurs between the occiput and the atlas or axis. In these cases a rigid tenderness is gradually developed, which is most distressing and persisting and evidently of an inflammatory character.

Or, as in Case 13, the paralysis may remain incomplete, being confined to the nerves which are connected with that part of the spine which is the seat of the wrench, one or other of their roots either having suffered lesion, or the nervous cord itself having been injured in its passage through the intervertebral foramen.

Lastly, as in Sir A. Cooper's case, a twist of the spine may slowly and insidiously be followed by symptoms of complete paraplegia, and eventually by death from extravasation of blood into the vertebral canal.

Case 13.—Miss ——, a lady, 28 years of age, was involved in the terrible catastrophe that occurred on the South Eastern Railway, at Staplehurst, on June 9, 1865, when in consequence of a bridge giving way a portion of a train was precipitated into a shallow stream. This lady lay for two hours and a half under a mass of broken carriages and débris of the bridge, another lady, a fellow-passenger, who had been killed, being stretched across her.

Miss ――― was lying in such a position that she could not move. Her head was forcibly twisted to the right side, and the neck bent forwards.

When extricated she was found to be a good deal cut about the head and face, and the left arm was extensively bruised, ecchymosed, and perfectly powerless.

Her neck had been so violently twisted or wrenched that for a long time Miss ――― lost completely all power of supporting the head, which she says felt loose. It used to fall on any side, as if the neck was broken, usually hanging with the chin resting on the breast.

Without going into an unnecessarily minute detail of all the distressing symptoms with which this young lady was affected, it suffices to say that she gradually recovered from all her general bodily sufferings, except these conditions, *viz.*, a weakened state of the neck, a loss of power in the left arm, and pain in the lower part of the back.

The neck had been so severely twisted and sprained that the ligamentous and muscular structures seemed to be loosened, so that in order to keep the head in position she was obliged to wear a stiff collar lest the head should fall loosely from side to side. At first it had a special tendency to fall forwards; but after a time the tendency was in a backward direction. When lying on her back she had no power whatever to raise her head, and was obliged to do so with her right hand put under it so as to support it. If she wished to get up when in bed, for instance, she was obliged to assume a most distressing action, being compelled to roll over on to her face, and then, pressing the forehead against the pillow, get upon her knees.

There was no pain in the cervical spine, nor could any irregularity of the vertebræ be detected. There was no pain in forcibly moving the head on the atlas, or rotating this bone on the axis. The looseness appeared to be in the lower part of the cervical spine.

The left arm had at first and for many weeks been completely powerless, all sensation as well as power of motion in it having been lost. Sensation gradually and slowly returned. But the whole of the nerves of the brachial plexus appeared to be partially paralyzed, so far as motor influence was concerned. The circumflex, the musculo-spiral, the median and the ulnar were all affected to such a degree as to occasion great loss of power to the muscles they respectively supplied. Thus she could not use the deltoid so as to raise the arm to the top of the head. She could not pick up a pin or even a quill between the thumb and forefinger. She could not hold a book. The power of grasping with the left hand and fingers was infinitely less than with the right, and there was some rigid contraction of the little and ring-fingers. The muscles of the left hand and of the ball of the thumb were wasted.

This crippled and partially paralyzed state of the left arm was a most serious and distressing inconvenience to the patient. Before the accident she had been an intrepid rider, a skilful driver, and an accomplished musician, playing much on the harp and piano. All these pursuits were necessarily completely put a stop to, and from being remarkable for her courage she had become so nervous as scarcely to be able to drive in a carriage.

Mr. Tapson had most skilfully and assiduously attended this very distressing case almost from the time of the accident, and the patient had occasionally had the advantage of Mr. Holmes Coote's advice. When I saw Miss —— in consultation with these gentlemen on April 20, 1866, ten and a half months after the accident, they told me that the condition of the neck had certainly, though very slowly, improved, but that the state of the left arm, which was such as has just been described, had undergone no change for several months.

The pain in the lower part of the back had increased during the last two months. There was no disturbance of the mind, and no sign of cerebral irritation. The bodily health generally was fairly good—as much so as could be expected under the altered circumstances of life that this accident had in so melancholy a manner entailed on this young lady.

The state of the cervical spine in this case was most remarkable. It was movable at its lower part in all directions as if it were attached to a universal joint, or had a ball-and-socket articulation, the weight of the head carrying it in all directions. It was almost impossible to conceive so great a degree of mobility existing without dislocation—but there was certainly neither luxation nor fracture, the vertebræ being apparently loosened from one another in their ligamentous connections and their muscular supports, so that the weight of the head was too great for the weakened spine to carry.

This loosening was most marked in the lower cervical region, and did not exist between the atlas and the occiput. It was clearly the direct result of the violent and long-continued wrench to which this part of the spine had been subjected.

The paralysis was confined to the left arm, no other part of the body having been affected by it. At first the paralysis was complete, the arm being perfectly powerless and sensation being quite lost. After a time sensation returned, but motion was still very imperfect, and no improvement had taken place in this respect for several months. As the nerves of the whole of the brachial plexus were implicated, and apparently to the same degree, it was difficult to account for this in any other way than by an injury inflicted upon them at their origin from the cord, or on their exit through the vertebral column. I think it most probable that this latter injury was the real cause of nervous weakness to the left arm, for the spine had been wrenched in the lower cervical region, in that part, in fact, which corresponds to the origin of the brachial plexus; and there was not at the time of my visit, nor did there appear to have been at any previous period, any disturbance in the functions of the spinal cord as a whole; the paralysis being entirely and absolutely localized to the parts supplied by the left brachial plexus, implicating these only so far as motor power was concerned, and affecting no other portion of the nervous system.

This lady brought an action for damages against the railway company at Guildhall in the spring of 1866. But as she had sustained no pecuniary loss by the accident, she was only awarded the wretched "compensation" of £1350. Mental sufferings, bodily pain, and disability, and complete annihilation of the prospects of a life, weigh lightly in the scales of justice, which are only made to kick the beam by the burden of the actual money loss entailed by the accident.

Case 14.—The following case, which I have seen several times in consultation with Dr. Russell Reynolds, under whose immediate care the patient was, and to whom I am indebted for its early history, affords an excellent illustration of some of the effects that may result from a severe twist or wrench of the spine.

Mr. G., about 23 years of age, a strong, well-formed, healthy young man, thrown from his horse on December 12, 1865. He fell on the back of his head, on soft ground, and rolled over. He got up immediately after the fall and walked to his house, a distance of about one hundred yards. He had no cerebral disturbance whatever, being neither insensible, delirious, concussed, nor sick. The head was twisted to the left side, and he felt pain in the neck He kept his bed in consequence of this pain in the neck till January 1st, 1866, and his room for a week longer. At this time he tried to write, but found great difficulty in controlling his right arm. He managed, however, to do so, and did write a letter. He was under surgical treatment in the country, and was not considered to have paralysis, as he could use his arms well for all ordinary purposes, and could walk without difficulty.

Towards the end of January, nearly six weeks after the accident, symptoms of paralysis very gradually and slowly began to develop themselves. The right arm became cold, numb, and was affected by creeping sensations. His right leg became weak, unequal to the support of the body, and he dragged his right foot.

He came to town on February 21st, when he was seen for the first time by Dr. Reynolds, who reports that at this period the paralysis of the right arm had become complete, and that of the right foot was partial, the patient walking with a drag of the foot. His limbs gave way under him, so that he had occasionally fallen. He had no pain in any part of the body; his mind was clear, but he was very restless.

On the 27th February, whilst stooping, he fell in his bedroom, struggled much, and was unable to rise. He was found, after a time, lying partly under his bed. On the following day it was found that the left side was partially paralyzed, the right side continuing in the condition already described. There was now considerable swelling and tenderness on the left side of the neck and about the third and fourth cervical vertebræ. He was seen shortly after this by Dr. Jenner, in consultation with Dr. Reynolds, and was ordered complete rest, with large doses of iodide of potass.

I saw him on March 3d, in consultation with Dr. Reynolds. I found him lying on his back in bed. The mind quite clear; spirits good. No appearance of anxiety or distress in the countenance; in fact I was much struck by the happy, cheerful expression of his countenance under the melancholy circumstances in which he was placed.

I found his condition much as has been described. There was complete paralysis of the right arm, partial of the right leg. The left arm was also partially paralyzed, and the left leg slightly so.

He was unable to stand. There was no affection of the bladder or of the sphincter ani. The skin was hot and perspiring; the pulse quick; urine acid.

He could not raise his head off the pillow, and lay quite flat on his back. On being raised up in the sitting posture, it was necessary to support his head with the hands; and when he was seated upright, he held the head firmly fixed, the spine being kept perfectly rigid. He was quite unable to turn or move the head.

The back of the neck was swollen, especially on the left side, and was tender on pressure. The swelling was less than it had been. The cervical vertebræ felt as if they were somewhat twisted, so that the head inclined towards the right side. It was doubtful whether this was really so. The patient continued the iodide of potass; and a gutta-percha case, extending from the top of his head to the pelvis, and embracing the shoulders and back of the chest, was moulded on him, so as to keep the head and spine motionless. He was ordered to lie on his back and not to move.

I saw the patient several times with Dr. Reynolds, and we were gratified to find that a steady improvement was taking place. On March 27th he had completely lost all symptoms of paralysis on the left side of the body; the right leg had recovered its power, and the paralytic symptoms had almost entirely disappeared from the right arm. He could raise it, grasp with his hand, and in fact use it for the ordinary purposes of life. He could stand, though in a somewhat unsteady way. This seemed rather owing to his having kept the recumbent position for so long a time than to any loss of nervous power in the legs.

The swelling of the neck had entirely subsided, and the cervical spine was straight, but it was rigid, and he could not turn the head. The support was habitually worn, and gave him great comfort.

This case is remarkable in several of its points. In the first place, the fact that the paralysis did not begin to show itself until several weeks—nearly six—had elapsed from the time of the accident is a matter of the greatest consequence in reference to these injuries. Then, again, the fact that although the brain was throughout unaffected, and the injury purely spinal, the paralysis was of a hemiplegic and not a paraplegic character, is also not without import. And lastly, the gradual subsidence of the very threatening symptoms with which the patient was affected, and the disappearance of the paralysis of the limbs in the inverse order to that in which it developed itself in them, should be observed.

That wrenches or twists of the spine may slowly develop paralytic symptoms, and may be attended eventually by a fatal result, is well illustrated by a case recorded by Sir Astley Cooper as occurring in the practice of Mr. Heaviside. It is briefly as follows: A lad, 12 years old, whilst swinging in a heavy wooden swing, was caught under the chin by a rope, so that his head and the whole of the cervical vertebræ were violently strained. As the line immediately slipped off, he thought no more of it. For some months after the occurrence he felt no pain or inconvenience, but it was observed that he was less active than usual, and did not join in the games of his schoolfellows. At that time it was found that he was really weaker than before the occurrence. He suffered from pains in the head and in the back of the neck, the muscles of which part were stiff, indurated, and very tender to external pressure. Movement of the head in any direction gave rise to pain, and there was diminution in voluntary power of motion in his limbs.

Eleven months after the accident the complaint and the paralytic affection of the limbs were gradually getting much worse, added to which he felt a most vehement and burning pain in the small of his back. His symptoms gradually became worse, difficulty of breathing set in, and he died exactly twelve months after the accident.

On examination after death the whole contents of the head were found to be perfectly healthy. There was no fracture or other sign of injury to the spine, but "the theca vertebralis was found overflowing with blood which was effused between the theca and the inclosing canals of bone. The effusion extended from the first vertebra of the neck to the second vertebra of the back, both included."[1]

This case is a most valuable one. It illustrates one of the important points in that last described, *viz.*, the very slow, gradual, and progressive development of paralysis in these injuries of the spine. And as it was attended by a fatal issue and the opportunity of a *post-mortem* examination, it also proves that this slow and progressive development of paralysis after an interval of "some months" may be associated with extensive and serious lesion

---

[1] Sir A. Cooper, Fractures and Dislocations, 8vo. ed., p. 530.

within the spinal canal, with the effusion, in fact, of a large quantity of blood upon the membranes of the cord,—the very condition that has already been shown (p. 38) to be the common accompaniment of many fatal cases of so-called "concussion of the spine."

Each of these three cases of twist of the spine is typical of a special group of these injuries. In the first case we have sudden and immediate paralysis of one arm produced by the wrench to which that portion of the spine that gives exit to the nerves supplying that limb had been subjected.

In the second case we have paralysis, resulting after an interval of some weeks, as a consequence of the pressure of the secondary inflammatory effusions that had been slowly produced by the injury to the spine and its contents,—that paralysis disappearing as these effusions were absorbed.

In the third case we have an instance of death resulting in twelve months after a wrench of the spine by the effects of hemorrhage into the spinal canal.

LECTURE THE FIFTH.

Period at which Symptoms begin to develop—Length of Time that often elapses—Concussion not associated with other Injury—Nature of Changes produced by Concussion—Early Symptoms of Railway Concussion—Detail of Symptoms of Railway Concussion—Symptoms of Railway Concussion—Interval between Accident and Symptoms—Pathology of Railway Concussion—Mr. Gore's Case.

ONE of the most remarkable circumstances connected with injuries of the spine is, the disproportion that exists between the apparently trifling accident that the patient has sustained, and the real and serious mischief that has occurred. Not only do symptoms of concussion of the spine of the most serious, progressive, and persistent character, often develop themselves after what are apparently slight injuries, but frequently when there is no sign whatever of external injury. This is well exemplified in Case 9, the patient having been partially paralyzed simply by slipping down a few stairs on her heels. The shake or jar that is inflicted on the spine when a person jumping from a height of a few feet comes to the ground suddenly and heavily on his heels or in sitting posture, has been well known to surgeons as not an uncommon cause of spinal weakness and debility. It is the same in railway accidents; the shock to which the patient is subjected in them being often followed by a train of slowly-progressive symptoms indicative of concussion and subsequent irritation and inflammation of the cord and its membranes.

But I may not only say that sudden shocks applied to the body are liable to be followed by the train of evil consequences that we are now discussing.. I may even go further, and say that these symptoms of spinal concussion seldom occur when a serious injury has been inflicted on one of the limbs, unless the spine itself has at the same time been severely and directly struck. A person who by any of the accidents of civil life meets with an injury by which one of the limbs is fractured or is dislocated, necessarily sustains a very severe shock, but it is the rarest thing possible to find that the spinal cord or the brain has been injuriously influenced by this shock that has been impressed on the body. It would appear as if the violence of the shock expended itself in the production of the fracture or the dislocation, and that a jar of the more delicate nervous structures is thus avoided. I may give a familiar illustration of this from an injury to a watch by falling on the ground. A watchmaker once told me that if the glass was broken, the works were rarely damaged; if the glass escapes unbroken, the jar of the fall will usually be found to have stopped the movement.

How these jars, shakes, shocks, or concussions of the spinal cord directly influence its action I cannot say with certainty. We do not know how it is that when a magnet is struck a heavy blow with a hammer, the magnetic force is jarred, shaken, or concussed out of the horse-shoe. But we know that it is so, and that the iron has lost its magnetic power. So, if the spine is badly jarred, shaken, or concussed by a blow or shock of any kind communicated to the body, we find that the nervous force is to a certain extent shaken out of the man, and that he has in some way lost nervous power. What immediate change, if any, has taken place in the nervous structure to occasion that effect, we no more know than what change happens to a magnet when struck. But we know that a change has taken place in the action of the nervous system just as we do in the action of the iron by the change that is induced in the loss of its magnetic force.

But whatever may be the nature of the primary change that is produced in the spinal cord by a concussion, the secondary effects are clearly of an inflammatory character, and are identical with those phenomena that have been described by Ollivier, Abercrombie, and others, as dependent on chronic meningitis of the cord, and subacute myelitis.

One of the most remarkable phenomena attendant upon this class of cases is, that at the time of the occurrence of the injury the sufferer is usually quite unconscious that any serious accident has happened to him. He feels that he has been violently jolted and shaken, he is perhaps somewhat giddy and confused, but he finds no bones broken, merely some superficial bruises or cuts on the head or legs, perhaps even no evidence whatever of external injury. He congratulates himself upon his escape from the imminent peril to which he has been exposed. He becomes unusually calm and self-possessed; assists his less fortunate fellow-sufferers, occupies himself perhaps actively in this way for several hours, and then proceeds on his journey.

When he reaches his home, the effects of the injury that he has sustained begin to manifest themselves. A revulsion of feeling takes place. He bursts into tears, becomes unusually talkative, and is excited. He cannot sleep, or, if he does, he wakes up suddenly with a vague sense of alarm. The next day he complains of feeling shaken or bruised all over, as if he had been beaten, or had violently strained himself by exertion of an unusual kind. This stiff and strained feeling chiefly affects the muscles of the neck and loins, sometimes extending to those of the shoulders and thighs. After a time, which varies much in different cases, from a day or two to a week or more, he finds that he is unfit for exertion and unable to attend to business. He now lays up, and perhaps for the first time seeks surgical assistance.

This is a general sketch of the early history of most of these cases of "concussion of the spine" from railway accidents. The details necessarily vary much in different cases.

There is great variation in the period at which the more serious, persistent, and positive symptoms of spinal lesion begin to develop themselves. In some cases they do so immediately after the occurrence of the injury, in others not until several weeks, I might perhaps even say months, had elapsed. But during the whole of this interval, whether it be of short or of long duration, it will be observed that the sufferer's condition, mentally and bodily, has undergone a change. His friends remark, and he feels, that "he is not the man he was." He has lost bodily energy, mental capacity, business aptitude. He looks ill and worn; often becomes irritable and easily fatigued. He still believes that he has sustained no serious or permanent hurt, tries to return to his business, finds that he cannot apply himself to it, takes rest, seeks change of air and scene, undergoes medical treatment of various kinds, but finds all of no avail. His symptoms become progressively more and more confirmed, and at last he resigns himself to the conviction that he has sustained a more serious bodily injury than he had at first believed, and one that has, in some way or other, broken down his nervous power, and has wrought the change of converting a man of mental energy and of active business habits into a valetudinarian, utterly unable to attend to the ordinary duties of life.

The condition in which a patient will be at this or a later period of his sufferings, will be found detailed in several of the cases that have been related, especially in Cases 5 and 6.

It may, however, throw additional light on this subject, if we analyze the symptoms, and arrange them in the order in which they will present themselves on making a surgical examination of such a patient, bearing this important fact in mind, however, that although all and every one of these symptoms may present themselves in any given case, yet that they are by no means all necessarily present in any one case. Indeed this usually happens, and we generally find that whilst some symptoms assume great prominence, others are proportionally dwarfed, or, indeed, completely absent.

The *countenance* is usually pallid, lined, and has a peculiarly careworn, anxious expression; the patient generally looking much older than he really is or than he did before the accident. I have seen one instance of flushing of face. This was marked in Case 11.

The *memory* is defective. This defect of memory shows itself in vaious ways; thus, Case 2 said that he could not recollect a message unless he wrote it down; Case 6 forgot some common words and misspelt others; Case 5 lost command over figures; he could not add up a few figures, and had also lost, in a great degree, the faculty of judging of weight, and of distance in a lateral direction; he forgot dates, the ages of his children, &c.

The *thoughts* are confused. The patient will sometimes, as in Case 11, break off in the middle of a sentence, unable to finish it; he cannot concentrate his ideas so as to carry out a connected line of argument; he attempts to read, but is obliged to lay aside the book or paper after a few minutes' attempt at perusal.

All *business aptitude* is lost, partly as a consequence of impairment of memory, partly of confusion of thought and inability to concentrate ideas for a sufficient length of time.

The *temper* often becomes changed for the worse, the patient being fretful, irritable, and in some way—difficult perhaps to define, but easily appreciated by those around him—altered in character.

The *sleep* is disturbed, restless, and broken. He wakes up in sudden alarm; dreams much; the dreams are distressing and horrible.

The *head* is usually of its natural temperature, but sometimes hot, as in Case 11. The patient complains of various uneasy sensations in it; of pain, tension, weight, or throbbing; of giddiness; of a confused or strained feeling in it. Frequently loud and incessant noises, described as roaring, rushing, ringing, singing, sawing, rumbling, or thundering. These noises vary in intensity at different periods of the day, but if once they occur, are never entirely absent, and are a source of great distress and disquietude to the patient.

The *organs of special sense* usually become more or less seriously affected. They become sometimes over sensitive and irritable, or are impaired in their perceptions, and at others perverted in their sensations. In many cases we find a combination of all those conditions in the same organ.

Vision is usually affected in various ways and in very different degrees. In some cases, though rarely, there is double vision and perhaps slight strabismus. In others an alteration in the focal length, so that the patient has to use glasses, or to change those he has previously worn. The patient cannot read for more than a few minutes, the letters running into one another. More commonly, muscæ volitantes and spectra, rings, stars, flashes, sparks— white, coloured, or flame-like—are complained of. This happened in Cases 5, 6, and 11. The eyes often become over sensitive to light, so that the patient habitually sits in a shaded or darkened room, turns his back to the window, and cannot bear unshaded gas- or lamp-light. This intolerance of light may amount to positive photophobia. It gives rise to a habitually contracted state of the brows, so as to exclude light as much as possible from the eyes. One or both eyes may be thus affected. Sometimes one eye only is intolerant of light. This intolerance of light may be associated with dimness and imperfection of sight. Perhaps vision is normal in one eye, but impaired seriously in the other. The circulation in the bottom of the eye is visible to some patients.

The hearing may be variously affected. Not only does the patient commonly complain of the noises in the head and ears that have already been described, but the ears, like the eyes, may be over sensitive or too dull. One ear is frequently over sensitive whilst the other is less acute than it was before the accident. The relative sensibility of the ears may readily be measured by the distance at which the tick of a watch may be heard. Loud and sudden noises are particularly distressing to these patients. The fall of a tray, the rattle of a carriage, the noise of children at play, are all sources of pain and of irritation.

Taste and *smell* are sometimes, but more rarely, perverted. Case 11 complained of occasional fetid smells, which were not perceptible to any one else.

The *sense of touch* is impaired. The patient cannot pick up a pin, cannot button his dress, cannot feel the difference between different textures, as cloth and velvet. He loses the sense of *weight*, cannot tell whether a sovereign or a shilling is balanced on his finger.

Speech is rarely affected. Case 12 stammered somewhat before the accident, but after it his speech became a most painful and an indescribably confused stutter that it was almost impossible to comprehend. The same phenomenon was observed in the Count de Lordat's case, p. 24. But it is certainly rare.

The *attitude* of these patients is usually peculiar. It is stiff and unbending. They hold themselves very erect, usually walk straight forwards, as if afraid or unable to turn to either side. The movements of the head or trunk, or both, do not possess their natural freedom. There may be pain or difficulty in moving the head in the antero-posterior direction, or in rotating it, or all movements may be attended by so much pain and difficulty that the patient is afraid to attempt them, and hence keeps the head in its attitude of immobility.

The movements of the trunk are often equally restrained, especially in the lumbar region. Flexion forwards, backwards, or sideways, is painful, difficult, and may be impossible; flexion backwards is usually most complained of.

If the patient is desired to stoop and pick anything off the ground, he will not be able to do so in the usual way, but bends down on the knee and so reaches the ground.

If he is laid horizontally and told to raise himself up without the use of his hands, he will be unable to do it.

The *state of the spine* will be found to be the real cause of all these symptoms.

On examining it by pressure, by percussion, or by the application of the hot sponge, it will be found that it is painful, and that its sensibility is exalted at one, two, or three points. These are usually the upper cervical, the middle dorsal, and the lumbar regions. The exact vertebræ that are affected vary necessarily in different cases, but the exalted sensibility always includes two, and usually three, at each of these points. It is in consequence of the pain that is occasioned by any movement of the trunk in the way of flexion or rotation, that the spine loses its natural suppleness, and that the vertebral column moves as a whole, as if cut out of one solid piece, instead of with the flexibility that its various component parts naturally impress upon its motions.

The movements of the head upon the upper cervical vertebræ are variously affected. In some cases the head moves freely in all directions, without pain or stiffness, these conditions existing in the lower and middle, rather than in the upper, cervical vertebræ. In other cases, again, the greatest agony is induced if the surgeon takes the head between his hands and bends it forwards or rotates it, the articulations between the occipital bone, the atlas, and the axis, being evidently in a state of inflammatory irritation. This happened in a very marked manner in Cases 5 and 6; and in both these it is interesting to observe that distinct evidences of cerebral irritation had been superadded to those of the more ordinary spinal mischief.

The pain is usually confined to the vertebral column, and does not extend beyond the transverse processes. But in some instances, as in Case 12, the pain extended widely over the back on both sides, more on the left than on the right, and seemed to correspond with the distribution of the posterior branches of the dorsal nerves. In these cases, from the musculo-cutaneous distribution of these nerves, the pain is superficial and cutaneous as well as deeply seated in the spine.

The muscles of the back are usually unaffected, but in some cases where the muscular branches of the dorsal nerves are affected, as in Case 12, they may be found to be very irritable and spasmodically contracted, so that their outlines are very distinct and marked.

The *gait* of the patient is remarkable and characteristic. He walks more or less unsteadily, generally uses a stick, or, if deprived of that, is apt to lay his hand on any article of furniture that is near to him, with the view of steadying himself.

He keeps his feet somewhat apart, so as to increase the basis of support, and consequently walks in a straddling manner.

As one leg is often weaker than the other, he totters somewhat, raises one foot but slightly off the ground, so that the heel is apt to touch. He seldom drags the toe, but walking flat-footed as it were on one side, the heel drags. This peculiar straddling, tottering, unsteady gait, with the spine rigid, the head erect, and looking straight forwards, gives the patient the aspect of a man who walks blindfolded.

The patient cannot generally stand equally well on either foot. One leg usually immediately gives way under him if he attempts to stand on it.

He often cannot raise himself on his toes, or stand on them, without immediately tottering forwards.

His power of walking is always very limited; it seldom exceeds half a mile or a mile at the utmost.

He cannot ride, even if much in the habit of doing so before the accident.

There is usually considerable difficulty in going up and down stairs—more difficulty in going down than up. The patient is obliged to support himself by holding on to the banisters, and often brings both feet together on the same step.

A sensation as of a cord tied round the waist, with occasional spasm of the diaphragm, giving rise to a catch in the breathing, or hiccup, is sometimes met with, and is very distressing when it does occur.

The *nervous power of the limbs* will be found to be variously modified, and will generally be so to very different degrees in the different limbs. Sometimes one limb only is affected, at others the arm and leg on one side, or both legs only, or the arm and both legs, or all four limbs, are the seat of uneasy sensations. There is the greatest possible variety in these respects, dependent of course entirely upon the degree and extent of the lesion that has been inflicted upon or induced in the spinal cord.

Sensation only may be affected, or it may be normal, and motion may be impaired; or both may be affected to an equal, or one to a greater and the other to a less, degree. And these conditions may happen in one or more limbs. Thus sensation and motion may be seriously impaired in one limb, or sensation in one and motion in another. The paralysis is seldom complete. It may become so in the more advanced stages after a lapse of several years, but for the first year or two it is (except in cases of direct and severe violence) almost always partial. It is sometimes incompletely recovered from, especially so far as sensation is concerned.

The *loss of motor power* is especially marked in the legs, and more often in the extensor than in the flexor muscles. The extensor of the great toe is especially apt to suffer. The hand and arm are less frequently the seat of loss of motor power than the leg and foot; but the muscle of the ball of the thumb, or the flexors of the fingers, may be so affected.

The loss of motor power in the foot and leg is best tested by the application of the galvanic current, so as to compare the irritability of the same muscles of the opposite limbs. The value of the electric test is, that it is not under the influence of the patient's will, and that a very true estimate can thus be made of the loss of contractility in any given set of muscles.

The loss of motor power in the hand is best tested by the force of the patient's grasp. This may be roughly estimated by telling him to squeeze the surgeon's fingers, first with one hand and then the other, or more accurately by means of the dynamometer, which shows on an index the precise amount of pressure that a person exercises in grasping.

It is in consequence of the diminution of motor power in the legs that those peculiarities of gait which have just been described are met with, and they are most marked when the amount of loss is unequal in the two limbs. The sphincters are very rarely affected in the cases now under consideration. Sometimes there is increased frequency of micturition, but I have rarely met with retention of urine or cases requiring the continued use of the catheter; nor have I observed in any case that the contractility of the sphincter ani had been so far impaired as to lead to involuntary escape of flatus or of feces.

Modification or diminution of sensation in the limbs is one of the most marked phenomena in these cases.

In many instances the sensibility is a good deal augmented, especially in the earlier stages. The patient complains of shooting pains down the limbs, like stabs, darts, or electrical shocks. The surface of the skin is sometimes over-sensitive in places on the back (as in Case 12), or in various parts of the limbs, hot, burning sensations are experienced in it. After a time these sensations give place to various others, which are very differently described by patients. Tinglings, a feeling of "pins and needles," a heavy sensation as if the limb was asleep, creeping sensations down the back and along the nerves, and formications, are all commonly complained of. These sensations are often confined to one nerve in a limb, as the ulnar, for instance, or the musculo-spiral.

Numbness, more or less complete, may exist independently of, or be associated with, all these various modifications of sensation with pain, tingling, or creeping sensations. Its extent will vary greatly; it may be confined to a part of a limb, may influence the whole of it, or may extend to several; its degree and extent are best tested by Brown-Séquard's instrument.

Coldness of one of the extremities, dependent upon actual loss of nervous power, and defective nutrition, is often perceptible to the touch, and may actually be established by the thermometer; but in many cases it is found that the sensation of coldness is far greater to the patient than it is to the surgeon's hand, and not unfrequently no appreciable difference in the temperature of two limbs can be determined by the most delicate clinical thermometer, although the patient experiences a very distinct and distressing sense of coldness in one of the limbs.

The condition of the limbs as to size, and the state of their muscles, will vary greatly.

In some cases of complete paraplegia, which has lasted for years, as in Case 3, it has been remarked that no diminution whatever had taken place in the size of the limbs. This was also the case in Case 2, where the paralysis was partial. It is evident, therefore, that loss of size in a limb that is more or less completely paralyzed is not the simple consequence of the disuse of the muscles, or it would always occur. But it must arise from some modification of innervation, influencing the nutrition of the limb, independently of the loss of its muscular activity.

In most cases, however, where the paralytic condition has been of some duration, the size of the limb dwindles; and on accurate measurement it will be found to be somewhat smaller in circumference than its fellow on the opposite side.

The state of the muscles as to firmness will also vary. Most commonly when a limb dwindles the muscles become soft, and the inter-muscular spaces more distinct. Occasionally in advanced cases a certain degree of contraction and of rigidity in particular muscles sets in. Thus the flexors of the little and ring fingers, the extensors of the great toe, the deltoid or the muscles of the calf, may all become the seats of more or less rigidity and contractions.

The *body* itself generally loses weight; and a loss of weight, when the patient is rendered inactive by a semi-paralyzed state, and takes a fair quantity of good food, which he digests sufficiently well, is undoubtedly a very important and a very serious sign, and may usually be taken to be indicative of progressive disease in the nervous system.

When the progress of the disease has been arrested, though the patient may be permanently paralyzed, we often see a considerable increase of size and weight take place. This is a phenomenon of such common occurrence in ordinary cases of paralysis from disease of the brain, that I need do no more than mention that it is also of not unfrequent occurrence in those forms that proceed from injury, whether of the cord or brain.

The condition of the *genito-urinary* organs is seldom much deranged in the cases under consideration, as there is usually no paralysis of the sphincters. Neither retention of urine nor incontinence of flatus and feces occurs. Sometimes, as in Case 5, irritability of the bladder is a prominent symptom. The urine generally retains its acidity, sometimes markedly, at others but very slightly so. As there is no retention, it does not become alkaline, ammoniacal, or otherwise offensive.

The sexual desire and power are usually greatly impaired, and often entirely lost. Not invariably so, however. The wife of Case 5 miscarried twice during the twelvemonth succeeding her husband's accident. I have never heard priapism complained of.

The *pulse* varies in frequency at different periods. In the early stages it is usually slow; in the more advanced it is quick, near to or above 100. It is always feeble.

The order of the progressive development of the various symptoms that have just been detailed is a matter of great interest in these cases, and each separate symptom comes on very gradually and insidiously. It usually extends over a lengthened period.

In the early stages, the chief complaint is a sensation of lassitude, weariness, and inability for mental and physical exertion. Then come the pains, tinglings, and numbness of the limbs; next the fixed pain and rigidity of the spine; then the mental confusion and signs of cerebral disturbance, and the affection of the organs of sense; the loss of motor power, and the peculiarity of gait.

The period of the supervention of these symptoms after the occurrence of the injury will greatly vary. Most commonly after the first and immediate effects of the accident have passed off there is a period of comparative ease, and of remission of the symptoms, during which the patient imagines that he will speedily regain his health and strength. This period may last for many weeks, possibly for two or three months. At this time there will be considerable fluctuation in the patient's state. So long as he is at rest, he will feel tolerably well; but any attempt at ordinary exertion of body or mind brings back all the feelings or indications of nervous prostration and irritation so characteristic of these injuries; and to these will gradually be superadded those more serious symptoms that have already been fully detailed, which evidently proceed from a chronic disease of the cord and its membranes. After a lapse of several months—from three to six—the patient will find that he is slowly but steadily becoming worse, and he then, perhaps for the first time, becomes aware of the serious and deep-seated injury that his nervous system has sustained.

Although there is often this long interval between the time of the occurrence of the accident and the supervention of the more distressing symptoms, and the conviction of the serious nature of the injury that has been sustained, it will be found, on close inquiry, *that there has never been an interval of complete restoration to health.* There have been remissions, but no complete and perfect intermission in the symptoms. The patient has thought himself and has felt himself much better at one period than he was at another, so much so that he has been tempted to try to return to his usual occupation, but he has never felt himself well, and has immediately relapsed to a worse state than before when he has attempted to do work of any kind.

It is by this chain of symptoms, which, though fluctuating in intensity, is yet continuous and unbroken, that the injury sustained, and the illness subsequently developed, can be linked together in the relation of cause and effect.

Having thus described the various symptoms that may arise from these shocks to and concussions of the spine, let us now briefly inquire into the pathological conditions that lead to and that are the direct causes of these phenomena.

I have already pointed out and discussed at some length, at p. 37 *et seq.*, the pathological conditions that are found within the spinal canal in those cases of paralysis, more or less complete, that result from direct and violent blows upon the back without fracture or dislocation of the bones entering into the formation of the vertebral column, and we have seen that in these cases the signs of spinal lesion are referable to extravasation of blood in various parts within the spinal canal, to rupture of the membranes of the cord, to inflammatory effusions, or to softening and disorganization of the cord itself.

In those cases in which the shock to the system has been general and unconnected with any local and direct implication of the spinal column by external violence, and where the symptoms, as just detailed, are less those of paralysis than of disordered nervous action, the pathological states on which these symptoms are dependent are of a more chronic and less directly obvious character than those above mentioned. They doubtless consist mainly of chronic and subacute inflammatory action in the spinal membranes, and in chronic myelitis, with those changes in the structure of the cord that are the inevitable consequences of a long-continued chronic inflammatory condition developed by it.

The only case on record with which I am acquainted, in which a *post-mortem* examination has been made of the spinal cord of a person who had actually died from the remote effects of concussion of the spine from a railway collision, is one that has very recently been related, and the parts exhibited to the Pathological Society by Dr. Lockhart Clarke. The patient, who had been under the care of Mr. Gore, of Bath, by whom the preparation was furnished, was a middle-aged man of active business habits. He had been in a railway collision, and, without any sign of external injury, fracture, dislocation, wound, or bruise, began to manifest the usual nervous symptoms. He gradually, but very slowly, became partially paralyzed in the lower extremities, and died three years and a half after the accident.

Mr. Gore has most kindly furnished me with the following particulars of the case. On the occurrence of the collision the patient walked from the train to the station close at hand. He had received no external sign of injury—no contusions or wounds, but he complained of a pain in his back. Being most unwilling to give in, he made every effort to get about in his business, and did so for a short time after the accident, though with much distress. Numbness and a want of power in the muscles of the lower limbs gradually but steadily increasing, he soon became disabled. There was great sensitiveness to external impressions, so that a shock against a table or chair caused great distress. As the patient was not under Mr. Gore's care from the first, and as he only saw the case for the first time about a year after the accident, and then at intervals up to the time of death, he has not been able to inform me of the precise time when the paralytic symptoms appeared, but he says that this was certainly within less than a year of the time of the occurrence of the accident. In the latter part of his illness some weakness of the upper extremities became apparent, so that if the patient was off his guard a cup or glass would slip from his fingers. There was no paralysis of the sphincter of the bladder until about eighteen months before his death, when the urine became pale and alkaline, with muco-purulent deposits. In this case the symptoms were not so severe as usual, there was no very marked tenderness or rigidity of the spine, nor were there any convulsive movements.

On examination after death, traces of chronic inflammation were found in the arachnoid and the cortical substance of the brain. The spinal meninges were greatly congested, and exudative matter had been deposited upon the surface of the cord. The cord itself was much narrowed in its anterior-posterior diameter in the cervico-dorsal region. The narrowing was owing to absorption of the posterior columns. These had not only to a great extent disappeared, but the remains were of a dark brownish colour, and had undergone important structural changes. This case is of remarkable interest and practical value, as affording evidence of the changes that take place in the cord under the influence of "concussion of the spine' from a railway accident. Evidences of chronic meningitis—cerebral as well as spinal—of chronic myelitis, with subsequent atrophy, and other organic changes dependent on mal-nutrition of the affected portion of the cord being manifest.

It is well known to pathologists that two distinct forms of chronic subacute inflammation may affect the contents of the spinal canal as the results of injury or of idiopathic disease, *viz.*, inflammation of the membranes, and inflammation of the cord itself.

In spinal meningitis the usual signs of inflammatory action in the form of vascularization of the membranes is met with. The meningo-rachidian veins are turgid with blood, and the vessels of the pia-mater will be found much injected, sometimes in patches, at others uniformly so. Serous fluid, reddened and clear, or opaque from the admixture of lymph, may be found largely effused in the cavity of the arachnoid.

Ollivier[1] states that one of the most constant signs of chronic spinal meningitis is adhesion between the serous lamina that invests the dura mater and that which corresponds to the spinal pia-mater. This he says he has often observed, and especially in that form of the disease which is developed as the result of a lesion of the vertebræ. He has also seen rough cartilaginous (fibroid?) laminæ developed in the arachnoid. Lymph also of a puriform appearance has often been found under the arachnoid, between it and the pia-mater.

In distinguishing the various pathological appearances presented by fatal cases of chronic spinal meningitis, Ollivier makes the very important practical remark—the truth of which is fully carried out by a consideration of the cases related in Lectures 2 and 3—that spinal meningitis rarely exists without there being at the same time a more or less extensive inflammation of the cerebral meninges, and hence, he says, arises the difficulty of determining with precision the symptoms that are special to inflammation of the membranes of the spinal cord.

When myelitis occurs, the inflammation attacking the substance of the cord itself, the most usual pathological condition met with is softening of its substance, with more or less disorganization of

---

[1] Vol. ii. p. 237.

its tissue. This softening of the cord as a consequence of its inflammation may, according to Ollivier, occupy very varying extents of its tissue. Sometimes the whole thickness of the cord is affected at one point, sometimes one of the lateral halves in a vertical direction is affected; at other times it is most marked in or wholly confined to its anterior or its posterior aspect, or the gray central portion may be more affected than the circumferential part. Then, again, these changes of structure may be limited to one part only,—to the cervical, the dorsal, or the lumbar. It is very rare indeed that the whole length of the cord is affected. The most common seat of the inflammatory softening is the lumbar region; next in order of frequency the cervical. In very chronic cases of myelitis, the whole of the nervous substance disappears, and nothing but connective tissue is left behind at the part affected.

Ollivier makes the important observation, that when myelitis is consecutive to meningitis of the cord, the inflammatory softening may be confined to the white substance.

But though softening is the ordinary change that takes place in a cord that has been the seat of chronic inflammation, yet sometimes the nervous substance becomes indurated, increased in bulk, more solid than natural, and of a dull white colour, like boiled white of egg. This induration of the cord may coexist with spinal meningitis, with congestion, and increased vascularization of the membranes. The case of the Count de Lordat (p. 24) is an instance of this induration and enlargement of the substance of the cord, and others of a similar nature are recorded by Portal, Ollivier, and Abercrombie.

It is important to observe, that although spinal meningitis and myelitis are occasionally met with distinct and separate from each other, yet that they most frequently coexist. When existing together, and even arising from the same cause, they may be associated with each other in very varying degrees. In some cases the symptoms of meningitis, in others those of myelitis, being most marked, and after death the characteristic appearances presenting a prominence corresponding to that assumed by their effects during life.

I have given but a very brief sketch of the pathological appearances that are usually met with in spinal meningitis and in myelitis, as it is not my intention in these lectures to occupy your attention with an elaborate inquiry into the pathology of these affections, but rather to consider them in their surgical relations.

I wish now to direct your attention to the symptoms that are admitted by all writers on diseases of the nervous system to be connected with and dependent upon the pathological conditions that I have just detailed to you, and to direct your attention to a comparison between these symptoms and those that are described in the various cases that I have detailed to you as characteristic of "concussion of the spine" from slight injuries and general shocks to the body.

The symptoms that I have detailed at pp. 74 to 83, arrange themselves in three groups:—

1st. The cerebral symptoms.
2d. The spinal symptoms.
3d. Those referable to the limbs.

In comparing the symptoms of "concussion of the spine" arising from railway and other accidents, as detailed in the cases I have related to you, with those that are given to and accepted by the profession as dependent on spinal meningitis and myelitis arising from other causes, I shall confine the comparison of my cases to those related by Abercrombie and Ollivier. And I do this for two reasons, first, because the works of these writers on diseases of the spinal cord are universally received as the most graphic and classical on the subject of which they treat in this country and in France; and, secondly, because their descriptions were given to the world before the railway era, and consequently could in no way have been influenced by accidents occurring as a consequence of modern modes of locomotion.

1. With respect to the cerebral symptoms. It will be observed that in most of the cases that I have related, there was more or less cerebral disturbance or irritation, as indicated by headache, confusion of thought, loss of memory, disturbance of the organs of sense, irritability of the eyes and ears, &c.;—symptoms, in fact, referable to subacute cerebral meningitis and arachnitis.

On this point the statement of Ollivier is most precise and positive. He says that it is rare to find inflammation of the spinal membranes limited to the vertebral canal. That we see at the same time a more or less intense cerebral meningitis. In the cases that he relates of spinal meningitis, he makes frequent reference to these cerebral symptoms—states that they often complicate the case so as to render the diagnosis difficult, especially in the early stages. In the *post-mortem* appearances that he details of patients who have died of spinal meningitis, he describes the morbid conditions met with in the cranium, indicative of increased vascularity and inflammation of the arachnoid. This complication of cerebral with spinal meningitis is nothing more than we should expect as a simple consequence of inflammation running along a continuous membrane. In both the fatal cases of meningitis of the spine recorded by Abercrombie, evidences of intracranial mischief are described.

2. The spinal symptoms that occurred in the cases of "concussion of the spine" which I have related, consisted briefly in pain at one or more points of the spine, greatly increased on pressure, and on movement of any kind, so as to occasion extreme rigidity of the vertebral column.

Ollivier says that one of the most characteristic signs of spinal meningitis is pain in the spine, which is most intense opposite the seat of inflammation. This pain is greatly increased by movement of any kind, so that the patient fearing the slightest displacement of the spine, preserves it in an absolute state of quiescence. This pain is usually accompanied by muscular rigidity. It remits, sometimes being much more severe than at others, and occasionally even disappears entirely. According to some observers, the pain of spinal meningitis is increased by pressure. But the correctness of this observation is doubted by Ollivier, who says that in chronic myelitis there is a painful spot in the spine where the pain is increased on pressure, and he looks upon this as indicative of inflammation of the cord rather than of the membranes.

3. The third group of symptoms dependent on concussion of the spine are those referable to the limbs. They have been described at pp. 81—83, and may briefly be stated to consist in painful sensations along the course of the nerves, followed by more or less numbness, tingling, and creeping; some loss of motor power affecting one or more of the limbs, and giving rise to peculiarity and unsteadiness of gait. No paralysis of the sphincters.

These are the very symptoms that are given by Ollivier and

others as characteristic of spinal meningitis, but more particularly of myelitis.

In spinal meningitis, says Ollivier, there is increased sensibility in different parts of the limbs, extending along the course of the nerves, and augmented by the most superficial pressure. These pains are often at first mistaken for rheumatism. There is often also more or less contraction of the muscles.

In myelitis the sensibility is at first augmented, but after a time becomes lessened, and gives way to various uneasy sensations in the limbs, such as formications, a feeling as if the limb was asleep (engourdissement). These sensations are first experienced in the fingers and toes, and thence extend upwards along the limbs.

These sensations are most complained of in the morning soon after leaving bed. They intermit at times, fluctuating in intensity, and in the early stages are lessened after exercise, when the patient feels better and stronger for a time, but these attempts are followed by an aggravation of the symptoms. Some degree of paralysis of movement, of loss of motor power, occurs in certain sets of muscles—or in one limb. Thus the lower limbs may be singly or successively affected before the upper extremities, or *vice versâ*. Occasionally this loss of power assumes a hemiplegic form. All this will vary according to the seat and the extent of the myelitis.

There is usually constipation in consequence of loss of power in the lower bowel. It is very rare that the bladder is early affected, the patient having voluntary control over that organ until the most advanced stages of the disease, towards the close of life, when the softening of the cord is complete.

Ollivier remarks, that in chronic myelitis patients often complain of a sensation as of a cord tied tightly round the body.

The gait (démarche) of patients affected with chronic myelitis is peculiar. The foot is raised with difficulty, the toes are sometimes depressed and at others they are raised, and the heel drags in walking. The body is kept erect and carried somewhat backwards.

If we take any one symptom that enters into the composition of these various groups, we shall find that it is more or less common to various forms of disease of the nervous system. But if we compare the groups of symptoms that have just been detailed, their progressive development and indefinite continuance, with those which are described by Ollivier and other writers of acknowledged authority on diseases of the nervous system, as characteristic of spinal meningitis and myelitis, we shall find that they mostly correspond with one another in every particular—so closely, indeed, as to leave no doubt that the whole train of nervous phenomena arising from shakes and jars of or blows on the body, and described at pp. 74 to 83 as characteristic of so-called "concussion of the spine," are in reality due to chronic inflammation of the spinal membranes and cord. The variation in different cases being referable partly to whether meningitis or myelitis predominates, and in a great measure to the exact situation and extent of the intra-spinal inflammation, and to the degree to which its resulting structural changes may have developed themselves in the membranes or cord.

LECTURE THE SIXTH.

DIAGNOSIS, PROGNOSIS, TREATMENT, OF CONCUSSION OF THE SPINE.

Diagnosis from Cerebral Concussion—From Rheumatism—From Hysteria—Prognosis of Concussion of the Spine—What is meant by Recovery—Probability of Recovery—Period of Fatal Termination—Treatment—Importance of Rest—Counter Irritation—Medical Treatment.

FROM the account that has just been given of the symptoms that may gradually develop themselves after concussion of the spine, I need say little about the diagnosis of this injury from other forms of cerebro-spinal disease. There are, however, three morbid states for one or other of which I have known it to be confounded, and from which it is necessary to diagnose it. These are, 1. The secondary consequences of cerebral concussion; 2. Rheumatism; and 3. Hysteria.

1. *From the secondary effects of cerebral concussion* it is not difficult to diagnose the consequences of concussion of the spine, in those cases in which the mischief is limited to the vertebral column. The tenderness and rigidity of the spine, the pain on pressing upon or on moving it in any direction, and the absence of any distinct lesion about the head, will sufficiently mark the precise situation of the injury.

The two conditions of cerebral and spinal concussion often coexist primarily. The shock that jars injuriously one portion of the nervous system, very commonly produces a corresponding effect on the whole of it, on brain as well as on cord; and, as has been fully pointed out in various parts of these lectures, the secondary inflammations of the spine, which follow the concussion, even when that is primarily limited to the vertebral column and its contents, have a tendency to extend along the continuous fibrous and serous membranes to the interior of the cranium, and thus to give rise to symptoms of cerebral irritation.

2. From *Rheumatism* the diagnosis may not always be easy, especially in the earlier stages of the disease, when the concussion of the spine and the consecutive meningitis have developed pain along the course of the nerves, and increased cutaneous sensibility at points. By attention, however, to the history of the case, the slow but gradually progressive character of the symptoms of spinal concussion, the absence of all fixed pain except at one or more points in the back, the cerebral complications, the gradual occurrence of loss of sensibility, of tinglings and formications, the slow supervention of impairment or loss of motor power in certain sets of muscles,—symptoms that do not occur in rheumatism,—the diagnosis will be rendered comparatively easy; the more so when we observe that in spinal concussion there is never any concomitant articular inflammation, and that although the urine may continue acid, it does not usually present evidences of a superabundance of lithates.

3. *Hysteria* is the disease for which I have more frequently seen concussion of the spine, followed by meningo-myelitis, mistaken, and it certainly has always appeared extraordinary to me that so great an error of diagnosis could so easily be made.

Hysteria, whether in its emotional or its local form, is a disease of women rather than of men, of the younger rather than of the middle-aged and old, of people of an excitable, imaginative, or emotional disposition rather than of hard-headed, active, practical men of business. It is a disease that runs no definite or progressive course, that assumes no permanence of action, that is ever varying in the intensity, in the degree, and in the nature of its symptoms; that is marked by excessive and violent outbreaks of an emotional character, or by severe exacerbations of its local symptoms, but that is equally characterized by long-continued and complete intermissions of its various phenomena.

Does this in any way resemble what we see in "concussion of the spine," or in the consecutive meningo-myelitis? In those cases in which a man advanced in life, of energetic business habits, of great mental activity and vigour, in no way subject to gusty fits of emotion, or to local nervous disquietudes of any kind,—a man, in fact, active in mind, accustomed to self-control, addicted to business, and healthy in body, suddenly, and for the first time in his life, after the infliction of a severe shock to the system, finds himself affected by a train of symptoms indicative of serious and deep-seated injury to the nervous system,—is it reasonable to say that such a man has suddenly become "hysterical," like a love-sick girl? Or is this term not rather employed merely to cloak a want of precise knowledge as to the real pathological state that has given rise to the alteration that is perceptible to the most casual observer in the mental state and bodily condition of the patient? To me, I confess, the sight of a man of forty-five, rendered "hysterical," not for a few hours or days even, by some sudden and overwhelming calamity that may for the time break down his mental vigour, but permanently so for months or years, would be a novel and a melancholy phenomenon, and is one that I have neither seen described by any writer with whose works I am acquainted, nor witnessed in a hospital experience of twenty-five years; and could such a condition actually be induced, it would certainly be to my mind an evidence of the most serious and disorganizing disease of the nervous system.

But, in reality, there can be but little difficulty in establishing the diagnosis between chronic meningo-myelitis and hysteria. The persistence of the symptoms, their slow development, their progressive increase in severity notwithstanding occasional fluctuations and intermissions in intensity, the invariable presence of more or less paralysis of sensation, or of motion, or both, will easily enable the surgeon to judge of the true nature of the case. That mental emotion is occasionally manifested by an unfortunate individual who has been seriously injured by an accident which tends to shake his whole nervous system, can scarcely be matter of surprise, the more so when, as commonly happens in these cases, he finds himself progressively and steadily deteriorating in health and strength, and sees in the future the gloomy prospect of a shattered constitution, of impaired mental vigour, and of loss of bodily activity; of the necessity of abandoning those occupations in which his life had been usefully spent, and on the continuance of which, probably, the support of himself and his family is dependent. That such an unhappy sufferer should occasionally be unnerved and give way to mental emotion is natural enough. It certainly appears to me that the term "hysteria," elastic as it is, can scarcely, with any regard to truth or justice, be strained so far as to embrace those feelings that naturally spring from the contemplation of so gloomy a prospect as this; and even if it be considered applicable to his mental state, it can in no way be looked upon as the cause of those bodily sufferings and disabilities which constitute the most important and serious part of his disease, and which have no analogy in development or progress with the ordinary physical phenomena of hysteria.

The prognosis of "concussion of the spine" and that of the consecutive meningo-myelitis is a question of extreme interest in a medico-legal point of view, and is one that is often involved in no little difficulty.

The prognosis requires to be made with regard, first, to the life, and, secondly, to the health of the patient. So far as life is concerned, it is only in those cases of severe and direct blows upon the spine, in which intra-spinal hemorrhage to a considerable extent has occurred (Case 4), or in which the cord or its membranes have been ruptured, that a speedily fatal termination may be feared.

In some of the cases of concussion of the spine, followed by chronic inflammation of the membranes and of the cord itself, death may eventually supervene after several, perhaps three or four, years of an increasingly progressive breaking down of the general health, and the slow extension of the paralytic symptoms. I have heard of several instances in which concussion of the spine has thus proved fatal some years after the occurrence of the accident. Mr. Gore, of Bath, who has had considerable experience in these injuries, writes to me in reference to the case related p. 84, that this is the third fatal case of which he has had more or less personal knowledge, the time from the injury to the occurrence of death varying from two and a half to five years.

In these cases, as in the one related at p. 84, the fatal result is the direct consequence of the structural changes that take place in the cord and its membranes. Indeed, this one case proves in the clearest and most incontestable manner the possibility of death occurring after a lapse of several years, from the progressive increase of those symptoms, which are dependent upon disease of the nervous system from concussion of the spine occurring from railway accidents, and attended by the usual symptoms of such injuries; the fatal termination being gradually induced by the slow and progressive structural changes which take place in the cord. This case establishes the fact beyond doubt that such a fatal termination is by no means impossible after an interval of several years, in cases of concussion of the spine in which deep-seated structural changes have developed in the cord.

The probability of such a melancholy occurrence is greatly increased if, after a year or two have elapsed from the time of the occurrence of the accident, the symptoms of chronic meningo-myelitis either continue to be gradually progressive, or, after an interval of quiescence, suddenly assume increased activity.

In fact, it is the excitation of this very form of disease, *viz.*, chronic inflammation of the spinal cord and its membranes, that constitutes the great danger in these injuries of the spine. When it has once gone on to the development of atrophy, softening, or other structural changes of the substance of the cord itself, complete recovery is impossible, and, ultimately, death is not improbable.

Ollivier states as the result of his experience, that although persons affected with chronic myelitis may live for fifteen or twenty years, yet that they more commonly perish within four years. This opinion as to the probable future of patients unfortunately affected by this distressing disease is perhaps too gloomy, so far as the fatal result is concerned, but it is an evidence of the very serious view that a man of such large experience in the diseases of the cord took of the probable issue of a case of chronic inflammation of that structure, and it is doubtless explicable by the fact that Ollivier's experience has necessarily been chiefly drawn from idiopathic or constitutional affections of that portion of the nervous system; and these may justly be considered to be more frequently fatal than those forms of the disease that arise from accident to an otherwise healthy man not predisposed to such affections.

Ollivier takes an equally unfavourable view of the ultimate result of spinal meningitis, and probably for the same reason. He says:[1] "Is spinal meningitis susceptible of cure? All observers agree in stating that death is the inevitable result." He qualifies this statement, however, by saying that he has found in one case after death from other disease, old thickening of the membranes of the cord, and that Frank relates another in which a fatal termination did not occur. The occurrence of convulsive movements is a most unfavourable sign. They indicate the existence of chronic myelitis, and are usually associated with deep disorganization of the structure of the cord. They are of a most painful character, and are apt to be excited by movements and shocks of the body, even of a very slight character. I have never known a patient recover who has been afflicted by them, progressive paralysis developing itself, and the case ultimately proving fatal. Mr. Gore, of Bath, informs me that he is acquainted with two cases which proved fatal at long periods of time after the accident, in both of which this symptom was present. One of these, a very healthy lad of nineteen, was injured on October 29, 1863, and died May 11, 1866. He suffered from convulsive attacks, with extreme pain in the spine, till the latter end of 1864, then the convulsions ceased, but the aching, wringing spinal pain continued; and his health broke down completely. Phthisis, to which there was no hereditary tendency, developed in the following spring, and he eventually died of that disease two years and a half after the injury.

From all this it is certain that concussion of the spine may prove fatal; first, at an early period by the severity of the direct injury (Case 4); secondly, at a more remote date by the occurrence of inflammation of the cord and its membranes (Ollivier); and, thirdly, after a lapse of several years by the slow and progressive development of structural changes in the cord and its membranes (Mr. Gore's Case, p. 84).

But though death may not occur, is recovery certain? Is there no mid-state between a fatal result, proximate or remote, and the absolute and complete recovery of the patient?

What is meant by the "recovery of the patient?" When you are asked, "In your opinion will this patient ever recover?" what are you to understand by that question? Is it meant whether there will be a mitigation of the symptoms—an amelioration of

[1] Vol. ii. p. 294.

health to some, perhaps even a considerable, extent—an indefinite prolongation of life, so that with care, by the avoidance of mental exertion and bodily fatigue of all kinds, the patient may drag on a semi-valetudinarian existence for fifteen or twenty years? Is this the meaning of the question? No, certainly not. If that question has any definite meaning, it is whether the patient will in time completely and entirely lose all the effects of the injury he has sustained—whether in all respects, mentally and bodily, he will be restored to that state of intellectual vigour and of corporeal activity that he enjoyed before the occurrence of the accident—whether, in fact, he will ever again possess the same force and clearness of intellect, the same aptitude for business, the same perfection of his senses, the same physical energy and endurance, the same nerve, that he did up to the moment of his receiving the concussion of his spine.

In considering the question of recovery after concussion of the spine, we have to look to two points, first, the recovery from the primary and direct effects of the injury, and, secondly, from the secondary and remote consequences of it.

There can be no doubt that recovery, entire and complete, may occur in a case of concussion of the spine when the symptoms have not gone beyond the primary stage, when no inflammatory action of the cord or its membranes has been developed, and more particularly when the patient is young and healthy in constitution. This last condition indeed is a most important one. A young man of healthy organization is not only less likely to suffer from a severe shock to the system from a fall or railway injury than one more advanced in life, but, if he does suffer, his chance of ultimate recovery will be greater, provided always that no secondary and organic lesions have developed themselves.

I believe that such recovery is more likely to ensue if the primary and direct symptoms have been severe, and have at or almost immediately after the occurrence of the accident attained to their full intensity. Case 1 is an instance of this, and many similar ones must present themselves to the recollection of most surgeons, and there are many such on record.

In these cases, under proper treatment the severity of the symptoms gradually subsides, and, week by week, the patient feels himself stronger and better, until usually in from three to six months at the utmost all traces of the injury have disappeared.

But incomplete or partial recovery is not unfrequent in these cases of severe and direct injury of the spine. Of this, Case 2 is an excellent illustration. The patient slowly recovers up to a certain point and then remains stationary, with some impairment of innervation in the shape of partial paralysis of sensation, or of motion, or both, usually in the lower limbs. The intellectual faculties or the organs of sense are more or less disturbed, weakened, or irritated, the constitution is shattered, and the patient presents a prematurely worn and aged look.

In such cases structural lesion of some kind, in the membranes, if not in the cord, has taken place, which necessarily must prevent complete recovery. When, therefore, we find a patient who, after the receipt of a severe injury of the spine by which the cord has been concussed, presents the primary and immediate symptoms of that condition, such as have been described in Case 1, we may entertain a favourable opinion of his future condition, provided we find that there is a progressive amelioration of his symptoms, and no evidence of the development of any inflammation, acute or chronic, of the membranes and the cord.

But our opinion as to his ultimate recovery must necessarily be very unfavourable if we find the progress of amendment cease after some weeks or months, leaving a state of impaired innervation. And this unfavourable opinion will be much strengthened if we find that subsequently to the primary and immediate effect of the injury, symptoms indicative of the development of meningo-myelitis have declared themselves. Under such circumstances of the double combination, of the cessation of improvement and the supervention of symptoms of intra-vertebral inflammatory action, partial restoration to health may eventually be looked for; but complete recovery is not possible.

When a person has received a concussion of the spine from a jar or shake of the body, without any direct blow on the back, or perhaps on any other part of the body, and the symptoms have gradually and progressively developed themselves, the prognosis will always be very unfavourable. And for this reason;—that as the injury is not sufficient of itself to produce a direct and immediate lesion of the cord, any symptoms that develop themselves must be the result of structural changes taking place in it as the consequence of its inflammation; and these secondary structural changes being incurable, must, to a greater or less degree, but permanently, injuriously influence its action.

The occurrence of a lengthened interval, a period of several weeks for instance, between the infliction of the injury and the development of the spinal symptoms, is peculiarly unfavourable, as it indicates that a slow and progressive structural change has been taking place in the cord and its membranes, dependent upon pathological changes of a deep-seated and permanently incurable character.

Abercrombie truly says: "Every injury of the spine should be considered as deserving of minute attention, and the most active means should be employed for preventing or removing the diseased actions which may result from it. The more immediate object of anxiety in such cases is inflammatory action; and we have seen that it may advance in a very insidious manner, even after injuries which were of so slight a kind that they attracted at the time little or no attention."

Well, then, when you see a patient suffering from the secondary effects of a slight injury of the spine, these effects having developed in an insidious but progressive manner, examine him with minute attention; and if you find evidence of inflammatory action in the cord and its membranes, as indicated by symptoms of cerebral irritation, spinal tenderness and rigidity, modifications of sensation, as pains, tinglings, and numbness in the limbs, and some loss of muscular or motor power, with a quick pulse and a shattered constitution, you must, at any period of the case, however early, give a most cautious prognosis. And if several months—from six to twelve—have elapsed without any progressive amelioration in the symptoms, you may be sure that the patient will never recover so as—to use the common phrase—"to be the same man" that he was before the accident. But if, instead of remaining stationary, a progressive increase in the symptoms, however slow that may be, is taking place, more and more complete paralysis will ensue, and the patient will probably eventually die of those structural spinal lesions that are described at p. 84.

I have purposely used the words "progressive amelioration" for this reason, that it often happens in these cases that under the influence of change of air, of scene, &c., a temporary amelioration takes place—the patient being better for a time at each new place that he goes to—or under every new plan of treatment that he adopts. Fallacious hopes are thus raised which are only doomed to disappointment, the patient after a week or two relapsing, and then falling below his former state of ill-health.

In forming an opinion as to the patient's probable future state, I believe that it is of less importance to look to the immediate or early severity of the symptoms than to their slow, progressive, and insidious development. Those cases are least likely to recover in which the symptoms affect the latter course.

The time that the symptoms have lasted is necessarily a most important matter for consideration. When they have been of but short duration, they may possibly be dependent on conditions that are completely, and perhaps easily, removable by proper treatment, as for instance, on extravasation of blood, or on acute serous inflammatory effusion (Case 14). But when the symptoms, however slight they may be, have continued even without progressive increase, but have merely remained stationary for a lengthened period of many months, they will undoubtedly be found to be dependent on those secondary structural changes that follow in the wake of inflammatory action, and that are incompatible with a healthy and normal function of the part. I have never known a patient to recover *completely and entirely so as to be in the same state of health that he enjoyed before the accident*, in whom the symptoms dependent on chronic inflammation of the cord and its membranes, and on their consecutive structural lesions, had existed for twelve months. And though, as Ollivier has observed, such a patient may live for fifteen or twenty years in a broken state of health, the probability is that he will die within three or four. There is no structure of the body in which an organic lesion is recovered from with so much difficulty and with so great a tendency to resulting impairment of function as that of the spinal cord and brain. And, with the exception probably of the eye, there is no part of the body in which a slight permanent change of structure produces such serious disturbance of function as in the spinal cord.

Treatment.—I have not much to say to you about the treatment of these injuries that we have been discussing. But I feel that my remarks on this subject would scarcely be complete were I to omit so important a matter from our consideration.

In the early stages of a case of "concussion of the spine," the first thing to be done is undoubtedly to give the injured part complete and absolute *rest.*

The importance of rest cannot be over-estimated in these cases. Without it no other treatment is of the slightest avail, and it would be as rational to attempt to treat an injured brain or a sprained ankle without rest, as to benefit a patient suffering from a severe concussion or wrench of the spine unless he is kept at rest. In fact, owing to the extreme pain in movement that the patient often suffers, he instinctively seeks rest, and is disinclined to exertion of any kind. It is the more important to insist upon absolute and entire rest in these cases, for this reason, that not unfrequently patients feel for a time benefited by movement—by change of air and of scene. And hence such changes are thought to be permanently beneficial. But nothing can be more erroneous than this idea, for the patient will invariably be found to relapse and to fall back into a worse state than had previously existed. In more advanced stages of the disease, when chronic meningitis has set in, the patient suffers so severely from any, even the very slightest movement of the body, from any shock, jar, or even touch, that he instinctively preserves that rest which is needed, and there is no occasion on the part of the surgeon to enforce that which the patient feels to be of imperative necessity for his own comfort.

In order to secure rest efficiently the patient should be made to lie on a prone couch. There are several reasons why the prone should be preferred to the supine position. In the first place, in the prone attitude the spine is the highest part of the body, thus passive venous congestion and determination of blood, which are favoured and naturally occur when the patient lies on his back, are entirely prevented, and that additional danger which may arise from this cause is averted so long as the prone position is maintained. Then again, the absence of pressure upon the back is a great comfort in those cases in which it is unduly sensitive and tender, and is a matter of additional safety to the patient, if he is paraplegic, by lessening the liability to sloughing from undue compression of the soft parts over the sacrum and nates. Lastly, the prone position presents this advantage over the supine, that it admits of the ready application of any local treatment that may be desired to the spine.

In some instances, as in Case 14, complete and absolute rest may be secured to the injured spine by the application of a gutta-percha case to the back, embracing the shoulders, nape, and back of the head, or, as in Case 13, by letting the patient wear a stiff collar so as to give the support that is needed to the neck.

But if rest is needed to the spine, it is equally so to the brain. I have repeatedly in these lectures had occasion to point out the fact that in cases of concussion of the spine the membranes of the brain become liable to secondary implication by extension of inflammatory action to them. The irritability of the senses—of sight and hearing, that is so marked in many of these cases—with perhaps heat of head, or flushings of the face, are the best evidences of this morbid action. For the subdual of this state of increased cerebral excitement and irritability, it is absolutely necessary that the mind should be kept as much as possible at rest, and that disquieting influences and emotions should, as far as practicable, be avoided. The patient, feeling himself unequal to the fatigue of business, becomes conscious of the necessity of relinquishing it, though not perhaps without great reluctance, and until after many ineffectual efforts to attend to it.

During the early period of concussion of the spine, much advantage will usually be derived from dry cupping along the back on either side of the vertebral column. In some cases I have seen good effects follow the application of ice-bags to the injured part of the spine.

At this period I believe that medicine is of little service beyond such as is required for the regulation of the general health on ordinary medical principles.

When the secondary effects of the concussion of the spine have begun to develop themselves, more scope presents itself for proper medical treatment, and much may often be done not only for the mitigation of suffering, but for the cure of the patient by carefully conducted local and constitutional treatment.

Rest as in the early stages must be persevered in, but in addition to this counter-irritation may now be advantageously employed. With this view the various forms in which this means is familiar to the surgeon—stimulating embrocations, mustard poultices, blisters, and setons or issues—may be successfully employed.

With regard to internal treatment, I know no remedy in the early period of the secondary stage, when subacute meningitis is beginning to develop itself, that exercises so marked or beneficial an influence as the bichloride of mercury in tincture of quinine or of bark. I have seen this remedy produce the most beneficial effects, and have known patients come back to the hospital to ask for the "bichloride" as the only medicine from which they had derived advantage. At a more advanced period, and in some constitutions in which mercury is not well borne, the iodide or the bromide of potass in full doses will be found highly beneficial, more especially when there are indications, as in Case 14, of the presence and the pressure of inflammatory effusion.

When all signs of inflammatory action have subsided—when the symptoms have resolved themselves into those of paralysis whether of sensation or of motion—but more especially in those cases in which there is a loss of motor power, with a generally debilitated and cachectic state, the preparations of nux vomica, of strychnine, and of iron may be advantageously employed. But I would particularly caution you against the use of these remedies, and more especially of strychnine, in all those cases in which inflammatory action is still existing, or during that period of any given case in which there are evidences of this condition. You will find that under such circumstances the administration of strychnine is attended by the most prejudicial effects, increasing materially and rapidly the patient's sufferings. But in the absence of this inflammatory irritation it will, if properly administered, be found to be a most useful remedy, more particularly in restoring lost motor power.

In those cases in which the strychnine may be advantageously administered, great benefit will also be derived from warm salt-water douches to the spine, and galvanism to the limbs.

At a more advanced period of the case, when general cachexy has been induced, and more or less paralysis of sensation and motion continues in the limbs, and nothing of a specific nature can be done in the way of treatment, our whole object should be to improve the general health on ordinary medical principles, so as to prevent as far as possible the development of secondary diseases, such as phthisis dependent on mal-nutrition and a generally broken state of health, and which may, after a lapse of several years, lead to a fatal termination.

THE END.

www.ingramcontent.com/pod-product-compliance
Lightning Source LLC
Chambersburg PA
CBHW020106010526
44115CB00008B/704